Girls Get Curves

Also by Danica McKellar

Math Doesn't Suck: How to Survive Middle School Math Without Losing Your Mind or Breaking a Nail

Kiss My Math: Showing Pre-Algebra Who's Boss

Hot X: Algebra Exposed

Girls Get
Curves

GEOMETRY TAKES SHAPE

Danica McKellar

HUDSON
STREET
PRESS

HUDSON STREET PRESS
Published by the Penguin Group
Penguin Group (USA) Inc., 375 Hudson Street, New York, New York 10014, U.S.A.
Penguin Group (Canada), 90 Eglinton Avenue East, Suite 700, Toronto, Ontario,
Canada M4P 2Y3 (a division of Pearson Penguin Canada Inc.)
Penguin Books Ltd., 80 Strand, London WC2R 0RL, England
Penguin Ireland, 25 St. Stephen's Green, Dublin 2, Ireland (a division of Penguin
Books Ltd.)
Penguin Group (Australia), 250 Camberwell Road, Camberwell, Victoria 3124,
Australia (a division of Pearson Australia Group Pty. Ltd.)
Penguin Books India Pvt. Ltd., 11 Community Centre, Panchsheel Park,
New Delhi – 110 017, India
Penguin Group (NZ), 67 Apollo Drive, Rosedale, Auckland 0632, New Zealand
(a division of Pearson New Zealand Ltd.)
Penguin Books (South Africa) (Pty.) Ltd., 24 Sturdee Avenue, Rosebank,
Johannesburg 2196, South Africa

Penguin Books Ltd., Registered Offices: 80 Strand, London WC2R 0RL, England

First published by Hudson Street Press, a member of Penguin Group (USA) Inc.

First Printing, August 2012
10 9 8 7 6 5 4 3 2 1

Copyright © Danica McKellar, 2012
Illustrations by Mary Lynn Blasutta
All rights reserved

PHOTO CREDITS: Page 33, © Kersti Reistad-Long; Page 85, © Tosin Emiola; Page 141,
© Puneet Batra; Page 196, © Melinda Huang; Page 197, © Julie M. Hadden; Page
239, © Peter A. Cragnolin; Page 300, © Mike Verta; Page 361, © Perry Cucinotta.

 REGISTERED TRADEMARK—MARCA REGISTRADA

CIP data is available.
ISBN 978-1-59463-094-1 (hc.)

Printed in the United States of America
Set in Stone Informal

I dedicate this book to young women everywhere who aren't afraid to stand up and tackle challenges—whether they be related to math, body image, or anything else. This is YOUR life, and you are stronger and smarter than you realize. I'm so proud of you!

Acknowledgments

Thank you to my parents, Mahaila and Chris, for always encouraging and believing in me! Thank you to my best friend in the whole world, my sister Crystal, and to the rest of my wonderful family, including (but not limited to) Chris Jr., Connor, Opa, Lorna (and kids!), Jimmy, Molly, and the Vertas. And to Grammy (1910–2010), who called me "clever" from my early childhood and was always proud of me, no matter what.

Thank you to the team at HSP—Clare Ferraro, Liz Keenan, Caroline Sutton, and Meghan Stevenson. It ain't easy, what we're doing! Thank you to my wonderful literary agent Laura Nolan at Paradigm and to my amazing publicist Michelle Bega at Rogers & Cowan. Thank you to my incredible lawyer Jeff Bernstein and assistant Bradley Polito, and to Cathey Lizzio, Pat Brady, and Matt Sherman for your support! Thanks so much to Kay Carlson, Jennifer Mallette, and Barbara Jacobson, and all else who proofread, including my mom, my dad, Brandy Winn, Damon Williams, the immensely gifted Jonathan Farley, and my amazing brother-in-law, Mike Scafati. Thank you to Nicole Cherie Jones, Brittany Pogue-Mohammed, and Bailey Thompson, and thank you Becky Batchelor for the line art. Thank you to all who shared their stories and quotes, including my goddaughter! Thank you to the educators who collected surveys, especially Dan Degrow, Jim Licht, Kris Murphy, Joanne Hopper, Pete Spencer, and the staff at St. Clair RESA, and also to Shirley Stoll, JoAnn Aleman, and Lisa Miller, and teachers Sue Olson, Denise Leonard, Matt Nimmons, Karen Robertson, and Rachael Sulkowski. And thank you to my late, great geometry teacher (and "onion" song performer), Mr. Maeder.

A special thank you to my mom for all the babysitting help! You're the best Tutana ever. And Olivia and Mitzy, your help has been invaluable. Finally, thank you to Mike for another amazing cover, and to my new baby boy, Draco, for opening my heart in ways I never knew possible. You mean the world to me!

What's Inside

PART 1: GEOMETRY BASICS

PART 2: INTRODUCTION TO PROOFS

PART 3: TRIANGLES, CONGRUENCE, AND MORE PROOFS

PART 4: PARALLEL LINES AND POLYGONS

PART 5: SIMILARITY, CIRCLES, AREA, AND VOLUME

Girls Get . . . What?

$\mathcal{M}$ath just got physical.

Look around—we live in a world of angles and curves. Geometry is responsible for the shape of the house you live in, the cars on the road, the shoes on your feet, and even the book in your hands. Diamond rings wouldn't be nearly so sparkly without the study of angles, and your favorite dress wouldn't fit nearly as well without the science of curves. But geometry does more than help us to master the physical world. Doing geometry—especially proofs—trains the logic center in our brains. And logic helps us stay clear and focused, which is useful in all parts of life!

For example, imagine you and your brother are in an infuriating argument. Your emotions get the best of you, and you hear yourself saying, "You're wrong because you're just—*wrong!*" . . . which only makes him laugh at you. Or imagine an interview for your dream job, where the boss looks at you and simply says, "So, why should I hire you? You're not as qualified as the other applicants." Your cheeks flush as adrenaline shoots through your body. How do you handle it? Well, here's where you could use logic to focus that surge of energy, charming her with a foolproof argument filled with passion and personality. But without that focus, you might get caught up in the assumption that she doesn't like you, find yourself flooded with desperation, and end up blurting out, "Because I . . . I really want it. A lot. Please?"

Geometry can help.

Just by *doing* geometry—especially proofs—we train our brains to think more logically, to avoid making assumptions, and to create airtight arguments. Logic will help you to be not only a better communicator, but also a more savvy citizen—one who is less prone to being lied to or taken advantage of. And that's always nice.

Look, it's no secret that girls are often labeled as "emotional" and "irrational." Of course we're emotional—it's what gives us energy and passion, and it's one of the most beautiful, powerful parts of being a woman. But we don't have to be irrational. No way! Being irrational happens when strong emotions cloud our minds, causing us to say things or act in ways that don't serve our goals. The logic we learn in geometry literally *focuses* our passion, giving us the power to keep our wits about us, even during emotional whirlwinds.

And this new power comes just in time.

Not only is the *math* changing and developing, but as young women, so are we. Our bodies go through some pretty massive changes, bringing all sorts of curves . . . and a whole new kind of power. This can be thrilling and scary at the same time—especially since our hormones are on a roller coaster of their own. Talk about an emotional whirlwind! These changes present us with a host of new challenges: How do we celebrate our curves without objectifying ourselves? And with the media plastering airbrushed images of "perfect" women everywhere, how do we find the balance of accepting our bodies while working to keep fit in healthy ways?

Believe it or not, *all* of these challenges can be made easier when we get good at geometry. Having logic at our fingertips keeps our energy and desires focused, which is crucially important not only for keeping a healthy self-image, but also for transforming our passions into success of all types.

Just as geometry is the tool humans have used to build the world around us since the beginning of time, it's the tool that will help you build the life you dream of, whatever shape it might take. Go get 'em!

FAQs: How to Use This Book

What Kinds of Math Will This Book Teach Me?

Geometry! The chapters of this book are filled with things like malls, ice cream, designer handbags, puppies, raspberry pie, lipstick, and cheerleading. By the time you finish reading them, however, you'll be a whiz at tons of geometry topics, including congruent triangles, circles, polygons, surface area, and volume. And of course, you'll be a seasoned expert at proofs!

Just to make sure you're *never* confused, every single (non-proof) problem has an answer at the back of this book. There are also fully worked-out solutions to all the problems, *including every single proof*, on the "solutions" page of GirlsGetCurves.com in case you get a different answer or want guidance on the proofs. I'm here for you, sister!

What Should I Already Know in Order to Understand This Book?

In geometry, most of the topics are brand new. But from time to time, we'll use pre-algebra and algebra concepts like fractions, solving for *x*, square roots, exponents, plotting points, etc. It's stuff you probably already learned, but everyone forgets things, right?

To make sure that you never feel lost, throughout the book I include just little bits of review and tons of footnotes that say stuff like, "To review such and such, see p.-- in *Hot X: Algebra Exposed* or *Kiss My Math* or even my first book, *Math Doesn't Suck*," so you can quickly flip to it. But if you don't own my first three books, that's fine, too. There are other places to review those topics (for example, by doing a search for your topic online). Either way, you're totally covered!

Do I Need to Read the Book from Beginning to End?

Nope! There are a few different ways to use this book:

- You can skip directly to the chapters that will help you with tonight's homework assignment or next week's test.

- You can skip to the math concepts that have always been problem areas to clear them up for good!

- Or you can, in fact, read this book from beginning to end and refer back to each chapter's "Takeaway Tips" for quick refreshers as you need them for assignments.

What's in This Book Besides Math?

In addition to the math I teach, look out for these fun extras, and more!

- Personality Quiz: What's Your Body Image? Find out now on p. 46!

- Confessionals: Stories from teens just like you!

- Real-life testimonials from gals who overcame their struggles in math and are now fabulously successful women: We've got everything from a pizzeria owner to a professional pool player to a journalist for *Elle* magazine. And yes, they all use math in their jobs.

  Let's do it!

Chapter 1

Hot Guys and Smelly Socks
Introduction to Logic and Reasoning

*L*et's say your best friend is planning to ask your latest crush to the dance, and it's making you crazy. You're not even sure if she knows how much you like him, but the whole thing is filling you with anger and jealousy. You're starting to see her as the enemy, which is also heartbreaking, because you've been friends for like 10 years!

Well, just hold on for a minute: Let's organize the facts. You said you're not sure if she knows how much you like him. After all, *if* she does know, *then* it might be time to find a new best friend, because she doesn't care about your feelings. However, *if* she doesn't know how much you like him (because you've been too embarrassed to admit it, perhaps?), *then* she's not the enemy at all! These are two different situations entirely, and you now realize that you need more information before you jeopardize a lifelong friendship.

It's really helpful to take the information we know and rewrite it into "if . . . then" statements, especially when the "if . . . then" logic is sort of hidden! These are called *conditional statements.*

What's It Called?

A **conditional statement** is a statement that implies a cause and an effect, and it can be written in the form "If A, then B." The "A" part is the cause (or hypothesis), and the "B" part is the effect (or conclusion). For example, "<u>If</u> he likes me, <u>then</u> my life will be complete." Instead of using "if . . . then" form, we can also use an arrow, like this: A => B, which can be read "If A, then B" or "A implies B" or "A means that B is true." For example, "He likes me => My life is complete."*

---

* See pp. 198–9 for dealing with obsession over guys . . . been there!

Let's practice writing conditional statements into "if . . . then" form:

General statement	The same statement in "if . . . then" form
If he's late, I'll be mad.	If he's late, *then* I'll be mad.
My nails break when I travel.	If I travel, *then* my nails break.
I love the weather when it's rainy.	If it's rainy, *then* I love the weather.
All ducks are cute.	If it's a duck, *then* it's cute.

For the breaking nails example, you might have been tempted to say, "If my nails break, then I travel," because that's the order the words were first written in. But of course that sentence doesn't have the same meaning as the original (and it's a pretty weird statement anyway). The *cause* should go after the "if," and the *effect* should go after the "then." Make sense?

And I know the "ducks" example might seem a little forced, but writing sentences into "if . . . then" form is a great warm-up for understanding logic. Let's practice!

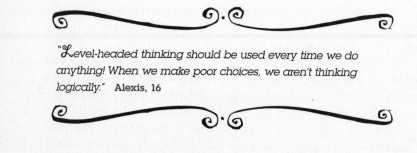

"*L*evel-headed thinking should be used every time we do anything! When we make poor choices, we aren't thinking logically." **Alexis, 16**

Doing the Math

Rewrite these sentences as "if . . . then" statements. Underline the "if" and the "then." I'll do the first one for you.

1. Squares are also rectangles.

Answer: If it's a square, then it's also a rectangle. *

2. I'm late for school when my sister takes forever in the shower.

3. I will scream if I hear that song one more time.

4. Eating more vegetables makes me feel healthy.

5. Odd numbers are not divisible by 2.

(Answers on p. 395)

Inductive vs. Deductive Reasoning

If your sister has been late for school every day this year, then you might guess that "your sister will always be late for school." Or if you look at a bunch of prime numbers, like 3, 7, 19, and 41, you might try to conclude that "all prime numbers are odd," which is totally false!

Inductive reasoning is when we make educated guesses (called **conjectures**)[†] based on a bunch of specific examples . . . but we haven't *proven* anything. Maybe tomorrow your sister's alarm will work and she'll end up being on time. And after all, 2 is a prime number, and the last time I checked, 2 wasn't odd.

Your sister being on time to school and the number 2 are **counterexamples** to the above conjectures. Coming up with counterexamples is like saying, "Ha ha, you're wrong, and here's the example to show how wrong you are." Look, inductive reasoning can be totally useful for understanding patterns, but it's a *bad idea* if you're trying to *prove* something, because a counterexample could always be lurking around the corner.

Deductive reasoning, on the other hand, means that we are told some general facts, called Givens, and we make specific conclusions based on them with a chain of "if . . . then" logic. This kind of reasoning

. . . .

* We'll meet our quadrilaterals (and a train-wreck reality TV family) in Chapter 16.

† Conjecture is pronounced con-jeck-sure, as in, "Suuuure, just try to act like your guess is a fact without looking like an idiot." That was sarcasm, in case you didn't notice.

is airtight (which means there could never be any counterexamples), and it's what we'll use when we do proofs.

For example, if we are given the facts "All baby ducks are cute" and "Annie loves all cute things," we can now <u>prove</u> that Annie loves baby ducks. First, we'll write both statements in "if . . . then" form.

Given: If it's a baby duck, then it's cute.

Given: If it's cute, then Annie loves it.

Therefore: If it's a baby duck, then Annie loves it.

Ta-da! We just did our first proof. That wasn't so bad now, was it? By the way, even if you don't agree that all baby ducks are cute (who are you?), because it was a "Given," you *have* to act as if it's true for the sake of this problem.

Let's make sure we remember the difference between induction and deduction, in case anyone asks (like on a test).

Deduction

Summary: We use an airtight chain of logic to prove conclusions based on given facts, often using "if . . . then" statements. Great for proofs!

First two syllables sound like "dee duck" as in "Look, dair's dee duck!"

Induction

Summary: We look at several specific examples and try to draw a general conclusion from them. Counterexamples could exist: *Bad idea* for proofs.

First two syllables sound like "in duck." Being "in" a duck? Another bad idea.*

* I don't know about you, but I'd rather not be "in" a duck, thank you very much. How would you even get in there? Don't answer that. Regardless, it's *a bad idea!*

Remember, the key to deduction is that we can only conclude things that *must* be true from the Givens—not things that *might* be true. Let's practice!

 Doing the Math

Given the statements below, use deduction to reach a new conclusion, if any. I'll do the first one for you.

1. Given: Bonzo has a moustache. All monkeys have moustaches.

<u>Working out the solution</u>: In this case, it might *seem* like we could conclude that Bonzo is a monkey, but we can't! You see, with this information, there's no airtight argument that leads us from A to B. I mean, Bonzo *might* be a monkey, but he might also be a giraffe who happens to have a moustache. See what I mean?

Answer: No conclusion possible

2. Given: All Barbies are dolls. All dolls are creepy.

3. Given: All puppies like bones. Sparky likes bones.

4. Given: All fruit grows on trees. All apples are fruit.

5. Given: All aliens speak Martian. Debbie speaks Martian.

(Answers on p. 395)

These types of deduction problems show up a lot on standardized tests. If you struggled with these (this is totally normal!), just hang in there and give yourself a break; you're training your brain in a whole new way. Remember the "if . . . then" statements we did a few pages ago? We'll soon be combining those with deduction . . . and smelly socks.

Hot Guys and Smelly Socks: The Converse, Inverse, and Contrapositive of a Statement

Let's say we're given *All hot guys wear smelly socks.* We could rewrite this as "<u>If</u> a guy is hot, <u>then</u> he wears smelly socks." Great. Now, if we're told that some guy (Zac) wears smelly socks, must he be hot? Nope! All sorts of people wear smelly socks, not just hot guys, after all. So the statement, "If a guy wears smelly socks, then he's hot," isn't necessarily true. In fact, we've swapped the cause and effect in the statement, and that's called the **converse**.*

The **converse**, **inverse**, and **contrapositive** of a statement are all ways to change an "if . . . then" statement by moving around its "cause and effect" parts in very specific ways.

Let's use p's and q's to create generic "if . . . then" statements, just to make everything shorter and easier to see. Later, we can substitute phrases like "has smelly socks" if we want.

Original statement: If p, then q.

Converse: If q, then p.

Inverse: If not p, then not q.

Contrapositive: If not q, then not p.

QUICK NOTE   We can actually think of p and q as *variables* that take the place of the original phrases. In fact, some people even use a negative sign instead of the word *not*. So for example, instead of "If not p, then not q," the inverse could be written, "If −p, then −q."[†]

Believe it or not, the contrapositive actually tells us the <u>same information</u> as the original statement. I mean, if a guy *doesn't* wear smelly

. . . .

* Do you know about Converse shoes? To remember the word *converse,* imagine you're doing a funky dance in your Converse shoes and your feet swap positions . . . just like the cause and effect do!

† Some textbooks use a wavy negative sign: ~p. I don't like it though. Looks kinda wimpy to me.

socks, then we know he *can't* be hot . . . because if he were hot, then he'd be wearing smelly socks!

Same information, written differently!

Original statement ⇔ Contrapositive

In our particular case, the p is hotness and the q is smelly socks, right? Check it out:

Logic Stuff

| Original Statement "If p, then q." | If a guy is hot, then he wears smelly socks. |
| | p q |

| Converse "If q, then p." | If the guy wears smelly socks, then the guy is hot. |
| | q p |

| Inverse "If not p, then not q." or "If –p, then –q." | If the guy isn't hot, then he doesn't wear smelly socks. |
| | –p –q |

| Contrapositive "If not q, then not p." or "If –q, then –p." | If the guy doesn't wear smelly socks, then the guy isn't hot. |
| | –q –p |

Do you see how the original statement and the contrapositive actually give the same information?

Watch Out!

When we talk about making part of a statement "negative," that just means it should be the *opposite* of whatever it was in the original statement. Just like with numbers, the negative of a negative is positive!* We'll do an example in #1 on p. 9.

How to Unfog Your Brain

When lots of negatives get involved, this stuff can get confusing—especially with contrapositives. Try this trick: If your brain is feeling cloudy, substitute different words for the confusing parts, so that "not" isn't needed.

For example, consider the statement: "If she likes that greasy donut, then she's crazy." Read that twice. Now, its contrapositive would be: "If she's not crazy, then she doesn't like the greasy donut." Um . . . say *what*? Brain fog! Replacing some words to get rid of all the negatives, we might get something like: "If she's smart, then she hates that greasy donut." Ah! Helps the meaning snap into place, doesn't it? It's just a little trick I like to use. The great news is, the replacement words don't have to be perfect in order for this trick to unfog the brain.

Doing the Math

Find the converse, inverse, and contrapositive of each statement below. Use the *p* & *q* chart on p. 7 to help you remember which is which. I'll do the first one for you. (Note: verb tenses may vary.)

. . . .

* For example, −(−5) = 5. See Chapter 3 in *Kiss My Math* for more on this.

1. If you lie, then you're not honest.

<u>Working out the solution:</u> So in this case, the p is you're lying, and the q is you're not being honest.

Converse: "If q, then p" becomes **If you're not honest, then you lie.**

Inverse: "If not p, then not q" becomes **If you don't lie, then you're honest.** (Notice how the negative "you're not" became the positive "you're".)

Contrapositive: "If not q, then not p" becomes **If you're honest, then you don't lie.**

2. If you tweet, then you're bored.

3. If she misses the bus, then she's not in a good mood.

4. If I stay up on Wednesday night, then I'm tired on Thursday.

(Answers on p. 395)

Converses in Life: Probably True or Not True?

- If the guy likes you, then he'll call you back.
- *Converse:* If the guy calls you back, then he likes you.

• Converse probably true . . . ?

- If you went to a concert last night, then you didn't get enough sleep.
- *Converse:* If you didn't get enough sleep, then you went to a concert last night.

• Converse probably *not* true!

- If the baby has a poopy diaper, then the baby's diaper smells.
- *Converse:* If the baby's diaper smells, then the baby has a poopy diaper.

- Converse probably true!

- If you are a square, then you have 4 sides.
- *Converse:* If you have 4 sides, then you are a square.

- Converse probably *not* true!

QUICK NOTE Arrows! On p. 1, we saw that we can write these conditional "if . . . then" statements by using this fancy little arrow: $\Rightarrow$. Notice that it goes in just one direction in order to show the cause and effect. So, "If she likes that greasy donut, then she's crazy" could be written as "She likes that greasy donut $\Rightarrow$ She's crazy." And our little "if p, then q" statement could be written like this: $p \Rightarrow q$. Make sure to still read this in your head as "p means q" or "If p, then q."

Beyond "If . . . Then": Biconditional Statements

If a statement is true and its converse is *also* true, then we can write both things in one compact form, using a little double-sided arrow. Here's an example: "If a number is even, then it's divisible by 2." Its converse is *also* true: "If a number is divisible by 2, then it's even." It turns out that we can save a whole bunch of space by writing one compact, biconditional statement, using a nifty little double-sided arrow.

"A number is even. $\Leftrightarrow$ A number is divisible by 2."

Biconditional rules are *two* rules, in disguise!*

. . . .

* Another way to think of the $\Leftrightarrow$ is to replace it with the phrase "is equivalent to."

Chains of Logic

Another tool we'll want under our belts as we move toward proofs is the idea of a *chain of logic*. This is where the "then" part of one statement becomes the "if" part of the next statement, creating a chain. They're kind of like chain belts, which can really tie an outfit together, don't you think?

For example, you know those girls who do everything the popular girls do? Let's say we know these "rules" exist in one of the cliques at school:

If Taylor wears pink, then <u>Lizzie wears pink</u>.

If <u>Lizzie wears pink</u>, then Jessica wears pink.

If Jessica wears pink, then Brittany wears pink.

And then we're *given* this new info:

OMG, Taylor is wearing pink.

Without that last line, we just have a bunch of conditional "rules." But when we find out that Taylor actually <u>does</u> wear pink, it sets off a chain reaction that causes not only Lizzie to wear pink but also Jessica and Brittany!

Notice that since *Lizzie wearing pink* is the "then" part of the first sentence and is also the "if" part of the next sentence, it links those sentences together like a chain and causes the logic to be airtight. We can literally create a chain out of these statements. It's much easier if we abbreviate things using variables.

T = **Taylor wears pink** J = **Jessica wears pink**

L = **Lizzie wears pink** B = **Brittany wears pink**

Let's say we were asked to prove that Brittany wears pink. No problem! We can stack our (abbreviated) sentences so that matching parts are lined up and then circle them for our chain belt:

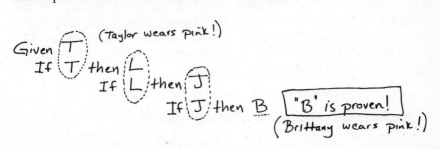

Ta-da! Our airtight chain belt proves Brittany is wearing pink.

QUICK NOTE   I like to use the word *rules* instead of conditional statements. They mean the same thing in this context.

Here's the step-by-step method for these chain belts!

Step By Step

Chain Belt Proofs (using deduction)

Step 1. I recommend rewriting the problem using variables—it just makes the chain belt neater and prettier. Then we write the conditional "rules" as "if . . . then" statements. Remember, biconditional rules are really two rules in disguise!

Step 2. Starting with the Given, we use the rules to make a chain link, matching the "if" and "then" parts at each stage. If it seems like we don't have the right rule, we can try using a contrapositive instead, because contrapositives give the same info.

Step 3. Repeat Step 2 until the "then" part of a rule we use is the thing we're supposed to prove. Done!

And... Action! Step By Step In Action

Given: Taylor doesn't wear pink. Prove: Jessica wears pink. Use the rules below:

1. Lizzie wears pink. ⇔ Jessica wears pink.

2. If Lizzie doesn't wear pink, then Taylor wears pink.

Step 1. To start, let's rewrite everything with variables like on p. 11 and rewrite the rules into "if . . . then" form, remembering to split the biconditional rule into its two parts. For example, "T" would mean that Taylor wears pink, and "not T" would mean that Taylor doesn't wear pink. So here's our problem, written more neatly:

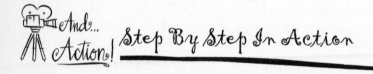

Given: Not T
Prove: J

Rules: { If L, then J
         If J, then L
         If not L, then T

With me so far? Make sure you see how we got these. Now, can we create a chain belt proof that starts with our Given, **Not T**, and ends up concluding **J**? Let's give it a shot!

Step 2. First, we'll write the Given: **Not T**. Hmm, which rule can link to this? Gosh, none of our "if . . . then" rules have **Not T** as the "if" part, which is what we'd need to make a chain. But one of the rules *does* have Taylor in it, so let's see if the contrapositive (see p. 7 for a refresher) is any more helpful. The contrapositive of "If not L, then T" is **If not T, then L**. Great! That becomes the next line, and we can circle the two **Not T** parts. Next, we need a match for **L**, so we can use **If L, then J** and circle the matching **L**'s.

Given (Not T) (Taylor doesn't wear pink!)
If (not T) then L
     If (L) then J ["J" is proven!]
                    (Jessica wears pink!)

Step 3. Lookie there, we chain-linked our way to the thing we were supposed to prove, which is **J**. So yep, it's proven: Jessica wears pink!

The key to making this chain belt was twofold: breaking up the biconditional rule into its two parts, and also using the contrapositive of one of the rules.

Watch Out!

We can't just decide to use the converse or inverse of a rule, because most of the time they won't be true! The reason we can use the contrapositive of any rule is because it actually gives the <u>same information</u> as the original rule.

QUICK NOTE   When things seem confusing, remember to start with the Given and see if an "if . . . then" rule can be written to have an "if" part that matches the Given. Just take things one step at a time.

Watch Out!

When circling chain belt links, make sure you really have a match. For example, <u>we could not make a link</u> from the below two rules, because even though it's tempting to circle the two q's, we can't! The "not q" *doesn't* match plain 'ol q. In fact, they're opposites, aren't they? Just pay attention and you'll do great.

$$\text{If } p \text{ then } q$$
$$\text{If not } q \text{ then } r$$

Not a chain link!

Health & Beauty Tip: Blink!

We all spend so much time staring at screens: computers, phones, TV, you name it. I recently learned that we blink up to 60% less often while looking at a screen than we do at other times! This leads to eyestrain and fatigue, which believe me, doesn't help when we're trying to study for a test or turn in an essay. Plus, it also leads to bloodshot eyes. Yikes! So whether you're studying at your computer, watching TV, or just messing around on your phone, remember to *blink*. I even wrote a little sign and taped it to my laptop: "Blink!" Personally, I sometimes also forget to breathe deeply, and oxygen is so important for staying alert, too. So, yep, I also have a little sign that says "Breathe!" Hey, whatever works, right?

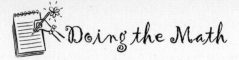

Doing the Math

Do the following chain belt proofs, using variables if needed. I'll do the first one for you.

1. *Rules:* **p ⇔ q**, and **If not q, then s**. *Given:* **Not p**. *Prove:* **s**.

<u>Working out the solution:</u> Um, what? Let's not panic! First, let's write the biconditional statement, p ⇔ q, into its two parts: **If p, then q,** and **If q, then p.** The other rule we've been told is: **If not q, then s.** Great, we have our rules in place. Now we need to use these rules to make a chain link from our *given*, <u>**not p**</u>, and somehow get to the thing we're supposed to prove, which is just "<u>**s**</u>," right? We can do this!

Starting with **not p,** can we make a link? Well, do any of the rules start with "If not p"? Nope, but the contrapositive of the rule "If q, then p" is **If not p, then not q.** Bingo! So we have our first link in the chain, and we circle the matching "not p" parts. Next, we can link to the rule **If not q, then s** and circle the matching "not q" parts. And hey, we've ended up with what we needed to prove! Nice.

Answer: See chain belt below.

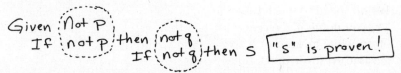

2. *Rules:* **If a, then c**, and **If b, then a**. *Given:* **b**. *Prove:* **c**.

3. *Rules:* If Paige is happy, then Rachel is happy. If Quinn is happy, then Paige is happy. *Given:* Rachel isn't happy. *Prove:* Quinn isn't happy. *(Hint: First, write everything with P, Q & R, and remember to consider contrapositives!)*

4. *Rules:* If Kala doesn't wear pink, then Lea wears pink. If Mindy doesn't wear pink, then Kala doesn't wear pink. Natalie wears pink. ⇔ Mindy wears pink. *Given:* Lea doesn't wear pink. *Prove:* Natalie wears pink. *(Hint: Write everything with K, L, M & N, and attack the chain one small step at a time, using contrapositives when needed. You can do it!)*

(Answers on p. 396)

Don't let your brain glaze over; stick with it! Doing these little chain belt proofs is making you *strong* and totally primed for the geometry proofs that most people shy away from. I'm so proud of you.

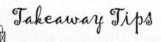

Deduction uses airtight logic to make conclusions based only on *given* information, including **conditional statements**, which we can call "rules." **Induction** is when we look at a bunch of specific examples and try to make a guess about a general conclusion.

If a conditional statement is written as $p \Rightarrow q$, then the **converse** is $q \Rightarrow p$ (think of Converse shoes switching positions!) and the **inverse** is $-p \Rightarrow -q$. The **contrapositive** is $-q \Rightarrow -p$, and it gives the <u>same information</u> as the original statement.

Using a chain of logic for a proof is when we link rules together, starting with the Given and ending with what we want to prove, matching the "if" and "then" parts of the rules along the way. It guarantees an airtight argument!

How to Look Like a Model in a Magazine Ad

Here's another conditional statement: "If you see a perfect-looking model in a magazine, then you can be sure it's been air-brushed." Want to know how to look like a model in a magazine ad? Have somebody take your picture, and then have someone airbrush it. Yep, magazines mostly use airbrushed, unrealistic photos—especially in the advertisements, which c'mon, make up most of the magazines. They create an impossible standard of beauty, and here's why: <u>Advertisements are designed to make us wish we looked different</u>. By using airbrushed images, companies can pretty much guarantee that no one could have hair as shiny, skin as flawless, a body as perfect, etc. You see, *it's in their best interest for us to believe we need their product to feel happy.* It's how they make money!

I can't remember the last mascara ad I saw that didn't have a model wearing obvious false lashes (and probably no mascara at all). Um, hello? Yeah, that mascara might make your lashes thicker, but how would you know if you're gonna use fake lashes anyway?

So remember that the "perfection" in advertisements is just a ploy to get us to spend money, and it's not fair for us to compare ourselves to those fake images. Can you imagine if a company used a real picture of a girl and said "You are good enough just the way you are. You don't need this lipstick at all, but it can be fun to wear sometimes." That would be so awesome. *(Check out p. 95 for how to outsmart misleading ads!)*

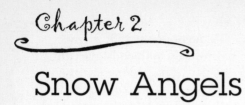

Chapter 2

Snow Angels

Geometry Basics

*H*ave you ever made a snow angel by lying in the snow and waving your arms up and down, making wings?* Ah, the simple joys of childhood that can be so easy to forget as our lives become increasingly complicated. Sometimes it's good to get back to the basics, y'know?

In this chapter, we're covering the basics of geometry, starting with those tiny little specks on the page we like to call **points.** Some of this stuff might seem obvious, but as we'll see in the next chapter, the less we *assume* we know about geometry, the better. So for now, pretend you're a little kid who doesn't know anything at all, perhaps playing in the snow on a quiet winter afternoon . . .

What Are They Called?†

Point, Line, Ray, Segment, Collinear

Points are represented by dots on the page. They have no "size" (they're infinitely small), and we label them with capital letters.

. . . .

* The key is to *get up* very carefully so you don't mess up the shape of your angel.

† These aren't really definitions. They are called "axioms," which basically means that somebody made them up a long time ago, and all of math has been built around them. They're the "starting rules" to the game. (Yes, math is the game!)

A **line** is an infinitely long, straight string of points that extends forever in both directions, so we put arrows on both ends! We can either label a line with a lowercase letter or by picking *any* two points on it and putting a tiny "line" with itty bitty arrows on top. The line below on the left could be called $\overleftrightarrow{XO}$ or $\overleftrightarrow{OX}$; it's up to us. The line on the right could be called *m*, or $\overleftrightarrow{CU}$, $\overleftrightarrow{CT}$, $\overleftrightarrow{CE}$, $\overleftrightarrow{UT}$, $\overleftrightarrow{ET}$, etc.* Those names all represent the *same line*.

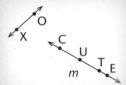

When points lie on the same (straight) line or segment, they are called **collinear** (pronounced: *co-LIN-ee-er*).† So, C, U, T, and E are collinear, but C, U, and O are not. With me so far?

Like a ray of sunshine, a **ray** is straight, has one endpoint, and goes forever in one direction only. We always label rays starting with the endpoint because that's where the ray starts (why didn't they call it a startpoint??) and then any other named point on the ray. Then we put a tiny little ray on top of the two letters, with the arrow going to the *right*. So the rays on the left and in the middle have only one name each: $\overrightarrow{OH}$ and $\overrightarrow{MU}$. (It seems like we should be able to call the middle ray $\overrightarrow{UM}$, but that's not allowed!) The ray on the right could be called $\overrightarrow{DU}$ or $\overrightarrow{DH}$, and that's it.

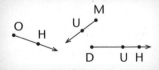

....
* We could also use $\overleftrightarrow{TE}$, $\overleftrightarrow{UE}$, $\overleftrightarrow{TU}$, $\overleftrightarrow{TC}$, $\overleftrightarrow{EU}$, $\overleftrightarrow{EC}$, or $\overleftrightarrow{UC}$. Phew!

† *Collinear* rhymes with "no skinnier," as in "Please *no*, don't get any *skinnier*." (Something you might say to a stick figure like the one on p. 21.)

A **segment** is straight, made up of points, and has two definite endpoints. We always name segments with their two endpoints, and then we put a little baby segment on top of them. Just like with lines, the order of the two letters doesn't matter, so the segment on the left could be called $\overline{MS}$ or $\overline{SM}$. And $\overline{WA}$ and $\overline{AY}$ are two different segments, but there is no segment called $\overline{WY}$ or $\overline{YW}$ pictured here. Also notice that $\overline{HO}$ is a segment, and its measure is smaller than $\overline{HT}$.

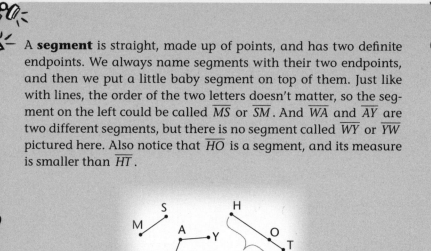

There's more than meets the eye when it comes to these diagrams. Just because something isn't labeled doesn't mean it doesn't exist, right? For example, in the MS WAY HOT diagram above, we could talk about the line $\overleftrightarrow{MS}$ or even $\overleftrightarrow{WY}$, even though they aren't pictured. We can *imagine* the line that goes through the points W and Y; we could even draw it!

And in the "DUH" ray on p. 19, we could talk about the line $\overleftrightarrow{DU}$ or the segment $\overline{UH}$. In fact, the *ray $\overrightarrow{UH}$* is technically shown; it just doesn't include any of the stuff to the left of the point U.

How many *segments* can you name within the *ray $\overrightarrow{DU}$*?*

QUICK NOTE   Every pair of points in the whole wide world is collinear. That's because, hey, take any two points, and we could draw a line through them.† Ta-da! We just showed they're collinear.

· · · · · · · · · · · · · · · · · · · ·

* There are three segments: $\overline{DU}$, $\overline{DH}$, and $\overline{UH}$. Um, duh? BTW, we could also call these same segments by their other names: $\overline{UD}$, $\overline{HD}$, and $\overline{HU}$.

† See p. 389 in the Appendix for more on drawing a line through two points.

Little M & M's

. . . or should I say *m* & *m*'s . . . ?

The segment $\overline{HT}$ in the HOT figure on p. 20 is labeled as having a measure of 4 units, so we'd say that $m\overline{HT}$ = 4. The little *m* stands for "the measure of." If you have a thing against little *m*'s, you could also simply write *HT* = 4, with *no* tiny little segment on top; they mean the same thing.* But we would never write $\overline{HT}$ = 4. Weird, huh?

Oh, and **betweenness** has a special meaning when it comes to points. In order for us to say that a point is **between** two other points, they all must be *collinear*. For example, in the MS WAY HOT diagram on p. 20, *O* is between *H* and *T*, but *A* is *not* between *W* and *Y* because those three points aren't collinear. Also, because *O* is between *H* and *T*, we know that *HO* + *OT* = *HT*. Got it? Great!

We'll practice some of this new vocab later. For now, how about a little snow?

Snow Angels . . . er, *Angles*

It's time to revisit our childhoods and make some snow angels. Except, um, flapping arms in the snow can result in some pretty cold armpits, so don't say I didn't warn you. Okay, actually, we're going to reverse two of the letters and make a snow **angle,** which we do by joining two *rays* together at their endpoints. Oh, and did I mention? There's no snow, either. I guess the *rays* of sunshine melted the snow.†

. . . .

* I'll do it both ways throughout the book, but I usually skip the *m*'s.

† Hmm, not quite as much like childhood as I'd imagined . . .

Angle, Sides, Vertex

An **angle** is made up of two rays that share the same endpoint. The two rays are the **sides** of the angle, and the shared endpoint is called a **vertex**.

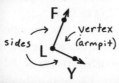

Notice on p. 21 that the vertex is at the "armpit" of our snow angel . . . the very *cold* armpit. In fact, cold armpits like these are enough to put anyone on the verge of moving to Texas, where it's easy to avoid the snow. To remember the name **vertex**, think about cold armpits and being on the "**ver**ge of moving to **Tex**as"—**vertex**!

The vertex (armpit) of the FLY angle is the point *L*, and its sides are the rays $\overrightarrow{LF}$ and $\overrightarrow{LY}$, and we could call the angle ∠*L*. Notice the little baby "angle" we put right before the *L*. We could also use three letters, like ∠*FLY* or ∠*YLF*, to describe this same angle. If we use three letters to label an angle, the middle letter has to be the vertex—and that actually seems right, doesn't it?

The angle in the middle is called ∠6; its vertex and sides haven't been given labels. On the right, the angle ∠7 can also be called ∠*SNO* or ∠*ONS*.

The big angle ∠*SNW* can also be called ∠*WNS*, but *none* of the angles could be called ∠*N*, because it wouldn't be clear which angle we were talking about. Three different angles use that same vertex, after all! The reason the angle on the left could be called ∠*L* is because there's only *one* angle at that vertex. Make sense?

QUICK NOTE   When looking at an angle like ∠SNO on p. 22, we could take a finger or a pencil and tap each corresponding letter in the diagram to more easily "see" which angle it represents.

When Angles Have Egos

Some people always have to be *right*. Others need their egos stroked all the time. Angles aren't so different, but we'll get to that . . .

The measure of an angle describes how "open" the "armpit" is, and it's usually measured as the smallest amount of rotation to get from one side of an angle to the other. In the SNOW diagram on p. 22, ∠ONW measures 50 degrees, and we can say ∠ONW = 50°.*

You might already know that there are 360 degrees in a circle. If you spin all the way around, you've spun 360°. Well, *half* of that is 180°, which is the "angle" of a straight line (this is called a **straight angle**), and 90° is a *quarter* turn, which makes up the all-important **right angle.** You'll make lots of right angles once you learn to drive and you know, turn *right* all the time (. . . or left, but I was making a point, okay?).

If you see a little baby square in the corner of an angle (like in SNOW on p. 22), this means we have a right angle, and we say that the two lines or segments are **perpendicular** to each other. We could even use an itty bitty picture of perpendicular lines to say $\overline{NS} \perp \overline{NW}$.

Angles smaller than 90° are called **acute,** obviously because they're so little and cute, and angles bigger than 90° are called **obtuse,** because they're um, obese?

What's Your Angle?

acute   right (90°)   obtuse   straight (180°)

....

* We have the option of using a little "*m*" for measure, and then we don't actually need the degree symbol; we could say *m*∠ONW = 50° or just *m*∠ONW = 50. I'll do some of each in this book, but I prefer ∠ONW = 50°, don't you?

QUICK NOTE   Whenever you see a , that means it's a Rule: a postulate, theorem, or definition that we'll use later in our proofs. (See p. 389 in the Appendix for those definitions.) You'll see that I use abbreviations like ∠ for angle or "seg" for segment; that's because it makes writing out proofs way easier. You'll see what I mean in a few chapters.

What's It Called?

Complementary and Supplementary angles

When two angles add up to 90° (a right angle), they are called **complementary angles**. These super-cocky angles like to compliment* each other on their hair or outfit, and they're always like, "We are so *right*." On p. 22, ∠7 and ∠ONW are complementary angles. And if we were told that ∠6 = 40°, then we'd know that ∠6 and angle ∠ONW were complementary too!

Complementary angles

If the sum† of two ∠'s is 90°, then they are comp.

<u>Alt:</u> **If the sum of two ∠'s is a right ∠, then they are comp.**

Read this last one as, "If the sum of two angles is a right angle, then they are complementary."

---

* Notice the difference in spelling: compl**e**mentary vs. compl**i**mentary.

† In case you forgot, a **sum** is what we get when we add two things together.

Two angles are called **supplementary angles** when they add up to a total of 180° (a straight line), which I think looks a lot like a page being turned in a book, like maybe a book from your school's *supplementary* summer reading list.

Supplementary angles

If the sum of two ∠'s is 180°, then they are supp.

<u>Alt:</u> **If the sum of two ∠'s is a straight ∠, then they are supp.**

(Read these out loud so you get used to the abbreviations!)

If we know that two angles are complementary and one of the angles measures 68° (like in the figure on the previous page), then we can easily find the other angle because we know $68° + y° = 90°$ has to be true. Solving this, we get $y° = 22°$. No sweat, right?

What if we know that $3x°$ and $x°$ are supplementary angles like in the figure above? Can we find out what they are? You bet! See, we know that $3x° + x° = 180°$, and to solve for x, we just combine like terms and then divide by 4, giving us $4x° = 180° \rightarrow x° = 45°$. That means the other angle is $3x° = 135°$, and yep, $45° + 135° = 180°$!

Let's practice these angles and dust off our algebra skills at the same time.*

QUICK NOTE   Sometimes angle variables will have a degree symbol after them (like x°), and sometimes they won't (like x). It can be different from book to book and from problem to problem. No biggie; in fact, it's good to be flexible!

....

* To brush up on basic equation-solving skills, check out Chapter 12 in *Kiss My Math*.

 *Doing the Math*

Find both angles described. I'll do the first one for you.

1. $\left(\frac{w}{2} + 5\right)^{\circ}$ and $(w + 1)^{\circ}$ are complementary.

<u>Working out the solution:</u> This means they add up to 90°, so: $\left(\frac{w}{2} + 5\right) + (w + 1) = 90$, and we solve for w! We drop the parentheses and combine 5 and 1 to get $\frac{w}{2} + w + 6 = 90 \rightarrow \frac{w}{2} + w = 84$. Let's multiply both sides by 2, and it becomes $w + 2w = 168 \rightarrow 3w = 168 \rightarrow \underline{w = 56}$. But that's not the answer! The problem wants the two angles, which are $\left(\frac{w}{2} + 5\right)^{\circ} = \left(\frac{56}{2} + 5\right)^{\circ} = \underline{33^{\circ}}$ and $(w + 1)^{\circ} = (56 + 1)^{\circ} = \underline{57^{\circ}}$. And yep, 33° + 57° = 90°!

Answer: 33° and 57°

2. $4y^{\circ}$ and y° are supplementary.

3. x° and $(x - 5)^{\circ}$ are supplementary.

4. $3g^{\circ}$ and $(g + 2)^{\circ}$ are complementary.

5. $\left(\frac{h}{5} - 7\right)^{\circ}$ and $(h + 1)^{\circ}$ are complementary.

(Answers on p. 396)

(C'mon, you can do it!)

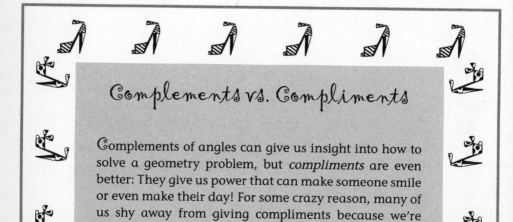

Complements vs. Compliments

Complements of angles can give us insight into how to solve a geometry problem, but *compliments* are even better: They give us power that can make someone smile or even make their day! For some crazy reason, many of us shy away from giving compliments because we're afraid of embarrassment somehow. But it's an illusion!

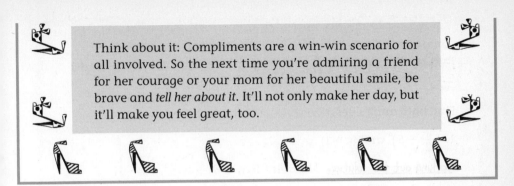

Think about it: Compliments are a win-win scenario for all involved. So the next time you're admiring a friend for her courage or your mom for her beautiful smile, be brave and *tell her about it*. It'll not only make her day, but it'll make you feel great, too.

Equal vs. Congruent

Hmm, that sounds like a legal case, or maybe a competition between artificial sweeteners . . . but it's not! In fact, these are two concepts (*equal* and *congruent*) whose distinction I used to find pretty annoying. I mean, if two angles have the same exact measurement, like if they are both 30°, then doesn't it seem like we should be able to say the two angles are *equal* to each other? Well, let's say that you and your best friend are exactly the same height. Does that make you the same person? Nope! Even truly identical twins are two separate people.

Have you ever heard the terms *con* and *con artist*? A con artist's "con" is usually pretty evil, like impersonating a museum curator in order to steal the royal jewels or something. He'd have to look just like the curator, though, right? Same height, weight, etc.

Much in the same way, two angles (or segments, or shapes of any type) are called **congruent** if their *measurements* are the same. Maybe these angles started off innocent enough, but once they realized they had a shot at the jewels, the con got bigger. Yes, the con grew. (Get it? *the con gru . . .*)

What's It Called?

Two things are **congruent** when all of their measurements are *equal* . . . and they could impersonate each other in a jewelry heist. We use the symbol ≅, which means "is congruent to." Looks kinda like a moustache from one of those funny disguises, if you ask me.

For example, if angles ∠ABC and ∠DEF both equal 55°, then the angles are *congruent* and we can use a moustache: ∠ABC ≅ ∠DEF.*

Remember this: _Numbers_ are equal; things† are congruent!

Chopstick Math . . . and Vertical Angles

Any time two chopsticks cross, they create two sets of congruent angles. Try it yourself: No matter how you cross a pair of chopsticks, or pens, or even two straight pinky fingers, you will *always* create two pairs of congruent angles. Do you see them? They look like two big V's opposite each other, and maybe that's why they're called **vertical angles**.‡ Well, guess what? We can *prove* that vertical angles are always congruent. All we need to know is a little something about supplementary angles! Hmm, how shall we start this proof? Well, let's first draw two random lines intersecting and label the lines and angles, just so it's easier to talk about them:

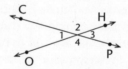

Looking at the diagram, sure, it's tempting to show off our new skills and say it's, like, "obvious" that ∠1 ≅ ∠3 and ∠2 ≅ ∠4. It sure seems like those angles could impersonate each other in a jewel heist. But we're going to actually *prove* they're congruent. Yeah, we're totally badass.

First, let's look at the line $\overleftrightarrow{CP}$ and just the angles ∠2 and ∠3, pretending the bottom half of the other line doesn't exist. (Use a finger to cover up ∠1 and ∠4.) Well gosh, now this looks a lot like a diagram of supplementary angles—and in fact, it is! You see, because $\overleftrightarrow{CP}$ is a straight line, that means **m∠2 + m∠3 = 180.** Now let's step back, and this time we'll look at the straight line $\overleftrightarrow{OH}$ (covering up angles ∠4 and ∠3) and notice that **m∠1 + m∠2 = 180.** Now we can do a little algebra to finish our proof:

. . . .

* We also have the option of writing an equals sign between the two *measures* of the angles: m∠ABC = m∠DEF.

† Like angles, segments, and later—shapes!

‡ In this case, vertical has nothing to do with "up and down," and there's no such thing as "horizontal angles." Weird, huh? Just remember the "V" shape that vertical angles make.

We've got a system of (underlined) equations, which we can now solve with substitution.* Rewriting the first equation as $m\angle 2 = 180 - m\angle 3$, we'll substitute that for $m\angle 2$ into the second underlined equation from p. 28, and we get:

$$m\angle 1 + \textbf{m}\angle\textbf{2} = 180 \quad \text{(the 2nd underlined equation}$$
$$\text{on p. 28)}$$

$$\rightarrow m\angle 1 + \textbf{(180} - \textbf{m}\angle\textbf{3)} = 180 \quad \text{(the substitution)}$$

$$\rightarrow m\angle 1 + 180 - m\angle 3 = 180 \quad \text{(dropped parentheses)}$$

$$\rightarrow m\angle 1 - m\angle 3 = 0 \quad \text{(subtracted 180 from both sides)}$$

$$\rightarrow m\angle 1 = m\angle 3$$

Well, how about that! We just totally proved that vertical angles have the same measurement, $m\angle 1 = m\angle 3$, which of course means that the angles themselves are <u>congruent</u>: $\angle\textbf{1} \cong \angle\textbf{3}$. Ta-da!

Can you figure out how to set up this same proof to show that $\angle 2 \cong \angle 4$? You bet you can! *(Hint: First cover up $\angle 4$ and $\angle 3$ to "see" some supplementary angles and get an equation, and then cover up $\angle 2$ and $\angle 3$.)* Give it a shot right now. It'll make your brain sturdy and sharp. I'm so proud of you.

Because there was nothing particularly special about our diagram—it truly represented any 'ol intersecting lines—we have indeed proven that **all vertical angles are congruent**. This is what's called a "paragraph proof."

What's It Called?

Vertical angles are the angles opposite each other, created from two lines (or segments or rays) intersecting like chopsticks. They make the shape of two big V's! And as we just proved, *vertical angles are always congruent.*

If $\angle$'s are vertical, then the $\angle$'s are $\cong$.[†]

Read this out loud as, "If angles are vertical, then the angles are congruent."

. . . .

* To brush up on solving systems of equations, including the "evil twins" method, see Chapter 12 in *Hot X: Algebra Exposed.*

† It totally annoys me when people overuse apostrophes, like in signs: *We Sell Banner's!* I want to say, "You sell banner's *what?* Banner's ink?" That being said, I use an apostrophe after the angle symbol because otherwise, $\angle$s looks like it could mean "angle S."

QUICK NOTE   If we pick any two angles that are created from two lines crossing, they'll either be congruent or supplementary. Kind of interesting, huh? Noticing this kind of thing will really help with diagrams and proofs later on.

Doodle Time!

As we get into working with diagrams, you'll see that congruency is kind of a big deal, and it's really helpful to be able to mark which things (segments or angles) are congruent to each other. Yep, we're going to be doodling!

The most common way to mark congruent angles (or segments) is with single, double, or triple tick marks, but we could use lots of things: circles, hearts, flowers, etc. I sorta like the double and triple tick marks because they look like kitty scratches.

The two triangles below on the right are called **congruent triangles,** because as you can see, corresponding parts all have the same measurement—they are mirror images of each other (and could totally impersonate each other to snag some jewels). We'll learn more about them in Chapter 8.

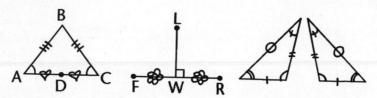

The following exercises will use stuff from the entire chapter, so flip back if you need refreshers!

"I love being smart and having knowledge: just knowing 'stuff.' I look at my math homework sometimes and am amazed I know what all 'that' is." Brenda, 15

Doing the Math

Use the diagram below on the right to answer these questions. Some will have more than one correct answer. I'll do the first one for you.

1. Find all pairs of congruent angles and state their congruence.

<u>Working out the solution</u>: Hmm, can we find congruent angles in this diagram? Yep! Those single tick marks mean that ∠R ≅ ∠T (there are a few names for these angles). But since we have chopsticks (intersecting segments), we have two pairs of vertical angles: ∠CAB ≅ ∠RAT and those wide ones, ∠CAR ≅ ∠BAT. (Notice that ∠A would be too vague, so we can't use it.)

Answer: ∠R ≅ ∠T, ∠CAB ≅ ∠RAT and ∠CAR ≅ ∠BAT

2. Find all pairs of congruent segments and state their congruence.

3. Give all possible names for the right angle in this diagram, *without* naming the points Q or T.

4. Name all the angles that appear to be obtuse *without* naming the point Q.

5. Name the two straight angles without naming the point Q.

6. How many angles appear to be acute?

7. Are the points B, A, and Q collinear?

8. Is the point A *between* the points B and Q?

9. Name the two perpendicular segments *without* naming the points A or Q.

10. What's the only other possible name for the ray $\overrightarrow{QA}$?

11. What are all other possible names for the angle ∠BAT? (*Hint: There are three others.*)

12. Name all pairs of supplementary angles *without* naming the point Q.

13. Find the pair of complementary angles. *(Hint: The degrees in a triangle always add up to 180°. You can do this!)*

(Answers on p. 396)

We'll do more with diagrams in the next chapter. Nice job!

Takeaway Tips

 Two things are congruent when their measurements are equal.

 The *vertex* of an angle is the "cold armpit" of the angle. Remember Texas!

 Right angles measure 90°; their sides are perpendicular (⊥).

 Two complementary ∠'s add up to 90°, and two supplementary ∠'s add up to 180°.

 Vertical ∠'s are the opposing V's made from crossed chopsticks and are always congruent.

TESTIMONIAL

Sara Reistad-Long

(New York, NY)
Before: Struggled to succeed in math
Now: Fashion, health, and lifestyle
journalist!

> "I wasn't going to let math stand in my way..."

I grew up in a very small town in Washington State. We were barely on the map. Even when I was a little girl, I loved the glossy pages of magazines and the stories of faraway places they told. At around age 11, I remember ordering my first "grown-up" magazine subscription—*Elle*—and imagining what it might be like to work at a place like that. My dreams of being a career journalist followed me to high school, and I knew if I wanted to succeed, I'd need to get into a great college. That meant I needed a strong GPA and high SAT scores, which included math. And while words always came pretty easily for me, math was another story.

I remember just sitting there, staring at my math homework and thinking, "This is a sink-or-swim moment. I can't give up. I'm going to beat this!" Time and time again, I had to dig down and come up with ways to make it interesting and find new ways to explain the concepts to myself. I also felt pressure *not* to succeed—even to dumb myself down. (No matter how old you are, it's always easier to seem "cool" by not caring.) But I couldn't let math—or my own insecurities—stand in my way. Although sometimes it felt like an uphill battle, I'm happy to say that all my hard work paid off! With my high scores and GPA, I got into Harvard University, which sure enough helped me get my first job...yep, at *Elle* magazine. And in fact, math is what got me my first regular column in the magazine: a money column! I was tasked with making finance relatable and interesting to a smart, fashionable audience.

Over the years, I've written for other fashion and beauty magazines such as *Glamour* and *Allure*, as well as big newspapers including the *New York Times* and the *Wall Street Journal*. And I love my job. Besides the fact that I've gone to some pretty cool celebrity parties, the work itself is always a fun challenge. For example, I might be assigned a story on how to get the *best* deals on beauty discount websites and then I'd go and figure out the actual numbers behind the discounts. Or if I'm told to write a story predicting next year's trends of healthy "superfoods"—like blueberries or carob—then I would read all the studies I could get my hands on to understand the research behind the health claims (and use deductive reasoning to extract the *real* story from the often-conflicting studies!) and also to estimate the probability of such trends starting, based on the statistics of past diet trends. Trends are all about math—percents, data charts, statistics, etc.—and *everything* we do in journalism is based on trends.

In a way, the more "fun" your desired profession is, the more important math and science are in your education. It shows employers that you're serious and well-rounded, and it can give you a real edge over others who want the same job. And of course, a solid background in math gives you more options if you change your mind. Following our dreams often takes some serious time and effort, but those are rewarding in their own way and make us feel good about ourselves. You can do *anything* you set your mind to!

Chapter 3

First Impressions

Working with Diagrams

As you've probably learned many times before, things—and people—aren't always what they seem. The cute guy who seems perfect might turn out to not respect you and totally not be worth your time. The quiet girl in the back of the room who never makes a peep might be an amazing singer, worthy of Broadway! If we're smart, we won't make too many assumptions based on appearances; they can get us into big trouble. (See pp. 44–5 for some stories.)

This is true in geometry, too! When we're working with diagrams, things may not always be what they seem. Before I show you what I mean, let's define some terms we'll be using in this chapter:

What Are They Called?

Plane, Coplanar, Parallel

A **plane** is a flat surface with no thickness, extending infinitely in two dimensions. (We can't draw that part, but just imagine that the plane *P* keeps spreading out.) If any two points on that surface are connected by a line, then all the points on that line must also be on the plane, because it's flat, after all!

Two lines (or points or segments, etc.) are called **coplanar** if they lie in the same plane. In our diagram, the lines *a* and *b* are coplanar, but *b* and *l* are not coplanar. Makes sense, right?

Two lines are called **parallel** if they are *coplanar* and they <u>never</u> intersect each other.* Parallel lines will often be marked with single or double arrows somewhere in the middle of them, and sometimes we'll be told lines are parallel with the symbol ∥, which looks like, well, tiny parallel lines!

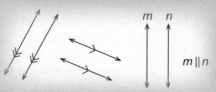

Above: three ways we can mark two parallel lines

"*I like geometry because I feel like I'm learning the secrets of how the real world works. It seems like the most 'useful' kind of math—people use it to actually build things!*" **Genevieve, 15**

What Are They Called?

Midpoint, Perpendicular Bisector, Angle Bisector

The **midpoint** of a segment is the point that divides the segment into two equal parts, because it's in the exact middle of the segment. On the next page, the point M is the **midpoint** of the segment $\overline{AY}$, and we can tell because $\overline{AM}$ and $\overline{MY}$ are marked as being congruent; in other words: $\overline{AM} \cong \overline{MY}$.

. . . .

* It's possible to have two lines that never intersect that also aren't parallel: They're called **skew lines,** and they live in three dimensions, not just on a 2-D plane. For example, on p. 35, we might be told that *b* and *l* are *skew lines*. Look at the "line" made from a wall meeting the floor. Now hold a nail file so it's vertical, and imagine both lines continue forever. Those are *skew lines*, too!

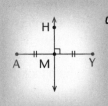

The **perpendicular bisector** of a segment is the line that is both *perpendicular* to the segment and that also passes through the segment's **midpoint**, so that it *bisects* that segment.* Here, $\overleftrightarrow{HM}$ is the perpendicular bisector of the segment $\overline{AY}$. We know because that little square tells us $\overleftrightarrow{HM} \perp \overline{AY}$ (the perpendicular part), and the kitty scratch marks tell us that $\overline{AM} \cong \overline{MY}$ (the bisector part).

 If a line (or seg) is a $\perp$ bisector, then it is $\perp$ to the seg and passes through the seg's midpoint.

 Alt: If a line (or seg) is a $\perp$ bisector, then it is $\perp$ to the seg and divides the seg into two $\cong$ seg's.

The **angle bisector** is the ray that divides an angle into two equal angles. Below, the ray $\overrightarrow{DI}$ is the angle bisector of $\angle WDE$, and the way we know is because the congruent kitty scratch marks tell us that $\angle WDI \cong \angle IDE$. We could say, "$\overrightarrow{DI}$ bisects $\angle WDE$."

 If a ray is an $\angle$ bisector, then it divides the $\angle$ into two $\cong$ $\angle$'s.

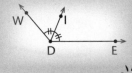

Watch Out!

In the diagram below, we *cannot* say that O is the midpoint of $\overline{NT}$, even though $\overline{NO} \cong \overline{OT}$. Why? Because O isn't even on $\overline{NT}$! In fact, $\overline{NT}$ would actually be the *straight* segment connecting N and T. ($\overline{NT}$ is not drawn, but we can imagine it; we'd end up making a triangle!) They like to keep us on our toes, don't they?

....

* See p. 389 in the Appendix for the Perpendicular Line Postulate.

Three-Way

We can divide things into <u>three</u> equal parts, too.

The points *R* and *E* **trisect** the segment $\overline{TS}$: They divide it into three smaller, congruent segments. In other words: $\overline{TR} \cong \overline{RE} \cong \overline{ES}$. And just like you might expect, two rays can **trisect** an angle, meaning that the two rays divide the angle into three equal parts. Here, the rays $\overrightarrow{SR}$ and $\overrightarrow{SO}$ divide the big angle $\angle TSI$ into three smaller, congruent angles. In other words: $\angle TSR \cong \angle RSO \cong \angle OSI$.

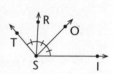

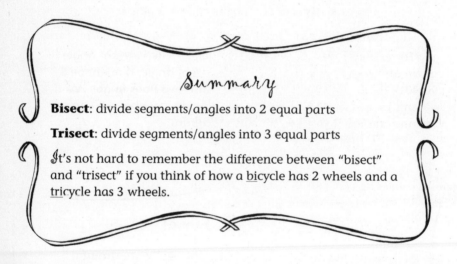

Summary

Bisect: divide segments/angles into 2 equal parts

Trisect: divide segments/angles into 3 equal parts

It's not hard to remember the difference between "bisect" and "trisect" if you think of how a <u>bi</u>cycle has 2 wheels and a <u>tri</u>cycle has 3 wheels.

QUICK NOTE   Lines (and rays) can't have midpoints or trisection points . . . because they're infinitely long!

"When someone tells you a rumor, it's best not to assume anything or jump to conclusions. There's a good chance the rumor's not even true, and it could cost you a friend." **Jenn, 15**

Looks Can Be Deceiving

As you'll see in the stories on pp. 44–5, it can be dangerous to make assumptions. When we work with diagrams, looks can be deceiving; something that seems "clear" from the picture might not be true at all! Check out the little arrow segments to the right. Would you believe that the center segment of *a* is actually longer than the center segment of *b*? It's true. Get out a ruler and see for yourself.

a.

b. >———<

And believe it or not, textbooks are allowed to draw their diagrams totally out of proportion, so even if we used a ruler to measure segments or a protractor* to measure angles, we really couldn't depend on the information we got.

For example, just because an angle *looks* like a right angle doesn't mean anything. Unless we're *told* it's a right angle, it might be 89.95° or 90.001°, for all we know! And just because two lines might *look* parallel doesn't mean they are. Here's a little guide for what we can typically assume (and not!) when looking at a diagram.

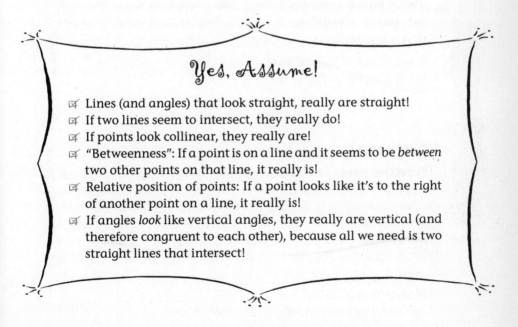

Yes, Assume!

- ☑ Lines (and angles) that look straight, really are straight!
- ☑ If two lines seem to intersect, they really do!
- ☑ If points look collinear, they really are!
- ☑ "Betweenness": If a point is on a line and it seems to be *between* two other points on that line, it really is!
- ☑ Relative position of points: If a point looks like it's to the right of another point on a line, it really is!
- ☑ If angles *look* like vertical angles, they really are vertical (and therefore congruent to each other), because all we need is two straight lines that intersect!

* See p. 387 for more on protractors.

Don't Assume Unless We Are TOLD!

- ☑ Just because an angle *looks* like a right angle, does NOT mean it is!
- ☑ Just because two segments (or angles) look congruent, does NOT mean they are!
- ☑ Just because one segment (or angle) looks bigger than another one, does NOT mean it is!
- ☑ Just because two lines look parallel, does NOT mean they are!
- ☑ Just because one segment (or angle) looks bigger than another, does NOT mean it is!
- ☑ There are tons more things we can't assume from looking at a diagram. Above are some common ones. Basically, if it isn't on the "Yes, Assume!" list, then *don't* assume it.

QUICK NOTE   We use a teeny, tiny triangle to mean "the triangle," so △ABC can be read as "the triangle ABC." Notice that we always use the three vertex points in the name of the triangle, and it doesn't matter what order we put them in.* (We'll formally define triangles in Chapter 6.)

Little Miss Perfect . . . or Not?

Is there a girl in your class who seems just perfect? She always seems to have the perfect clothes, the perfect house, and the perfect thing to say? You might assume her life *feels* perfect, too, but people often have insecurities and problems they'd never admit, and deep down, they feel more like you than you realize!† It's never a good idea to *assume* . . .

Take a look at this diagram of what *seems* to be a "perfect dress." Okay, so it's really a couple of triangles, but we can use our imagination, right?

....

* Unless we're talking about congruent or similar triangles—then the order *does* matter. More on that on p. 120 and p. 286.

† The truth? *None* of us has it all figured out. We just do our best, so let's be nice to ourselves, okay? See p. 78 for more on so-called perfect girls.

Here's a partial list of what we can (and can't!) assume from looking at it. Read carefully and make sure you understand each one.

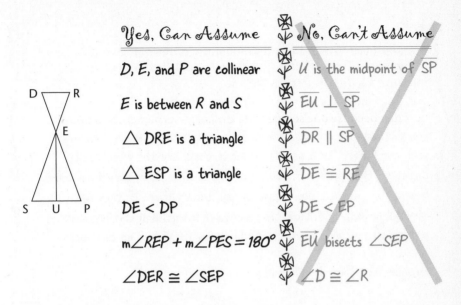

Yes, Can Assume	No, Can't Assume
D, E, and P are collinear	U is the midpoint of $\overline{SP}$
E is between R and S	$\overline{EU} \perp \overline{SP}$
△ DRE is a triangle	$\overline{DR} \parallel \overline{SP}$
△ ESP is a triangle	$\overline{DE} \cong \overline{RE}$
DE < DP	DE < EP
m∠REP + m∠PES = 180°	$\overrightarrow{EU}$ bisects ∠SEP
∠DER ≅ ∠SEP	∠D ≅ ∠R

Now, we might learn that the things in the "No, Can't Assume" column are true after we get more information. But at this point, we certainly mustn't *assume* they are.

It takes some focus to find all those letters and make sure you follow what each thing means, but it's worth it! It'll make you an expert diagram navigator in no time.

What's the Deal?
Parallel . . . Segments?

I know you're lying awake at night wondering things like, "Wait. Two segments could be coplanar and never intersect—and still not be parallel! Then how can we possibly define *parallel* for segments and/or rays?"

Never fear; there is a way to define *parallel* for segments and rays. We say that two segments are parallel if the lines they are *contained within* are parallel. So for any two segments, imagine if we extended them into lines. If the lines are parallel, then the segments are parallel. If the lines aren't parallel, then their segments aren't parallel. Done!

 Doing the Math

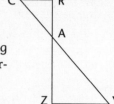

Based on the CRAZY diagram, label the following statements "True," "False," or "Not Enough Information." I'll do the first one for you.

1. $\angle ACR \cong \angle AYZ$

Working out the solution: First, let's locate these angles. Tracing with our finger (or a pencil), it seems these are the two "pointy" angles at the upper left and lower right. Hmm, they sure seem congruent, but we just don't have enough information. If they were vertical angles, we'd know they were congruent, and if they were supplementary, we'd know they add up to 180°, but no such luck!

Answer: Not Enough Information

2. R, A, and Z are collinear.

3. $CY > RZ$

4. $m\angle CAR + m\angle RAY = 180°$

5. R, A, and Y are collinear.

6. $RA > RZ$

7. $\overline{CR} \parallel \overline{ZY}$

8. $m\angle CAY = 180°$

9. $m\angle RZY = 90°$

10. $\angle CAR \cong \angle ZAY$

(Answers on p. 396)

"Drawing" Conclusions . . . Literally!

Okay, so you have this great friend who's always been right there by your side. Let's call her Tamara. Then one day this other girl, Brittany, tells you that Tamara wants to date your boyfriend, Zac. You look over right then and see Tamara talking to Zac.

Whoa, head rush. Alright, let's not jump to conclusions. What do we actually know? Despite how it looks, here are the only facts: 1. Tamara is talking to Zac. 2. Brittany *said* Tamara wants to date Zac.

Instead of ruining a lifelong friendship, you might spend a few seconds coming up with another scenario that *also satisfies the facts*. Like, for example, maybe Brittany just saw Tamara talking to Zac and jumped to conclusions herself! Same facts, different possible conclusion.

Something similar works in geometry: A great way to separate what we *know* about a diagram from what we are *assuming* about it is by drawing a second diagram that still satisfies all the facts we've been given. For example, say we're given $\angle KSI \cong \angle ISR \cong \angle RST$, with the SKIRT diagram to the right.

It might seem like those little segments along the bottom should be congruent, $\overline{KI} \cong \overline{IR} \cong \overline{RT}$, but we can't assume it! Here's a diagram that we could draw ourselves, which satisfies the Givens (and everything we can assume from the diagram). Now it's pretty clear that those little segments aren't necessarily congruent just because the angles are!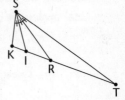

To see more of these examples and try "drawing your own conclusions" yourself, check out the free worksheet at GirlsGetCurves.com!

Takeaway Tips

 Parallel lines are coplanar lines that never intersect.

 Segments and angles can be bisected (divided into two equal parts) or trisected (divided into three equal parts).

A perpendicular bisector bisects a segment and is also $\perp$ to that segment.

Looks can be deceiving! When working with diagrams, we need to pay attention to the difference between what we *know* and what we are tempted to assume.

Danica's Diary
CONFESSIONAL: ASSUMPTIONS GONE WRONG!

*L*earning how to make conclusions based on actual facts, and not assumptions, is important in more than just geometry. Sure, it's a skill that good lawyers use every day when crafting a good argument, but it's a life-savvy skill that we all need. For example, when you're asked to sign a contract of any sort—whether it's for your first credit card application, to open a bank account, or even for an appearance on television—being able to see what's actually on the page can be crucially important so that you don't end up getting taken advantage of! Believe me, most companies will try to get away with whatever they can, because they figure most people don't take the time to read the small print, thinking, "Oh, I'm sure this bank doesn't charge fees for a free checking account," or "I'm sure the production company wouldn't use embarrassing footage that I didn't approve of," etc.

I've personally avoided both of these situations, and many more, thanks to the skill of seeing what's really on the page and not making assumptions. Did I mention that assumptions can be devastating in social settings, too? You know what can happen when we "assume": It makes an "ass" out of "u" and "me"!

Has there ever been a time when you made a fool of yourself because you made an assumption you shouldn't have? This is what you said!

"*I* heard that my boyfriend was cheating on me, so I went up to him and BLEW UP. He was like, 'What the heck are you talking about?' Turns out, it was just a rumor started by a girl who doesn't like me, and I just assumed it was true!" —Amanda, 16

"*W*hen I was in the 8th grade, we had a test coming up in world history that I assumed would be easy, because all the other tests had been easy. So I didn't study, and I failed it. That totally sucked." —Megan, 14

"*Last* year, I walked up to my friends and they all stopped talking. This kept happening; I assumed they were talking behind my back, and I felt really hurt. But it turns out they were planning a surprise birthday party for me all along!" —Wendy, 16

"*If* your friends are fighting, you shouldn't jump to conclusions about who's right or wrong; it's best to stay neutral." —Allie, 15

"*I* once judged somebody by her looks, and I assumed we wouldn't have chemistry. But after I got to know her, I liked her so much that I asked her out, and now we've been together for two years." —Dale, 16

"*I* remember one time there was this girl in my gym class and she was really quiet and wore all black. Everyone—including me—made fun of her. A couple weeks later my friends and I got in a fight and we weren't on speaking terms, so I ended up being all by myself. The next day that girl came and sat by me. She turned out to be the nicest person I'd ever met, and I felt like a fool for assuming who she was based on appearances." —Darcy, 14

"*D*on't assume other people don't like you! There is a girl who had always seemed very distant toward me, and she recently told me that she always thought I hated *her*! Now we're great friends." —Bailey, 17

The next time you have a strong emotional reaction to a situation, try sitting down and writing out the "Givens"—the things you actually know for a *fact*. Then write down the things you're *assuming* but don't know for sure, and finally write down the conclusion that's been upsetting you. You might find that you've been reacting to things that aren't necessarily real!

And always remember this: We may not be able to control the world around us, but *our reactions are a choice*. When something happens that might make us feel upset, we can choose to stay really mad about it, or we can decide to take a breath, calmly look at the facts, and look for a different perspective on the matter. It takes practice, for sure, but realizing that we have a *choice* in how we feel can be very empowering.

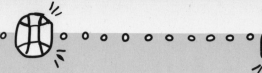

QUIZ:
What's Your Body Image?

*T*ake this quiz by expert psychologist Dr. Robyn Landow and contributor Bailey Thompson, and see how you fare!

1. When you pass a mirror, you usually:
- **a.** Avoid it like the plague. You'd rather not think about how you look because it'll just depress you.
- **b.** Sneak a look just long enough to remember to stand up straight and suck in your stomach.
- **c.** Stop to admire yourself. You're lookin' goood . . .

2. You've always wanted to try scuba diving, and a certified scuba instructor is visiting your school next month for free lessons! The only catch? You'll have to wear a skintight wet suit, and your body's not exactly in ideal shape. You've got four weeks. You:
- **a.** Decide this is the extra motivation you've needed to start waking up early for a 20-minute jog every morning. It'll feel great!
- **b.** Freak out. You eat almost nothing for two weeks, trying to lose weight, then (because you're starving) pig out on sweets and end up putting on even more weight. Same old pattern.
- **c.** Who the heck cares how your wet suit fits?

3. You're walking down the school's hall when you notice a very pretty girl. Your initial feeling is:
- **a.** Envy/Admiration. You pledge to practice better health habits, like only drinking water (no more soda!) so that your skin can clear up and get that "glow."
- **b.** Confidence. Sure, she may be gorgeous in her own way, but she's no match for you.
- **c.** Depression/Jealousy. You immediately compare yourself to her, noting all the reasons why you aren't as beautiful and never could be.

4. Your general attitude when it comes to your body is this:
- **a.** You're constantly dissatisfied with how you look but feel powerless to change things.
- **b.** You do your best to stay fit and healthy. Feeling good and looking good often go hand-in-hand, after all!

c. You don't really exercise much, and you usually eat fast food because it's so easy.

5. When you put on your favorite jeans, you notice a "muffin top" happening. You:
   a. Throw on a dress that hides the extra fat. Hmm, maybe it's time to cut down on the greasy fries. After all, you'd like those jeans to be flattering again!
   b. Break down in tears and cancel your plans. You just feel too fat to be seen in public.
   c. Go out as is. Who cares if you've got a muffin top? Lots of people do.

6. Your best friend just asked you to join the local tennis club. You:
   a. Accept the offer instantly. There's nothing bad about a little extra exercise, and learning a new sport is definitely fun!
   b. Turn her down. You already get plenty of exercise.
   c. Decline immediately. You just don't like the idea of everyone watching you run around. I mean, what are the chances that there won't be jiggling in some part of your body? How embarrassing!

7. A few friends take you to an all-you-can-eat buffet for dinner. There's a huge variety of healthy and not-so-healthy fare. You:
   a. Fill your plate with healthy foods that will make you feel energized, not overly full.
   b. Keep your plate small. There are guys around here, and you certainly wouldn't want anyone to think you're fat.
   c. Dig in, and pile on the fried shrimp! One night of eating like this can't hurt anyone, right?

8. One of your girlfriends checks her weight nearly every day. While on the scale, she explains how good it feels and suggests that you check your weight often, too. You:
   a. Check every day, too. It feels good to see the number go down, but it's a huge panic whenever it goes up.
   b. Have no interest in your weight; it's just a number. What matters is how you look and feel.
   c. Check every few weeks, just because you're curious.

9. A guy friend you've been crushing on invites you to the beach. You:
   a. Say you're busy. You don't want to be seen in a bathing suit but don't want to have to explain your reasons.
   b. Don't want to miss the chance to hang out, but you suggest a different spot, hoping he'll go for it!

c. Accept his offer. You just got a new bikini anyway, so you might as well try it out. If he doesn't like who you are—all of you—then it's not meant to be anyway.

10. You find yourself flipping through the latest style magazine, looking at the models. What do you think of them?
   a. They are the ultimate in perfection, and you wish you looked like them.
   b. They are unrealistic. You know all of the photos are totally air-brushed, and you remind yourself that real is more beautiful than the vacant-eyed ditzes they seem to be.
   c. Some of them look unhealthy, but you secretly wish you looked like them anyway.

Scoring:

1. a = 4; b = 2, c = 1	4. a = 4; b = 2; c = 0	7. a = 2; b = 4; c = 1
2. a = 2; b = 4; c = 0	5. a = 1; b = 4; c = 0	8. a = 4; b = 1; c = 2
3. a = 2; b = 0; c = 4	6. a = 2; b = 0; c = 4	9. a = 4; b = 3; c = 1
		10. a = 4; b = 2; c = 3

If you scored 6–15:

You are confident and non-apologetic. You're either in great shape and you know it, or you're not in great shape, and you're totally okay with it. Either way, you accept the way you look and don't feel you should change for anyone. It's great to be that confident . . . but remember, it's important to put some effort into health and fitness! Especially if you scored close to 6, you might want to ask yourself where your "I don't care" attitude is coming from. It can be easy to look at all the unhealthy habits of people around us and allow our own expectations to be lowered. Yikes! It's your job (and no one else's) to take care of your body, and that means making healthy food choices wherever possible, getting at least some light exercise (try yoga!), and drinking plenty of water. While it's extremely dangerous to obsess over appearance, complacency isn't much better. There's a huge difference between having an eating disorder and simply working hard to be the healthiest we can be. Caring about how we look is actually a positive thing! It's all about balance.

If you scored 16–29:

You're neither obsessed with your body, nor are you complacent: You care about how you look, but sometimes you feel insecure, and that's totally normal! Most of us want to put our best foot forward, whether for our first day of school, a night out with the girls, or heck, just checking ourselves out in the mirror. We want to be proud of how we come across in this world and feel great at the same time. And as you know, a much better sign of your body's fitness than some number on a scale is how you look in clothes—and especially how you feel (light and free, or sluggish?). You don't think your body is "perfect" (whatever that means, anyway!), but you don't feel you need to be, either. There's a big difference between a healthy desire to be fit and an unhealthy fixation on what you perceive to be imperfections. It can be easy—for all of us—to fall prey to the advertisements full of "perfect" models that surround us 24/7, whether in magazines, on billboards, or on TV. (See p. 17 for more on that.) Just stay the course: Healthy food/beverage/exercise choices will keep you energized and fit, and those great habits will guide you toward a lifetime of confidence and happiness!

If you scored 30–40:

Geez, girl, go easy on yourself! Nobody's perfect, after all. Everybody, and I mean everybody, has a thing or two she wishes were different about her face or body, or feet, or hair. (*Danica says: See p. 236 for mine!*) But it sounds like your body may have become an unhealthy obsession for you, and it's time to seek out help. Please talk to a guidance counselor, the nurse at school, or another trusted, objective adult. The sooner you can start changing your thinking patterns, the better off you'll be. Also, check out pp. 339–41 for a spotlight feature on a teenager who is struggling with these very same issues. Remember, you are not alone!

Chapter 4

Why, Mommy? Why?

Introduction to Two-Column Proofs

You know how little kids love to ask the question "Why?" but somehow never get satisfied? Does this sound familiar from, say, 10 years ago or so? Imagine you, all dressed for school with your little lunch box, and your mom, scooting you toward the door.

"Hurry up and get in the car." "Why?" "We don't want to be late for school." "Why?" "Because going to school makes you smart, and it's important to be smart." "Why?" "Because if you're smart, you can have a good career someday, which you might want." "Why?" "Because having a good career can help you be a positive member of society, which is important." "Why?" "Being a positive member of society will lead to, you know, inner peace." "Why?" "WILL YOU JUST GET IN THE CAR ALREADY!!"

As far as the little kid is concerned, there's no end; there will always be another "Why?" to ask. I imagine that moms find this sort of thing a little bit cute and also pretty annoying. It doesn't matter how many answers she gives, there's always another "Why?" The mom has to justify every single thing she says!

If you've ever babysat a little kid, you've probably experienced this sort of thing yourself. When you're doing a geometry proof, imagine you're babysitting a little kid and that you can't do anything without an adorable little voice saying "Why?" The great part about proofs though, is that there IS an end to them. You will prevail!

Now let's talk for a moment about Rules . . . and becoming Queen.

☺

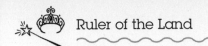

Math is like a puzzle—a big game with Rules. Yes, I like to call them *Rules* with a capital "R." And these Rules will make up most of the links in the chain belts we'll build for our proofs, just like we did in Chapter 1.

The more Rules we know, the more powerful we are, because we can prove more things! Like if you became Queen, wouldn't you want to know all the Rules of the land? That way the citizens would respect you more, and then you would have more power.*

Do we know any Rules yet? Sure! Every definition we've seen is a Rule. And definitions are an especially handy type of Rule because their converses are always true.

For example, the definition of *angle bisector* is "a ray that divides an angle into two equal parts." Here's how that looks as an "if . . . then" Rule:

"*If* a ray divides an angle into two equal parts, *then* it is the **angle bisector** of that angle."

And because this is a definition, its converse is also true:

"*If* a ray is the **angle bisector** of an angle, *then* it divides the angle into two equal parts."

This is true for all definitions! For more on this nifty fact, see pp. 66–7.

Besides **definitions**, any other Rule we come across will probably be labeled as a **property, postulate,** or **theorem**. There are differences among these types of Rules, but nothing we need to worry about at the moment.† So for now, let's lump them all together, call them "Rules," and be done with it.

Here's what I want you to remember about Rules: They can all be written in "if . . . then" form, so they can all be used for chains of logic during proofs. I'll mark Rules‡ with this Queenly little crown.

Do we know any Rules besides some definitions? Yep! Remember the chopsticks from pp. 28–9 and how vertical angles are always congruent to each other? That gave us this Rule: *If* two angles are vertical, *then* the angles are congruent. Abbreviated, this looks like:

 If two ∠'s are vertical, *then* the ∠'s are ≅.

. . . .

* I haven't spent any time thinking about what it would be like to become Queen. None at all.

† See p. 389 in the Appendix for more on this.

‡ Even though some definitions won't get this little crown, they can be used as Rules, too.

Notice that the converse of that Rule *isn't* true. It would be: **If two ∠'s are ≅, *then* the ∠'s are vertical.** Lots of congruent angles aren't vertical, after all!

Here's another one we know, whose converse *is* true:

 If two lines are ⊥, *then* they create right ∠'s.

 If two lines create a right ∠, *then* they are ⊥.

And here's one of my favorite Rules of all time . . .

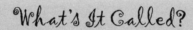

What's It Called?

The Transitive Property

This is something you might have seen before, but with *equal signs*: If $a = b$ and $b = c$, then $a = c$. The fact that $a = b$ and $b = c$ actually "forces" a and c to be equal, too. See what I mean? And the same type of logic applies to congruent segments (or angles). So we get these Rules:

 Transitive Property for Segments: If two segments are ≅ to the same segment, then they are ≅ to each other.

 Transitive Property for Angles: If two ∠'s are ≅ to the same ∠, then they are ≅ to each other.

For example, looking at the angles in the sparkly diamonds below, if we are told that ∠1 ≅ ∠2, and we're also told that ∠2 ≅ ∠3, can you figure out what new piece of information we can now conclude?

You guessed it! Because of the Transitive Property, knowing that $\angle 1 \cong \angle 2$ and $\angle 2 \cong \angle 3$ means we can conclude that $\angle 1 \cong \angle 3$.

Here's another way to look at it:

$$a \underset{\cong}{\overset{b}{}} c$$

It's like we have a circle of congruency with one "link" left empty, and we get to fill in the gap! In fact, you might come across the four-element version of this property, too, which can be looked at like this:

$$a \cong d$$
$$b \cong c$$

See, if two *congruent* things (*a* and *b*) are congruent to two *other* things (*d* and *c*), then those two *other* things must also be congruent! This is really just the regular Transitive Property applied twice, so we don't need to make a big deal out of it.

We're ready for the next phase of Queen training . . .

Proving Yourself: Baby Steps Toward Proofs

So how do we *use* these Rules in proofs? Here's how it will usually go:

Diagram & Givens + Rules → Conclusion

We see the diagram, we are provided with some Givens, and we already know some Rules. We apply those Rules on the diagram and the Givens, and we come up with a new conclusion. It's just like the chain belts from Chapter 1!

But Why, Mommy, Why? The Two-Column Proof

Here's how proofs look: In the left-hand column, we write true Statements about our diagram—specific facts we are given and facts we can conclude. The last Statement is always the thing we wanted to prove, and some people like to add a little ∴ right before it.* Nothing says "this is the final statement" like three little dots, right? Hmm.

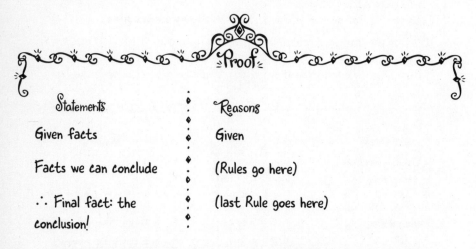

Proof

Statements	Reasons
Given facts	Given
Facts we can conclude	(Rules go here)
∴ Final fact: the conclusion!	(last Rule goes here)

And in the right-hand column, we put our Reasons; this is where we justify each Statement we made, and we use Rules to do it.

Just imagine that a little kid is watching you write the proof, and she keeps saying "Why?" after each Statement you make. In an attempt to satisfy her, you calmly provide a Reason, and you put these "Reasons" in the right-hand column.

QUICK NOTE   Everything we write in the Reason column will be either "Given" or an "if . . . then" Rule.†

....

* Some people call ∴ the "therefore" symbol.
† You'll see later on (pp. 122–3) that this won't always be true. But for now, it is!

Remember our pink-wearing gals from Chapter 1? Let's do the proof from p. 11. (I'll include a third column just to make things easier.)

Here are the "Rules" we had:

 If Taylor wears pink, then Lizzie wears pink.

 If Lizzie wears pink, then Jessica wears pink.

 If Jessica wears pink, then Brittany wears pink.

And here's the problem:

Given: Taylor wears pink. Prove: Brittany wears pink.

Let's do it!

Proof

Statements (Boldly declare something!)	Here's WHY we can make these Statements →	Reasons
1. Taylor wears pink.	(But WHY? How do we know?)	**1. Given** (Oh, right; we were told this in the problem!)
2. Lizzie wears pink.	(Why can we be so bold and say that?)	**2. (because) If Taylor wears pink, then Lizzie wears pink.** (Oh, I see why we could say that!)
3. Jessica wears pink.	(How do we know that Jessica wears pink?)	**3. (because) If Lizzie wears pink** (which we now know), **then Jessica wears pink.** (This is one of our Rules!)
4. ∴ Brittany wears pink.	(But why, Mommy, why?)	**4. (because) If Jessica wears pink, then Brittany wears pink.** (And that's the thing we wanted to prove!)

This can be a lot to digest! Try reading this proof like a book: left to right and then down to the next row. So you'd read, "Taylor wears pink. *But why? How do we know?* Given. *Oh, right; we were told this in the problem.* Lizzie wears pink. *Why can we be so bold and say that? Because* if Taylor wears pink, then Lizzie wears pink. *Oh, I see why we could say that!* Jessica wears pink . . ." etc.

QUICK NOTE   In the example on p. 55, I used creative underlining styles to show which phrases match up with other phrases in both columns. If it helps, notice that in the right-hand column, if we could circle the parts with matching underlines, it would be the same circling we did on p. 11 to create an airtight chain belt.

> In a two-column proof, each Rule's "if" part matches previous Statements, and each Rule's "then" part matches the current Statement.

Don't worry about memorizing the thing in the black box; just look back at the pink-wearing-gals proof, and *notice* how it's true on each line. This is worth your time, I promise!

Here's the step-by-step method, which works whether we're talking about fashion trends or, you know, geometry.

☺

Step By Step

Two-Column Proofs:

Step 1. Look at the givens and the thing we want to prove. See if you can figure out a game plan: Are there any helpful things from the diagram we can assume (like vertical angles)? What Rules (definitions, theorems, etc.) do we know that might help us get to our desired conclusion? Jot them down!

Step 2. Set up the Statements and Reasons columns. Fill in the first line(s) with what we're given, and write "Given" for the Reason(s).

Step 3. If you can see what the next Statement and Reason should be, great! Fill 'em in, making sure the Reason answers the question, "But WHY can we say that?" If you're <u>not</u> sure how to proceed, consider the "if . . . then" Rules that you know. Which one has a matching "if" part to one of our *previous Statements* and will bring you closer to the thing you want to prove? <u>The "then" part of this Rule will be the Statement.</u> Fill in that Statement and Rule (Reason). Use creative underlines to show the matching parts and guarantee that the logic is airtight.

Step 4. Repeat Step 3 until the new Statement is the thing we wanted to prove. Done!

Here's a simplified way to look at the two-column format:

The Two Columns

Statements
It's the progression of the Statements leading up to the conclusion: the thing we want to prove.

Reasons
These are all the Reasons—the answers to "Why, Mommy, why?" at every step of the way.

Let's see how this proof business works in a real live geometry example.

CHAPTER 4: WHY, MOMMY? WHY?   57

Let's do a proof using the Transitive Property and our favorite kind of take-out math (chopsticks)!

Given: $\angle 1 \cong \angle 3$. **Prove:** $\angle 2 \cong \angle 3$.

Let's jump in!

Step 1. Okay, here's our strategy: Because chopstick angles are congruent ($\angle 1 \cong \angle 2$), and we're given that $\angle 1 \cong \angle 3$, we can take these two "ingredients" and apply the Transitive Property to show $\angle 2 \cong \angle 3$. So that's our game plan. What Rules will we use? The ones about vertical angles and transitivity, of course! **If two $\angle$'s are vertical, then the $\angle$'s are $\cong$**, and the **Transitive Property** for angles (we'll write it in "if . . . then" form with the actual angles themselves).

Step 2. Let's set up the columns! (See the next page.) Our first Statement is just our Given, and we can write "Given" for its Reason. Nice. Another thing we're given (because we can assume it from looking at the diagram) is that $\angle 1$ and $\angle 2$ are vertical angles.

Step 3. Next, we can use our "**If two $\angle$'s are vertical, then the $\angle$'s are $\cong$**" Rule, which takes us from the Statement "$\angle 1$ and $\angle 2$ are vertical angles" to the Statement "$\angle 1 \cong \angle 2$." See what I mean? The Rule takes us from the "if" part to the "then" part.

Step 4. Now that our two ingredients, $\angle 1 \cong \angle 2$ and $\angle 1 \cong \angle 3$, are established, the next Rule we use will be the Transitive Property to reach our final conclusion: $\angle 2 \cong \angle 3$. That Rule (with our specific angles) is: **If $\angle 1 \cong \angle 2$ and $\angle 1 \cong \angle 3$, then $\angle 2 \cong \angle 3$.** Nice.

Following is the actual two-column proof, with a center column added to help keep our brains on. Oh, and be sure to check out the underlines—the "if" parts always match up with the previous Statements, and the "then" parts match up with the current Statements, so we know our logic is airtight!

"*Logic is great for outsmarting siblings.*" Soraya, 16

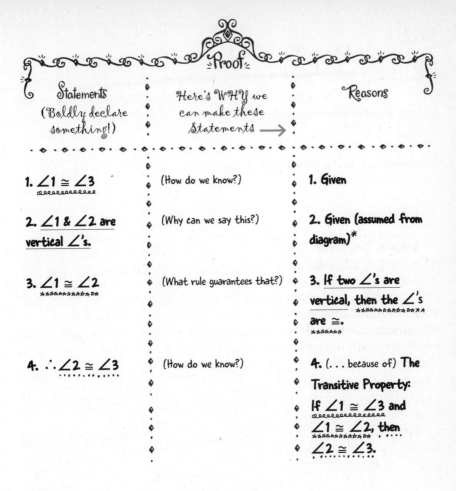

Proof

Statements (Boldly declare something!)	Here's WHY we can make these Statements →	Reasons
1. $\angle 1 \cong \angle 3$	(How do we know?)	1. Given
2. $\angle 1$ & $\angle 2$ are vertical $\angle$'s.	(Why can we say this?)	2. Given (assumed from diagram)*
3. $\angle 1 \cong \angle 2$	(What rule guarantees that?)	3. If two $\angle$'s are vertical, then the $\angle$'s are $\cong$.
4. $\therefore \angle 2 \cong \angle 3$	(How do we know?)	4. (... because of) The Transitive Property: If $\angle 1 \cong \angle 3$ and $\angle 1 \cong \angle 2$, then $\angle 2 \cong \angle 3$.

And because $\angle 2 \cong \angle 3$ is what we wanted to prove, we're done!

    Read that proof out loud a few times, and let it sink in. Reading it out loud will help keep your brain from wandering. And don't worry if you wouldn't have known how to create it. If you already knew this stuff, you wouldn't need to be in school, would you? Just follow the logic and let it sink in. You're doing awesome!

....
* Advanced footnote: Eventually we'll see that this step is optional (see footnote on p. 129), but for now we'll definitely include it!

Notice that I've been using the symbols ≅ for "congruent" and ∠ for "angle." Here is a (partial) list of easy ways to cut down on the writing during proofs. Sounds good to me!

Congruent: ≅	Angle: ∠
Segment: **seg**	Parallel: ‖
Perpendicular: ⊥	Triangle: △
Complementary: **comp.**	Supplementary: **supp.**

We don't have to use these, or we can choose to only use some. It's up to us!*

Mini-Proofs: Proofs That Seem Like They Don't Need Proving

Let's step back for a moment and do some super-short proofs, sometimes called "mini-proofs." These tiny proofs can be maddening. "Why do I need to prove this? It's obvious! I don't know how to prove something so obvious." (Actually, you will know how in just a moment.)

Here's the key: After the "Given" line, these mini-proofs usually only require a single definition. Those are the only "Reasons"! Look for a key word in the problem (either in the Given or the conclusion we're supposed to prove) to tell us *which* definition might be helpful. That definition should link us from the Given to the conclusion. And that's really it! This'll make more sense in an example.

Watch Out!

Because "if . . . then" definitions can be written in either order (because their converses are always true), when we use them as Rules in proofs, it can be easy to write the definition in the wrong order! We have to make sure the "if" part matches with a *previous* Statement in our proof, and the "then" part matches the *current* Statement, and we'll be golden.

....

* Every teacher (and textbook) has a different opinion on how much stuff needs to be written out, so check with yours to make sure you don't get any points taken off.

Take Two: Another Example

Given: E is the midpoint of $\overline{GM}$.

Prove: $\overline{GE} \cong \overline{EM}$.

$\overset{\textstyle\bullet\;\;\;\;\;\bullet\;\;\;\;\;\bullet}{\text{G}\quad\;\text{E}\quad\;\text{M}}$

We might be thinking, "Wait, how do we prove this? That's like the same thing as the Given!" But . . . it's really not the *same* Statement, is it? After all, one involves the word *midpoint*, and one involves the congruent symbol, $\cong$. It looks like we need a Rule to link the two Statements together. Let's follow the steps:

Step 1. This seems like a mini-proof, so what definition might help us? How about the definition of a midpoint? **If a point divides a segment into two $\cong$ segments, *then* the point is the midpoint.** Let's see if this game plan works.

Step 2. We fill in the "Given" line; no problem.

Step 3. We want to use the definition of *midpoint,* but we know that the "if" part of our Rule must match a previous Statement in our proof, right? The only previous Statement is, "*E* is the midpoint of $\overline{GM}$." That means the "if" part better have the word *midpoint* in it. Oops! We need to make an adjustment to the game plan: It looks like we'll need the *converse* of the definition of *midpoint,* which is: **If a point is a midpoint, *then* it divides a segment into two $\cong$ segments.** And not only does the "if" part match a previous Statement, but the "then" part matches the final thing we want to prove. Fantastic! Let's underline the matching parts. . .

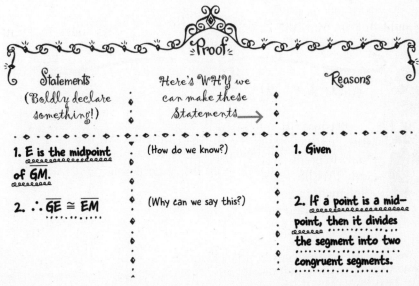

Proof

Statements (Boldly declare something!)	Here's WHY we can make these Statements →	Reasons
1. E is the midpoint of $\overline{GM}$.	(How do we know?)	1. Given
2. $\therefore \overline{GE} \cong \overline{EM}$	(Why can we say this?)	2. If a point is a midpoint, then it divides the segment into two congruent segments.

Ta-da!

I know it can feel sort of pointless to do these, because they're so short. But truthfully, these mini-proofs are great practice for longer proofs because they train us how to correctly link one statement to the next, in an airtight way. We'll see that longer proofs often require one or more small links that don't seem to need to be stated, but without them, we'd lose our airtight "if . . . then" chain of logic.

I mean, if you wanted to learn how to make chain belts, you'd start out by learning how to make just *one link* first, right?

QUICK NOTE   When we use creative underlines to keep track of the chain of logic, notice that every "if . . . then" Rule will use at least <u>two</u> styles of underlines, every time! Also, the Reasons "Given" and "Assumed from diagram" <u>don't</u> need to be underlined.

FYI, I've been putting a center column in these proofs to help show you how the logical flow works, but you probably won't see that extra column anywhere else. So from this point on, most of the time I won't include it—and neither should you!

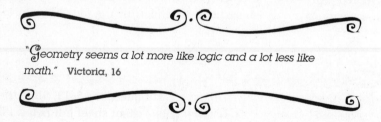

"*Geometry seems a lot more like logic and a lot less like math.*" Victoria, 16

Here's a partial list of Rules that we can use as a "cheat sheet" for the exercises below. You haven't been Queen for very long, so a little extra help is acceptable, don't you think?

Rules Cheat Sheet

Vertical Angles: **If two ∠'s are vertical,** *then* **the ∠'s are ≅.**

Right angle definition: **If an ∠ measures 90°, then it is a right ∠.**

Right angle definition (converse): **If an ∠ is a right ∠, then it measures 90°.**

Perpendicular definition: **If two lines (or segments) create a right ∠, then they are ⊥.**

Perpendicular definition (converse): **If two lines (or segments) are ⊥, then they create right ∠'s.**

Angle bisector definition: **If a ray divides an ∠ into two ≅ ∠'s, then the ray is the ∠ bisector.**

Angle bisector definition (converse): **If a ray is an ∠ bisector, then it divides the ∠ into two ≅ ∠'s.**

Midpoint definition: **If a point divides a segment into two ≅ segments,** *then* **the point is the midpoint.**

Midpoint definition (converse): **If a point is a midpoint,** *then* **it divides a segment into two ≅ segments.**

Transitive Property: We should write **"Transitive Property"** *and fill in the following statement with specific angles or segments for a, b, c during proofs: If a ≅ b and b ≅ c, then a ≅ c.*

Doing the Math

For questions 1–4, do the mini-proofs. For questions 5–13, fill in the gaps for the longer proofs. Use the above cheat sheet if it helps. I'll do the first one for you!

1. Given: ∠A = 90°. Prove: ∠A is a right angle.

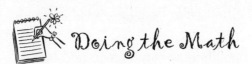

Working out the solution: It can be tempting to say, "But there's nothing to prove. That's the definition of a right angle!" And that's a good thing, because it's exactly the *Reason* we'll use in the proof: the definition of a right angle!

Answer: (see proof below)

Statements | Reasons

1. ∠A = 90°
   <u>ₒₒₒₒₒₒₒₒₒₒₒₒₒₒₒₒ</u>

1. Given

2. ∴ ∠A is a right angle.
   <u>★★★★★★★★★★★★★★★★★★★★★★</u>

2. If an angle measures 90°, then it
   <u>ₒₒₒₒₒₒₒₒₒₒₒₒₒₒₒₒₒₒₒₒₒₒₒₒₒₒₒₒₒₒ</u> <u>★★★★★★★★</u>
   is a right angle.
   <u>★★★★★★★★★★★★★★★★★★★★★★</u>

2. See HOP. Given: $\overline{HO} \cong \overline{OP}$. Prove: O is the midpoint of $\overline{HP}$.

3. See SKIP. Given: $\overrightarrow{SI}$ is the angle bisector of ∠KSP. Prove: ∠KSI ≅ ∠ISP.

4. Use the JUMP diagram below. Given: $\overline{JM} \perp \overline{UP}$. Prove: ∠JMP is a right angle.

Questions 5–8: Complete the fill-in-the-blank proof, and include the underlines.

(Ignore the Givens from #4)

Given: $\overline{JM} \cong \overline{UM}$, and M is the midpoint of $\overline{UP}$.
Prove: $\overline{JM} \cong \overline{MP}$.

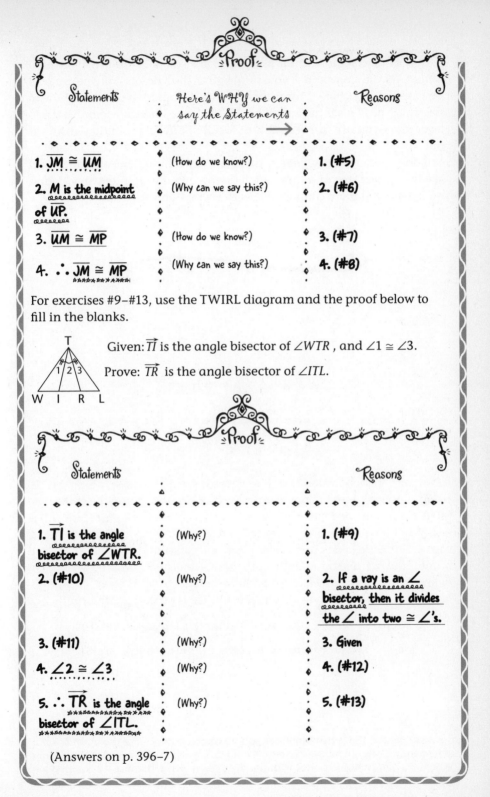

Proof

Statements	Here's WHY we can say the Statements →	Reasons
1. $\overline{JM} \cong \overline{UM}$	(How do we know?)	1. (#5)
2. M is the midpoint of $\overline{UP}$.	(Why can we say this?)	2. (#6)
3. $\overline{UM} \cong \overline{MP}$	(How do we know?)	3. (#7)
4. ∴ $\overline{JM} \cong \overline{MP}$	(Why can we say this?)	4. (#8)

For exercises #9–#13, use the TWIRL diagram and the proof below to fill in the blanks.

Given: $\overrightarrow{TI}$ is the angle bisector of ∠WTR , and ∠1 ≅ ∠3.

Prove: $\overrightarrow{TR}$ is the angle bisector of ∠ITL.

Proof

Statements		Reasons
1. $\overrightarrow{TI}$ is the angle bisector of ∠WTR.	(Why?)	1. (#9)
2. (#10)	(Why?)	2. If a ray is an ∠ bisector, then it divides the ∠ into two ≅ ∠'s.
3. (#11)	(Why?)	3. Given
4. ∠2 ≅ ∠3	(Why?)	4. (#12)
5. ∴ $\overrightarrow{TR}$ is the angle bisector of ∠ITL.	(Why?)	5. (#13)

(Answers on p. 396–7)

Proofs take some getting used to. Don't worry if you feel frustrated at first! In the beginning, you might find that proofs fall into one of two categories: either, "This is too obvious. I would have no idea how to *prove* this," or "This is too complicated. I have no idea how to prove this!" Never fear—you are fully capable of tackling any 'ol proof they throw at you.

Look for key words, match them up with the Rules you know, and take things one step at a time. We'll do more in the chapters ahead, and you'll get much more comfortable with them, I promise!

What's the Deal?
Why Should the Converses of Definitions Always Be True?

In Chapter 1, we learned that just because an "if . . . then" statement is true, it doesn't mean its converse is also true. So why should the converses of all definitions be true? Well, what is a definition in the first place? It's really just a *name* we give to something so it's easier to talk about. Let's say we think erasable purple pens are so special that they need their own name, and we decide to call them "purplemeisters." There, we've just defined the word *purplemeister*. So let's write it in "if . . . then" form and also write down its converse, which you'll see is automatically true!

If a pen is purple and erasable, *then* it is a **purplemeister**.

If a pen is a **purplemeister**, *then* it is purple and erasable.*

. . . .

* And of course, those two sentences can be combined to create the biconditional statement: "A pen is a purplemeister. ⟺ A pen is purple and erasable." But during proofs, it's more helpful to stick with the "if . . . then" form, because it's easier to follow the chain of logic.

And the same thing is true for *any* geometry term we've defined, like **right angle**:

> *If* an angle measures 90°, *then* it is a **right angle**.

> *If* an angle is a **right angle**, *then* it measures 90°.

See what I mean? Because we've defined the term, there's no getting around the fact that the converse must be true! We'll use definitions constantly in proofs, and now you totally understand why their converses are true. Nice.

Takeaway Tips

 Proofs: Diagram & Givens + Rules → Conclusion

 In a two-column proof, the Reasons answer the question "But Why?" for each Statement. Think of your babysitting job!

 In a two-column proof, for every "Reason," we'll write "Given" or an "if . . . then" Rule.

 Every Reason's "if" part should match with Statements from a *previous* line, and its "then" part should match the Statement on the *same* line. We can use creative underlines to help keep track!

 Every definition's *converse* is true and can be used as an "if . . . then" Rule.

Mini-proofs teach us how to go from one statement to another in an airtight way. Doing these "single links" will totally help us build longer proofs later on.

Danica's Diary
LOGIC AND LISTENING TO YOUR GUT

In geometry proofs, logic rules the day. In
life, logic is an amazing tool, but we also
need to pay attention to something that often
seems to defy logic: our gut instincts.

As we get older, we begin to face big
decisions—things that will affect the quality
of our lives as young women, and even for the rest of
our lives! These decisions range from the college
we choose to the guys we date. It becomes more and
more important for us to see how our instincts can
guide us through big and (seemingly) small decisions,
and to recognize the danger of *wanting* to want things.

Often, we "want to want" things in order to make
life easier. Maybe your parents want you to go to a
particular college, and you know it would make them
happy and proud . . . so maybe you *want* to want to go
there, but you don't actually *want* to.

Or maybe you're about to get into an elevator, but
there's a really strange man inside who's giving you
the creeps. It would be weird and rude for you to not
get in the elevator with him, right? So even though you
don't want to get into the elevator, you *want* to want to
get in.* Logic might say, "Oh, what are the chances
that he's dangerous? It would just be easier to get in."
But your instincts may have picked up something your
brain hadn't noticed—his nervous breathing pattern or
the way he looked at you when the doors opened. That
inner voice actually guides us all and protects us from
bad decisions, and we'd be wise to listen to it! When we
"want to want" things, we're fighting that voice and
ignoring our instincts that keep us true to ourselves.

My second serious boyfriend was a guy named Jeff.
Okay, not really, but let's just use that name. Things
were so great—for awhile. He was smart and fun and
gorgeous and treated me great. But after awhile, I
started feeling that things weren't quite right. I was

. . . .

* In this situation, make any excuse you have to, but do not get into the elevator! He
could easily pull the "stop" button and trap you inside. We have instincts for a
reason.

totally in love, and I didn't want to believe that he wasn't "the one." But my gut *knew* something was wrong. As it turns out, he was taking illegal drugs—yikes! I mean, I never saw him take the drugs; he never did it around me, but his behavior was just becoming . . . unpredictable and moody. I couldn't believe it: This guy I was so crazy about was doing something seriously stupid to himself. Even then, I was tempted to stick it out with him, because he told me he was trying to get sober. And he *was* trying, just not very successfully. It's funny; near the end of our relationship, I would make sure never to ride in a car with him just in case he wasn't sober, and yet I believed he'd be better any day. I was in some sort of denial, believing his words more than his actions: I was *wanting* to believe him, even though deep inside I knew otherwise.

I remember going on a field trip for my oceanography class at UCLA, and as I was riding in a boat, jumping across the choppy waters with the wind in my hair, I suddenly felt so free and light—a feeling I hadn't had since I found out about Jeff's drug use. I knew then that I was "wanting to want" to be with Jeff, which is different from actually *wanting* to be with him. I realized that I was trying to force something that wasn't meant to be. Shortly after that, I ended things. It was hard. I cried a lot in the following few weeks, but it was the right thing, and I knew it. My gut told me so.

Our guts are smart, and often "see" more than we can with our conscious minds. We should listen to their warnings! If you want to understand more about how your instincts can keep you safe (and how to tell the difference between instincts and paranoid fears), check out the book *The Gift of Fear* by Gavin DeBecker. It's pretty awesome.

Chapter 5

Lounge Chairs and Leftover Pizza

The Addition, Subtraction, Multiplication, & Division Properties . . . and More Proofs!

As proofs get more involved, there will be more than one Rule used and more than one chain link created . . . which means we'll need more Rules! Shall we sit by the pool as we learn them?

The Addition Properties

Sitting by the pool in chairs is nice, but it's even better when we can put our legs up, don't you agree? Have you ever seen an ottoman footstool, which can extend the length of a chair, sort of creating a "lounge chair"? Let's say that you and your BFF are sitting in identical chairs but you only have one ottoman, so you're sharing it (pushed up against the chairs). Then your new "lounge chairs" would also have identical lengths to them, right? You would have just as much room for your legs as your BFF would. So that's nice and fair.

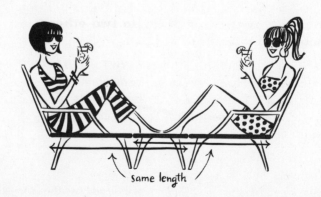

same length

In geometry, there's a very similar Rule: If we have two segments we know are congruent, and we add the *same segment* to both of them, we'll end up with two congruent (longer) segments.

 If a segment is added to two ≅ segments, *then* the sums* are ≅.

In BOYS, Given $\overline{BO} \cong \overline{YS}$, we can conclude: $\overline{BY} \cong \overline{OS}$. That little middle segment adds to both of them, and that means those overlapping longer segments must be congruent, just like our lounge chairs, right?

And of course, if we add two congruent segments to *two other* congruent segments, we end up with two (longer) congruent segments.

 If two ≅ segments are added to two other ≅ segments, *then* the sums are ≅.

In OFTEN, Given $\overline{OF} \cong \overline{EN}$ and $\overline{FT} \cong \overline{NT}$, we can conclude: $\overline{OT} \cong \overline{ET}$. It's not hard to see that when the pieces are all congruent, the longer segments must also be congruent.

The same Rules apply to angles:

 If an angle is added to two ≅ angles, *then* the sums are ≅.

In FLIRT, *Given* ∠FTL ≅ ∠ITR, we can conclude: ∠FTI ≅ ∠LTR. We're just adding that middle angle to both, and so the overlapping "medium-sized" angles must be congruent!

 If two ≅ angles are added to two other ≅ angles, then the sums are ≅.

In WITH, *Given:* ∠TWH ≅ ∠THW and ∠IWT ≅ ∠IHT *(these are marked as congruent angles on the diagram), we can conclude:* ∠IWH ≅ ∠IHW. Yep! If the two sets of smaller angles are congruent, then the bigger angles they make up should also be congruent.

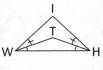

. . . .
* In case you forgot, a **sum** is what we get when we add two things together.

Now imagine the lounge chair situation in reverse: The BFFs are still using the same ottoman, but this time we're only told that the two big "lounge chairs" are congruent. Well, when we take away the shared ottoman, we know the two short chairs *must* also be congruent, right? Sure sounds like subtraction to me!

The Subtraction Properties

  If a segment is subtracted from two ≅ segments, then the differences* are ≅.

In *CUTE*, Given $\overline{CT} \cong \overline{UE}$, we can conclude $\overline{CU} \cong \overline{TE}$. We're just subtracting $\overline{UT}$ from each of the longer, congruent segments, after all! Imagine we're told that those longer, overlapping segments both measure 7 cm. ($CT = UE = 7$ cm), and that $UT = 2$ cm. I bet you can see how CU and TE would each <u>have</u> to be 5 cm. So yep, we'd get congruent segments.

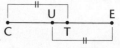

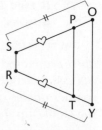 **If two ≅ segments are subtracted from two other ≅ segments, then the differences are ≅.**

In *SPORTY*, Given $\overline{SO} \cong \overline{RY}$ (kitty scratches) and $\overline{SP} \cong \overline{RT}$ (hearts), we can conclude that $\overline{PO} \cong \overline{TY}$. Subtracting the "hearts" from the "kitty scratches," it makes sense that the little leftover segments $\overline{PO}$ and $\overline{TY}$ would *have* to be congruent.

. . . .

* In case you forgot, a **difference** is what we get when we subtract one thing from another.

If an angle is subtracted from two ≅ angles, then the differences are ≅.

In SMART, Given ∠STA ≅ ∠MTR, we can conclude that ∠STM ≅ ∠ATR. After all, what happens when we subtract that middle angle, ∠MTA, from each of the two big, overlapping congruent angles, ∠STA and ∠MTR? We get two smaller congruent angles: ∠STM and ∠ATR.

 If two ≅ angles are subtracted from two other ≅ angles, then the differences are ≅.

In GIRLS, Given ∠GIL ≅ ∠ILS and ∠RIL ≅ ∠RLI, we can conclude that ∠GIR ≅ ∠RLS. Here, we want to subtract the two congruent angles *inside* the triangle, ∠RIL and ∠RLI, from those two big, wide congruent angles, ∠GIL and ∠ILS, respectively. What do we get? We get two *smaller* outer angles, ∠GIR and ∠RLS. Can you see how they must also be congruent?

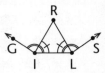

 QUICK NOTE   If we're asked to find a particular value and we don't know how to do it, we can start by asking ourselves, "Given what we know, what else CAN we conclude?" And often that answer will lead us much closer to the final answer we're looking for.

"*I know the benefits of being smart, but I think the biggest reason that I want to be smart is because I have blond hair. My friends kid around with me about me being blond, especially when I have a moment where I can't remember something really trivial. I want to be smart so I can help to completely destroy the 'blonde' stereotype.*" **Ashley, 13**

Let's practice some of this stuff!

![Doing the Math notepad icon] *Doing the Math*

Use the addition and subtraction properties for segments and angles to answer the following questions. I'll do the first one for you.

1. Given: $\angle GIL \cong \angle ILS$, $\angle RIL \cong \angle RLI$, $m\angle GIR = (9x - 2)°$, and $m\angle RLS = (11x - 18)°$. Find $m\angle GIR$.*

<u>Working out the solution:</u> Hmm, this seems like a monster problem! But never fear, we have logic on our side. Because $\angle GIL \cong \angle ILS$ and $\angle RIL \cong \angle RLI$, we can just subtract two congruent things from two other congruent things, and get two new congruent things: $\angle GIR \cong \angle RLS$, which is another way of saying $m\angle GIR = m\angle RLS$. Great! We've been told that $m\angle GIR = (9x - 2)°$, and $m\angle RLS = (11x - 18)°$. Now that we know they are equal to each other, we can say: $(9x - 2)° = (11x - 18)°$. It's pretty clear this means the stuff *inside* the parentheses has to be equal: $9x - 2 = 11x - 18$. Subtracting $9x$ from both sides, we get: $-2 = 2x - 18$, and adding 18 to both sides, we get: $16 = 2x \rightarrow x = 8$. But this isn't the final answer! The problem asked for $m\angle GIR$, which we know equals $(9x - 2)°$, right? So we can just plug in x's value: $m\angle GIR = (9x - 2)° = (9[8] - 2)° = (72 - 2)° = 70°$.

Answer: $m\angle GIR = 70°$

2. Use the SMART diagram. Given: $\angle STA \cong \angle MTR$, $m\angle ATR = 23°$, $m\angle MTA = 81°$. Find $m\angle STR$.

....

* We'll use the little *m*'s for these exercises, just to make sure you can be flexible!

3. Use the SPORTY diagram. Given: $\overline{SP} \cong \overline{RT}$, $\overline{SO} \cong \overline{RY}$, $PO = 3x$ cm, $TY = (x + 4)$ cm. Find TY.

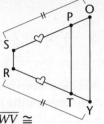

4. Use the WAVE diagram to the right. Given: $\overline{WV} \cong \overline{AE}$. Which other segments can we say are congruent?

5. Part a. In the "ice cream" diagram to the right, if $\angle IRE \cong \angle ERM$ and $\overrightarrow{RE}$ bisects $\angle CRA$ (meaning that $\angle CRE \cong \angle ARE$), name the other two sets of angles at the vertex R (that's 4 angles total) that must be congruent.

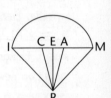

Part b. If $m\angle IRA = 65°$, find $m\angle MRC$.

(Answers on p. 397)

Now let's do some proofs using these properties! They might be minis. Remember, it's all about creating links between each step in an airtight way. Use the steps on p. 57 if it helps.

QUICK (REMINDER) NOTE   Each Rule in the Reasons column links **from** one (or more) of the previous Statements **to** the current Statement.

Doing the Math

Do the following two-column proofs, using the addition/subtraction Rules from pp. 71–3 and referring to the list of Rules from p. 63. I'll do the first one for you.

1. In SPARKLY, Given: $\angle SYA \cong \angle LYR$. Prove: $\angle PYA \cong \angle RYK$.

<u>Working out the solution</u>: First, let's find these angles on the diagram. Got 'em? Good. So, we know those bigger angles are congruent, and they have the ones we're supposed to prove *inside* them. Sure seems like a subtracting angles problem! Hmm. We'd need $\angle SYP$ and $\angle LYK$ to be congruent to each other, because if we subtracted those off the big angles, then we could show that $\angle PYA \cong \angle RYK$. Make sense? But is it even true that $\angle SYP \cong \angle LYK$? Yep, I see chopsticks (vertical angles), so they *have* to be congruent! Sounds like we have a game plan: Vertical angles means congruent angles ($\angle SYP \cong \angle LYK$), and then we subtract those off of $\angle SYA$ and $\angle LYR$ to conclude $\angle PYA \cong \angle RYK$. Let's do it, and we'll underline matching parts to make sure our logic is airtight.

Statements	Reasons
1. $\angle SYA \cong \angle LYR$	1. Given
2. $\angle SYP$ and $\angle LYK$ are vertical angles.	2. Given (assumed from diagram)*
3. $\angle SYP \cong \angle LYK$	3. If two $\angle$'s are vertical, then the $\angle$'s are $\cong$.
4. $\therefore \angle PYA \cong \angle RYK$	4. Subtraction Property: If two $\cong$ $\angle$'s ($\angle SYP \cong \angle LYK$) are subtracted from two $\cong$ $\angle$'s ($\angle SYA \cong \angle LYR$), then the differences are $\cong$.†

. . . .

* Any step whose reason is "Given (assumed from diagram)" can be skipped, but including it can make things clearer while we're still getting used to proofs! (See the footnote on p. 129 for more.)

† Depending on your teacher, you might eventually be allowed to simply say "Subtraction Property."

2. Mini-proof! In EARING, Given: $\angle 1 \cong \angle 3$ and $\angle 2 \cong \angle 4$. Prove: $\angle EAR \cong \angle ING$.

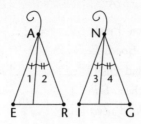

3. Mini-proof! In DARK, Given: $\overline{DR} \cong \overline{AK}$. Prove: $\overline{DA} \cong \overline{RK}$.

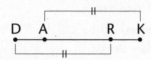

For #4–7, fill in the blanks for this proof using BLUE. Given: L is the midpoint of $\overline{BU}$ and U is the midpoint of $\overline{LE}$. Prove: $\overline{BU} \cong \overline{LE}$.

B   L   U   E

Proof

Statements	Reasons
1. L is the midpoint of BU.	1. Given
2. $\overline{BL} \cong \overline{LU}$	2. (#4)
3. U is the midpoint of LE.	3. Given
4. (#5)	4. If a point is a midpoint, then it divides the seg into two $\cong$ seg's.
5. $\overline{BL} \cong \overline{UE}$	5. (#6)
6. $\therefore \overline{BU} \cong \overline{LE}$	6. (#7)

(Answers to #4–7 on p. 397; proof solutions at GirlsGetCurves.com)

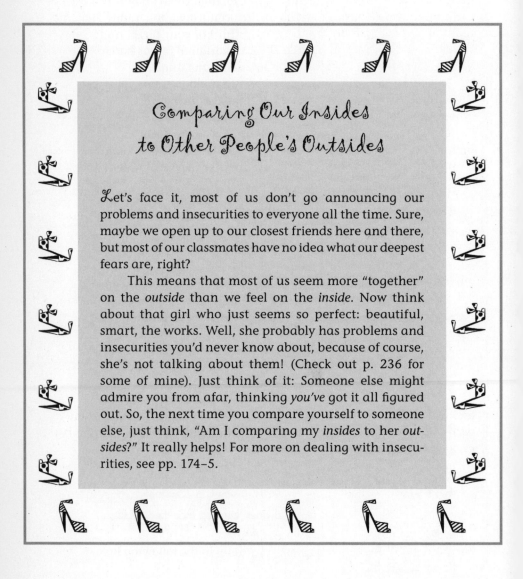

Comparing Our Insides to Other People's Outsides

Let's face it, most of us don't go announcing our problems and insecurities to everyone all the time. Sure, maybe we open up to our closest friends here and there, but most of our classmates have no idea what our deepest fears are, right?

This means that most of us seem more "together" on the *outside* than we feel on the *inside*. Now think about that girl who just seems so perfect: beautiful, smart, the works. Well, she probably has problems and insecurities you'd never know about, because of course, she's not talking about them! (Check out p. 236 for some of mine). Just think of it: Someone else might admire you from afar, thinking *you've* got it all figured out. So, the next time you compare yourself to someone else, just think, "Am I comparing my *insides* to her *outsides*?" It really helps! For more on dealing with insecurities, see pp. 174–5.

Leftover Pizza . . . and the Multiplication Properties

Imagine that you and your sister go to different pizza restaurants and both come home with leftover pizza—exactly half a pizza each. If those half-pizzas are congruent (identical in size), then you'd know for sure that the original whole pizzas must have been congruent, too, right? The next night, if you each brought home only a third of your pizza, and if those thirds were congruent, you'd *also* know the original pizzas had to be congruent. The same thing works with segments and angles!

In OMG YEA, let's say that M and E are midpoints on their segments, which means they bisect their segments; in other words, they divide their segments in half.

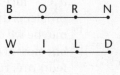

Now let's say we know that $\overline{OM} \cong \overline{YE}$. Those are each half of their respective segments, and since the halves are congruent, the entire segments must be congruent, too—just like the pizza! And that means $\overline{OG} \cong \overline{YA}$.

By similar logic, if in BORN WILD on the right we're told that O, R, I, and L are trisection points, and we're also told that $\overline{BO} \cong \overline{WI}$ (the first little part of each segment—these are like the pizza thirds), we can totally conclude that $\overline{BN} \cong \overline{WD}$. The long segments are three times the lengths of the smaller congruent segments.

The same is true for angles (which actually look much more like pizza).* So in the "THINK ABOUT" diagram below, if we are given that $\angle 1 \cong \angle 4$, and also that $\overrightarrow{TI}$, $\overrightarrow{TN}$, $\overrightarrow{AO}$, and $\overrightarrow{AU}$ are all angle trisectors (that means they divide their big angles into three equal parts), then we know that the two *big* angles, $\angle HTK$ and $\angle BAT$, must be congruent. They're just three times the little congruent angles $\angle 1$ and $\angle 4$, after all!

And below in the MATH MORE diagram, if we're told that the rays $\overrightarrow{HA}$ and $\overrightarrow{MR}$ are angle bisectors (so they cut their big angles in halves), and if we're also given that $\angle 8 \cong \angle 9$, then we can totally conclude that the big angles must be congruent: $\angle MHT \cong \angle OME$.

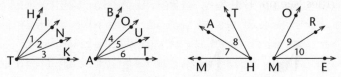

Pant, pant! All of this can be summed up in the following Rule:

 Multiplication Property: If segments (or angles) are $\cong$, then their like multiples are $\cong$.

....

* Some New York City pizza restaurants sell enormous "slices" that we might very well cut into thirds or fourths and still not finish. Picture that for these angle examples!

The Division Properties

The leftover pizza logic works in reverse, too: If you know that you and your sister started out with congruent pizzas, then if you each cut your pizzas in half, all your half-pizzas would have to be congruent to each other, right? So in the MATH MORE diagram on p. 79, if we're instead told that $\angle MHT \cong \angle OME$ and that $\overrightarrow{HA}$ and $\overrightarrow{MR}$ are angle bisectors, then we know that the smaller angles are all <u>half</u> the size of whatever the angle measure is for the bigger (congruent) angles, right? And that means all those little half-angles are congruent to each other, just like the pizza! In other words: $\angle 7 \cong \angle 8 \cong \angle 9 \cong \angle 10$.

And I bet you can see where this is going for the angle trisectors, too: In the THINK ABOUT diagram on p. 79, if we know that the rays $\overrightarrow{TI}$, $\overrightarrow{TN}$, $\overrightarrow{AO}$, and $\overrightarrow{AU}$ are all angle trisectors, then they divide the big angles into three equal parts each, right? And so if we're told those big angles, $\angle HTK$ and $\angle BAT$, are congruent, then all the little ones must be congruent to each other, too. In other words: $\angle 1 \cong \angle 2 \cong \angle 3 \cong \angle 4 \cong \angle 5 \cong \angle 6$.

The same goes for segments. In OMG YEA on p. 79, if we know that $\overline{OG} \cong \overline{YA}$ and that M and E are midpoints, then we can conclude that all the halves are congruent: $\overline{OM} \cong \overline{MG} \cong \overline{YE} \cong \overline{EA}$. And in BORN WILD, if we're told that O, R, I, and L are trisection points and that $\overline{BN} \cong \overline{WD}$, then we can conclude that all those little thirds are also congruent: $\overline{BO} \cong \overline{OR} \cong \overline{RN} \cong \overline{WI} \cong \overline{IL} \cong \overline{LD}$.

Here's how we sum up all this division stuff:

 Division Property: If segments (or angles) are $\cong$, then their like divisions are $\cong$.

QUICK NOTE Why do we have to say "*like* multiples" and "*like* divisions"? Because that way, we know the pizzas were cut into equal slices. For example, if you cut your pizza into halves and your sister cut hers into thirds, and if one of your slices is congruent to one of hers, your "whole" pizzas wouldn't be congruent at all! They'd have to be *like divisions* of the pizzas.* Got it? Good!

....

* By the way, these Multiplication and Division Properties work for the same reason that if we know $x = y$, then we could conclude things like $3x = 3y$ and $\dfrac{x}{2} = \dfrac{y}{2}$. Multiples and divisions . . .

Bonus Rules—for Supplementary and Complementary Angles

Here's one more set of Rules, just for angles. If we have two angles that are supplementary to two congruent angles, then the other set of angles must be congruent! The same goes for complementary angles. Here are the actual Rules:

 If ∠'s are supp. to ≅ ∠'s, then they are ≅.

 If ∠'s are comp. to ≅ ∠'s, then they are ≅.*

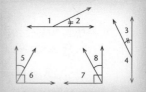

For example, in the diagram to the right, if we're given that ∠2 ≅ ∠3, then since they are supplementary to ∠1 and ∠4, respectively, do you see how we can totally conclude that ∠1 ≅ ∠4? Just imagine that ∠2 and ∠3 each equal 30°. Then ∠1 and ∠4 would have to each be 150°! Similarly, if we know that ∠5 ≅ ∠8, we can conclude that ∠6 ≅ ∠7 (notice the right angles). They're *forced* into being congruent.

Makes sense, right?

Notice that since we've been told that ∠2 ≅ ∠3, if we have a statement involving ∠2, we can write the *same statement* using ∠3 in its place. So in the diagram, we can say, "∠1 is supplementary to ∠2." By sticking ∠3 in its place, we can now say, "∠1 is supplementary to ∠3."

In fact, we've just used another property, called the Substitution Property.

 The Substitution Property: If two angles[†] are ≅ and we have a statement involving the measure of one ∠, we can replace it with the other ∠.

Nice.

---

* We can substitute "**to ≅ ∠'s**" with "**to the same ∠**" in both statements and get more true Rules!

† Instead of angles, we could also use this for segments, triangles, or really, any geometric object!

Brain Boosting Afternoon Snacks

What we eat affects not only how we look and feel, but also how the brain functions! Here's a list of easy, healthy snacks from my friend and nutritionist Jen Drohan. They'll keep your body *and* brain energized. (For more tips, see p. 154.) Thanks, Jen!

* apple slices with almond butter
* a smoothie made with frozen blueberries, orange juice, and a banana
* a handful of walnuts with some raisins or dried cranberries
* celery and cucumber slices with hummus
* carrots with salsa
* avocado with chopped cucumber and tomatoes mixed in lemon and olive oil
* a baked sweet potato with sliced avocado on the side (Sounds weird, but it's a delicious combo!)

(Visit Jen at HappyHealthyBelly.com!)

QUICK NOTE   It's always easy to *start* a proof; we just write down what we were given, and in the Reasons column, we write "Given." Nice! Then we have to think about what Rules we know that might help us link from what we *know* to what we're supposed to prove.

QUICK NOTE   The wording of the Multiplication/Division Rules (and many Rules to come) can make it difficult to do our underlines in any straightforward way. But the underlines have been like training wheels to help us follow the logic, which we don't need as much now anyway! So from this point, we'll stop actually underlining parts of our "if . . . then" Rules. The important thing is to make sure we're creating an airtight argument from one line to the next.

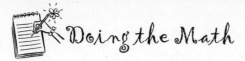

Doing the Math

Use the Multiplication & Division Properties and Bonus Rules from p. 79–81 to do these problems. I'll do the first one for you.

1. Refer to FAMOUS. Given: $\overrightarrow{SA}$ & $\overrightarrow{SM}$ trisect $\angle FSO$, and $\overrightarrow{SM}$ & $\overrightarrow{SO}$ trisect $\angle ASU$, $\angle FSO \cong \angle ASU$, and $\angle 4$ is supplementary to $\angle 5$. Prove: $\angle 1$ is supplementary to $\angle 5$.

<u>Working out the solution:</u> Let's see: We have big congruent angles, and we're being asked about smaller angles. Sounds like the Division Property! This time, the big congruent angles are overlapping ($\angle FSO \cong \angle ASU$). Hmm, we eventually want to prove that $\angle 1$ is supplementary to $\angle 5$, right? We know that <u>$\angle 4$ is supplementary to $\angle 5$</u>, so if we could just prove that $\angle 1 \cong \angle 4$, then we could substitute $\angle 1$ for $\angle 4$ and conclude that <u>$\angle 1$ is supplementary to $\angle 5$</u>, right? And we can prove that $\angle 1 \cong \angle 4$, because they're *like divisions* (thirds) of two big congruent (overlapping) angles. Nice.

Statements	Reasons
1. $\overrightarrow{SA}$ & $\overrightarrow{SM}$ trisect $\angle FSO$.	1. Given
2. $\overrightarrow{SM}$ & $\overrightarrow{SO}$ trisect $\angle ASU$.	2. Given
3. $\angle FSO \cong \angle ASU$	3. Given
4. $\angle 1 \cong \angle 4$	4. Division Property: If two $\angle$'s are $\cong$ ($\angle FSO$ & $\angle ASU$),* then their like divisions (thirds) are $\cong$.
5. $\angle 4$ is supp. to $\angle 5$.	5. Given
6. ∴ $\angle 1$ is supp. to $\angle 5$.	6. Substitution Property (using #4 & #5)

....

* We didn't *have* to say "($\angle FSO$ & $\angle ASU$)," but it sure makes it easier to keep track of what's going on.

2. In the FAMOUS diagram exercise for #1 on the previous page, if ∠5 = 150°, what is ∠FSU? *(Hint: We now know that ∠1 is supp. to ∠5.)*

3. For HOWDY, Given: O & W trisect $\overline{HY}$, and D is the midpoint of $\overline{WY}$. If DY = 2, what is HY?

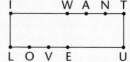

4. For IWANTLOVEU, Given: $\overline{WT} \cong \overline{LE}$, W is the midpoint of $\overline{IT}$, and E is the midpoint of $\overline{LU}$. Prove: $\overline{IT} \cong \overline{LU}$.

5. Again, for IWANTLOVEU, Given: A & N trisect $\overline{WT}$, O & V trisect $\overline{LE}$, and $\overline{WT} \cong \overline{LE}$. Prove: $\overline{LO} \cong \overline{AN}$.

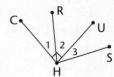

6. For CRUSH, Given: $\overline{CH} \perp \overline{UH}$, ∠1 ≅ ∠3. Do a paragraph proof to explain why ∠2 is complementary to ∠3. *(Hint: Use the definition of ⊥ from p. 63 to figure out the measurement of ∠CHU, then say something about the relationship between ∠1 and ∠2, and then use the Substitution Property.)*

(Answers to #2–3 on p. 398; proofs at GirlsGetCurves.com)

"*Sometimes I get sooo frustrated with geometry proofs it makes me want to quit, but I know they're making me smarter, and I definitely want to be smarter.*" Iris, 15

Takeaway Tips

 If we add or subtract congruent segments (or angles), we'll get more congruent segments (or angles). Remember the short pool chairs and ottomans.

 If two segments (or angles) are congruent, then their like multiples are congruent, and their like divisions are congruent, too. Just think of leftover pizza!

The Substitution Property says that if two segments (or angles) are congruent, then if we have a statement that uses one of them, we can stick the other one in its place.

TESTIMONIAL

Tonya Sims

(Chicago, IL)
<u>Before</u>: Insecure outcast
<u>Now</u>: Confident, fabulous entrepreneur and software developer!

My awkward phase was longer than most, lasting roughly from kindergarten through the ninth grade. I wore enormous glasses, I was always taller than everyone (6'2'' by the seventh grade), I was extremely skinny, and I had a severe overbite that earned me the nickname "Bucky Beaver." The teasing was so bad that I became mute and was terrified to speak or show my teeth and gums. I was finally allowed to get contact lenses after sitting my parents down in the kitchen one night and using sheets of white paper to present points to them of why getting rid of the glasses would help my confidence, so that helped a little. And that same year I also got braces; I was so relieved! But that only brought on a new name: "Brace Face." I couldn't win.

I started sabotaging my grades in the hopes that "not caring" would somehow make me seem "cool" and maybe I'd finally fit in. But dumbing myself down didn't bring me confidence or popularity, just lower grades. So I started studying again, and I'm glad I did. It turns out the key to my happiness wasn't fitting in; it was standing out! My height helped me make the varsity basketball team as a freshman. I led the team in scoring, we made it to the state tournament, and my confidence soared. It's funny how life works, and you just never know how the stuff that makes you feel like an outcast can actually be gifts in disguise. And that year, I also discovered a love of geometry that would lead to a dream career that combines logical thinking with my passion for sports!

"My awkward phase was longer than most."

Today I'm an entrepreneur and a software developer: I recently started a company that makes software to help connect high school athletes with college coaches looking for recruits. The software code that I write may not "look" like the proofs we do in geometry, but it's based on the same "if . . . then" logic and deductive reasoning. For example, let's say a college coach is looking for a basketball player who: Is 6'0 or taller, averages 15 points per game or more, averages 6 rebounds per game or more, has a high school GPA of 3.0 or higher, has an ACT score of 25 or higher, and lives within a 200-mile radius of the college. She'd type in these criteria, and the software would use my algorithm to find the right recruits:

```
If (player >= 6'0) && (player.pointspergame >= 15) && (player.reboundspergame >= 6) && (player.gpa >= 3.0) && (player.act >= 25) && (player.radius <= 200) return player; end
```

The software would create a list of girls that fit all of these requirements, the coach could contact them directly, and maybe the next basketball star would be discovered! The thing I love most about software development is having the power and creative control at my fingertips to help people reach their goals, and maybe even change the world. It gives me chills just thinking about it.

And I'm truly grateful for all my high school experiences. Even the awkwardness helped prepare me for today, in more ways than one. First, most entrepreneurs are deemed as outcasts. Most people just don't get how or why someone would want to start their own business. We have to keep moving forward even if someone doesn't like us or thinks that we're weird. Second, what kind of people change the world?

Certainly not anyone who's afraid to challenge
the status quo, doesn't want to stand out or look
different, or won't voice an opinion because it's
unusual! If someone calls you weird, rejoice! If you
learn to embrace your differences, those same quirky
things can someday define who you are—and show you
what you have to offer the world that is truly
special.

Chapter 6

Ski Lessons

Introduction to Triangles

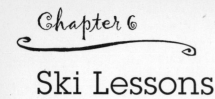

𝒥 like mountains, don't you? They're big, and trees make the air fresh. Just looking at a mountain makes me want to take a nice, deep breath. Some adventurous people like to ski down their steep slopes, or even *scale* the sides of big mountains. Less adventurous people might prefer to just *lean* against them (that doesn't sound very fun, though).

And who's ever seen a mountain whose sides were perfectly symmetrical? Not me. Well, there were these two hikers, Maddy and Leeza, and Maddy said she swore she saw a mountain whose sides were the same length. Of course, Leeza didn't believe her, and Maddy said, "I *saw* so, Leez!"

In this chapter, we'll learn all about triangles, practice some proofs, and all of this mountain talk is about to help us remember some new vocab.

What Are They Called?

scalene, isosceles, equilateral, base, legs, base angles

A **scalene triangle** is a triangle that has no congruent sides. The lengths are all different from each other, just like with real-life mountain peaks. *Scalene* is pronounced like "scale-lean." You know, two things a mountain hiker might do.

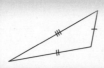

An **isosceles triangle** is a triangle in which at least two sides are congruent. *Isosceles* is pronounced, "I saw so, Leez!" The two congruent sides are called the **legs**, and the other side is called the **base**. The two angles opposite the legs are called the **base angles**.

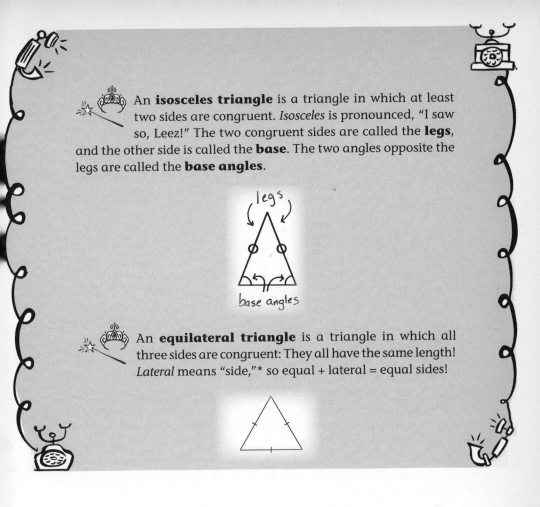

An **equilateral triangle** is a triangle in which all three sides are congruent: They all have the same length! *Lateral* means "side,"* so equal + lateral = equal sides!

I'm Ready for My Close-Up: Sides and Angles

There's an easy relationship between the lengths of the sides of a triangle and the size of the angles opposite those sides. The bigger the angle, the bigger the opposite side will be. We see this on movie sets all the time. The lights have metal sides called "barn doors" that open and close, and as the angle of the doors changes, so does the size of the spotlight on the opposite wall. The bigger the angle of the barn doors, the bigger the spotlight.

----

* Like how our "lat" muscles are the ones extending along the <u>sides</u> of our bodies.

We can imagine that each vertex of a triangle is the location of a movie set light! Then it makes sense that for triangles, the *biggest* side will be opposite the *biggest* angle, the *smallest* side will be opposite the *smallest* angle, and the *medium* side will be opposite the *medium* angle. It follows that if two sides of a triangle are congruent (as happens in isosceles and equilateral triangles), then the angles opposite those sides will be congruent, too—and vice versa!*

Because of this side/angle relationship, the three types of triangles we saw on pp. 88–9 can actually be defined by their *angles* instead of their *sides*:

 A **scalene triangle** is a triangle in which no *angles* are congruent.

 An **isosceles triangle** is a triangle in which at least two *angles* are congruent.

 An **equilateral triangle** is a triangle in which all three *angles* are congruent.†

So if we're asked to prove that a triangle is isosceles, for example, we could either prove that two sides are congruent or that two angles are congruent—whichever one happens to be easier to do. Nice!

. . . .

* We'll do more with this concept on p. 148 when we learn the "If sides, then angles" Rule and its converse.

† We could also call them *equiangular* triangles. Some people do!

We've just seen how to describe triangles based on whether any of their sides and/or angles are congruent, right? Well, we can also describe triangles based on how *big* their angles are. Remember acute, obtuse, and right angles from p. 23? Triangles use these same names!

What Are They Called?

Acute triangle, Obtuse triangle, Right triangle

An **acute triangle** is a triangle in which *all* angles are acute (less than 90°).

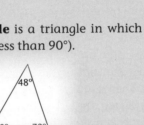

An **obtuse triangle** is a triangle in which one angle is obtuse (more than 90°).

A **right triangle** is a triangle that has one right angle (90°). The side opposite the right angle is called the **hypotenuse**. (More on that in Chapter 11!)

   If you look at the definitions carefully, you'll see that every triangle is either acute, obtuse, or right; it can't be more than one of those. After all, if a triangle has an obtuse angle or a right angle, then it's automatically disqualified from the "acute" category, right? And if an obtuse triangle also had a right angle, well, then it wouldn't be a triangle anymore! It would look something like this: ⌐/. The two sides couldn't reach each other. I bet now you can see why the definition of obtuse triangle says that only *one* of its angles is obtuse.

QUICK NOTE   Scalene and isosceles triangles can be acute, obtuse, or right. But equilateral triangles can only be acute! (We'll see why on p. 104.) They are pretty "cute," I guess.

Purple Frogs: "Always, Sometimes, Never"

We've all seen true/false questions; we're given a statement like "Frogs can be purple. True or false?" and we'd say "true."* Well, sometimes we're given a statement and we have to say how *often* it's true. Is it *always* true, *sometimes* true, or *never* true?

For example, if the statement were "Frogs are purple," then we'd say "sometimes," because we know purple frogs exist, but we also know that frogs can be, you know, green. Sometimes they're purple, and sometimes they're not.

In geometry, these exercises tend to show up when we're dealing with definitions and classifications of shapes. For example, if the statement were "Right triangles are acute," we'd answer "never" because right triangles have one angle that measures 90°, which means it can't be acute (*all* angles must be less than 90° in an acute triangle).

Depending on the topic, this type of exercise can be easy, or it can cause total brain lock. They're good logic builders and they also pop up on standardized tests, so it's never a bad idea to practice them! Here's a game plan for tackling some of the trickier ones . . .

Step By Step

A strategy for "Always, Sometimes, Never" geometry exercises:

Step 1. If the statement itself is confusing, try rewriting the statement in the conditional "if . . . then" form, like we practiced on pp. 2–3, and make sure you can picture what it's saying.

. . . .

* I'm not kidding; look it up. Purple frogs exist!

Step 2. Try drawing a sketch of the situation the statement describes, which will help with Step 3.

Step 3. Is the statement *ever* true? In other words, is there any specific example that satisfies the condition? Look for specific examples and/or counterexamples to the statement, and:

If there are no possible specific examples → **Never**

If there are specific examples and no possible counterexamples → **Always**

If there are specific examples and counterexamples → **Sometimes**

QUICK NOTE   You know you've found a counterexample to a conditional statement if you've thought of a situation where the "if" part is satisfied, but the "then" part is <u>not</u> satisfied.

And... Action! Step By Step In Action

Is the following statement always, sometimes, *or* never *true?*

Equilateral triangles are isosceles triangles.

Step 1. Rewriting this in "if . . . then" form, we could say, "If a triangle is equilateral, then it is isosceles." Another way is, "If it's an equilateral triangle, then it's an isosceles triangle."

Step 2. To sketch this, we'll sketch an equilateral triangle, and we'll see if it has to be isosceles.

Step 3. Yep! This triangle we drew is indeed isosceles, because it has at least two congruent sides (in fact, it has three). So the answer *isn't* "never." Now the question is, can we find a counterexample? In other words, can there be an equilateral triangle that *isn't* isosceles? Nope! There is no possible counterexample, because by definition, equilateral triangles must have three congruent sides, which means they *must* have at least two congruent sides. With no possible counterexamples, the answer must be "always." Done!

Answer: Always

With all the new definitions in geometry, it can be easy to glaze over some of the important details, and doing "Always, Sometimes, Never" exercises is actually a great way to solidify this stuff.

Doing the Math

Answer "always," "sometimes," or "never" to the following statements. I'll do the first one for you.

1. Isosceles triangles are obtuse triangles.

<u>Working out the solution</u>: We can write this as, "If a triangle is isosceles, then it's obtuse." Can we draw an example of an isosceles triangle that is also obtuse? Sure! Let's sketch one. 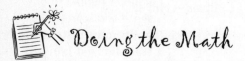 We just need one angle to be bigger than 90°, after all. On the other hand, we know that just because a triangle is isosceles doesn't mean it has to be obtuse; we've seen skinny (acute) isosceles triangles, like the one on p. 89. We have an example and a counterexample, so we know the answer must be "sometimes." Done!

Answer: Sometimes

2. Isosceles triangles are acute triangles.

3. Isosceles triangles are equilateral.

4. Equilateral triangles are scalene.

5. Scalene triangles are isosceles.

6. Isosceles triangles are right triangles.

7. A triangle has two obtuse angles.

8. A triangle has three acute angles.

9. A triangle has at least one acute angle.

(Answers on p. 398)

Reality Math:
"False Logic" Advertising?

The whole point of advertising is to persuade people to buy products or services, right? And it makes sense that companies would want to be as persuasive as possible. In most instances, there are regulations to keep companies from flat out lying in their ads.* But it should be no surprise that companies often find ways to mislead the public . . .

A bottle of moisturizer says, "30% more than the competition!"

The advertisers want us to think, "Wow, 30% more, for the same price!", but they're not actually saying *anything* about the price, are they? Sure, maybe they're giving us 30% more moisturizer, but they might be charging 40% more! We'd need to check out the volumes and prices, and figure it out for ourselves.†

A sign outside a store says, "EVERYTHING IN THE STORE: up to 75% OFF!"

They want us to think, "Hey, everything in this store is totally discounted! Let's go in!" But because it says "up to" 75% off, most of the items could be 20% off or 10% off or even not on sale at all. What this is really guaranteeing is that nothing is discounted *more* than 75%. Tricky, huh?

A website says, "Now through Sunday, only $19.99!"

They want us to think, "Wow, I better buy this soon, because I don't want to miss the sale!" Well . . . I mean, the sale's last day *might* be Sunday, but the sale might actually keep going through Monday and beyond. Notice the placement of the comma: If it had said, "Now through Sunday only, $19.99!" then it *would* mean that Monday is too late. We'd have to check it out for ourselves.

If we really pay attention to the *wording* in advertisements, we'll see that fewer actual facts are given than first meet the eye. But we'll keep our brains turned on, we'll use logic, and we won't be fooled by misleading ads ever again!

. . . .

* Unless if they're *obviously* lying. For example, have you ever noticed how many restaurants claim to have the "World's Best Burger"? Yeah . . .

† To brush up on unit prices (a great way to compare two products!), check out Chapter 17 in *Math Doesn't Suck*.

Proof Practice!

Now, let's do a few proofs using our new definitions from this chapter. And here's a cheat sheet of some Rules from *previous* chapters:

 If two ∠'s are supp., then the sum of their measures is 180°.

 If two ∠'s are vertical, then they are ≅.

Transitive Property: If $a \cong b$ and $b \cong c$, then $a \cong c$. (And we fill in specifics.)

Remember how we rewrote sentences into "if . . . then" form on pp. 2–3? Well, rewriting some of our definitions like this is a great habit to get into with proofs. For example, we could rewrite the definition of an isosceles triangle from p. 90: **If two ∠'s of a △ are ≅, then the △ is isosceles.** Or we could rewrite the definition of an obtuse triangle: **If one ∠ of a △ is > 90°, then the △ is obtuse.*** Your teacher might not insist on using the "if . . . then" format for all your "Reasons," but it's a great way to make sure the logic is airtight.

Here's something else to think about as you do more proofs on your own. There's often more than one way to do them. Our goal isn't necessarily to figure out the exact proof that someone else thought of (like a teacher or a textbook writer). Our goal is to figure out an airtight proof that *works*. Usually that's all your teacher will be looking for anyway.

---

* Technically, the ">" symbol means "<u>is</u> greater than," so we could just write ". . . of a △ > 90°", but it's easier to read with the extra "is." These abbreviations are meant to make life easier during proofs, not more confusing, so let's use an extra "is" or "are" wherever it makes sense!

 Doing the Math

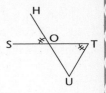

Complete the two-column-style proof below, using the Step By Step on p. 57 if it helps. First, I'll show you a full proof.

1. Given: $\angle HOS \cong \angle OTU$. Prove that $\triangle OUT$ is isosceles.

<u>Working out the solution:</u> To prove a triangle is isosceles, we can either prove that it has two congruent sides or two congruent angles. Because we're given info about its angles, let's pick that one! A game plan? Hmm, since vertical angles are congruent, it looks like we can use vertical angles to prove that $\angle HOS \cong \angle TOU$, then the Transitive Property tells us that $\angle TOU \cong \angle OTU$, and then the (angles) definition of an isosceles triangle will finish it off:

Statements | Reasons

Statements	Reasons
1. $\angle HOS \cong \angle OTU$	1. Given
2. $\angle HOS$ & $\angle TOU$ are vertical angles.	2. Given (assumed from diagram)*
3. $\angle HOS \cong \angle TOU$	3. If two $\angle$'s are vertical, then they are $\cong$.
4. $\angle TOU \cong \angle OTU$	4. Transitive Property: If $\angle HOS \cong \angle OTU$ and $\angle HOS \cong \angle TOU$, then $\angle TOU \cong \angle OTU$.
5. $\therefore \triangle OUT$ is isosceles.	5. If two $\angle$'s of a $\triangle$ are $\cong$, then the $\triangle$ is isosceles.

. . . .

* This is actually an optional step (see the footnote on p. 129).

For #2–8, fill in the blanks of the proof below:

Given: $\overline{LT} \cong \overline{TS}$, and $\overline{LT}$ is the perpendicular bisector of $\overline{AS}$. Prove that $\triangle ATL$ is a right isosceles triangle.

Proof

Statements	Reasons
1. $\overline{LT} \cong \overline{TS}$	1. (#2)
2. $\overline{LT}$ is the $\perp$ bisector of $\overline{AS}$.	2. (#3)
3. (#4)	3. If a segment is a $\perp$ bisector, then it divides a segment into two $\cong$ parts. (partial definition of $\perp$ bisector)
4. $\overline{AT} \cong \overline{LT}$	4. (#5)
5. $\triangle ATL$ is isosceles.	5. (#6) (Good progress!)
6. (#7)	6. If a segment is a $\perp$ bisector, then it creates a right angle. (partial definition of $\perp$ bisector)
7. $\therefore$ (#8)	7. If a $\triangle$ is isosceles and has a right angle, then it is a right isosceles $\triangle$.

(Answers on p. 398)

Ski Lessons and the Triangle Inequality

Imagine you're taking your first ski lesson, and your overzealous teacher has put you at the top of a very steep mountain. Flattering, sure, but just a bit terrifying. Wouldn't it be great if you could make the mountain less steep? Imagine if we could push that top vertex down, but keep the other points (vertices) where they are.

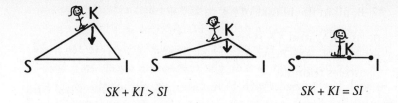

$$SK + KI > SI \qquad\qquad SK + KI = SI$$

Okay, maybe we didn't need to make the mountain totally flat (that would be um, pretty boring), but notice something: If the bottom length, *SI*, stays the same, then as the point *K* gets closer to the ground, <u>*SK* and *KI* get shorter and shorter</u>, until *K* is on the flat ground, and in fact, collinear with *S* and *I*, at which time it's pretty clear that *SK* + *KI* = *SI*, right?

But in order for *K* to be off the ground even the tiniest bit (in other words, even if △*SKI* is the skinniest little triangle in the world) then those two segments (*SK* and *KI*) have to get just a tiny bit longer than they were when it was flat. Then we *have* to replace the equals sign with a > sign, and we get: *SK* + *KI* > *SI*. And this kind of inequality exists for every triangle! It's called the Triangle Inequality.

 Triangle Inequality: The sum of the lengths of any two sides of a triangle is always greater than the length of the third side.

So we can take any two sides of any triangle, add up their lengths, and we *know* the sum will be bigger than the length of the third side.

For example, in the triangle to the right, we know for sure that $1.5 + 2 > x$, right?

Simplifying, we get $3.5 > x$; in other words, $x < 3.5$.* But this works for *any* two sides, so we could also say with confidence that $1.5 + x > 2$, which simplifies to $x > 0.5$. Combining the two inequalities, we now know that: **$0.5 < x < 3.5$**. So there's only a certain range of values that *x* could be in order for this to be a triangle. In fact, if *x* were smaller than 0.5 or bigger than 3.5, then two of the sides of the triangle simply couldn't reach each other, no matter how hard they tried.

Can't you just hear those little arms straining to reach each other? It's sad, really.

. . . .

* To brush up on inequalities, check out Chapter 14 in *Kiss My Math*.

QUICK NOTE   Notice that in the previous example, we only used two inequalities to get the restrictions on x. There are actually *three* true statements of inequality that we can make for any triangle, but one of the statements usually isn't very helpful. For example, with that same triangle, we also could have made this statement: x + 2 > 1.5. In other words, x > −0.5. But we already knew that x has to be a positive number, because it represents a *length*. So even though it's a true statement, let's face it, "x > −0.5" doesn't help us very much.

Watch Out!

For the Triangle Inequality, we have to use > ("is greater than") and not ≥ ("is greater than *or equal to*"). If the sum of two sides of a triangle were *equal* to the length of the third side, that would mean there's no triangle in the first place—just three points hanging out on the same line (collinear points), like a totally flat mountain. Yeah, no fun at all.

Step By Step

Working with the Triangle Inequality, where one side is labeled "x."

Step 1. If the triangle hasn't been labeled, use variables along with the Givens to label each side.

Step 2. Pick any two sides of the triangle. Their sum will be greater than the length of the third side, so write that inequality down, and simplify it so we get "*x* < something" or "*x* > something."

Step 3. Repeat this with the other two sides. At least one of the resulting inequalities won't be very helpful; that's okay!

Step 4. If possible, combine the two (helpful) inequalities into one conjunction,* like this: "something < *x* < something else." Do a reality check: Do these restrictions make sense for the triangle?

Step 5. If that's what the problem was asking for, we're done. If not, we can use this information to get the final answer. Voilà!

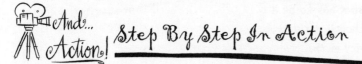

And... Action! Step By Step In Action

Given: △GEM *is isosceles, with GE = EM and GM = 6. What are the possible values for ME?*

Step 1. Let's label the two congruent legs of this isosceles triangle with "*d*" (for, um, diamond). Great, so our triangle has lengths *d*, *d*, and 6.

Step 2. Let's create our first inequality. We can pick any two sides we want; how about *d* and *d*? Because this is a triangle, we can say for sure that $d + d > 6$. Simplifying, we get $2d > 6$ → **$d > 3$**. Great! We already know something about what *d* has to be.

Step 3. Picking another two sides, let's do: $d + 6 > d$. Simplifying, we can subtract *d* from both sides and we get $6 > 0$. Yeah, this must be one of those unhelpful inequalities. Moving on to the third inequality, we get $6 + d > d$. But wait, that's the same as the one we just did! Yep, because it's isosceles, we only get one helpful inequality and two identical, unhelpful inequalities. Right on.

Step 4. The only helpful inequality is **$d > 3$**. Does this make sense? Sure! Imagine we pushed the top of the mountain straight down; the two legs of our triangle (the *d*'s) would both end up getting squashed down to equal 3 each. But if we pulled up the top of the mountain, well, the sky's the limit! The *d*'s could get as big as we can imagine (which is pretty big).

Step 5. Remembering that the original problem asked about the length of $\overline{ME}$, we should write the answer in those terms, and *d* = ME, after all, so it's not hard to make that adjustment!

Answer: *ME* > 3

. . . .

* To read more about inequalities, conjunctions, and control freaks, check out pp. 102–5 in *Hot X: Algebra Exposed*.

Doing the Math

Answer the questions below, using the Triangle Inequality. I'll do the first one for you!

1. Find the range of possible values for *x* in the above triangle.

<u>Working out the solution:</u> You might be thinking, wait! How can we find x's values, when there's an "a" hanging around? Well, that just means we find x's range of possible values in <u>terms of a.</u>[*] Let's do it! The first inequality could be: $a + x > 2a$. To isolate x, we subtract a from both sides and get: **x > a.** The next inequality could be: $a + 2a > x$; in other words: **x < 3a.** The third (unhelpful) inequality is: $2a + x > a \rightarrow x > -a$. Yep, not helpful, since we know a must be positive. So duh, of course x is bigger than −a. If it weren't, then x would be a negative number! Combining the two helpful inequalities into a conjunction, we get: **a < x < 3a.**

Answer: $a < x < 3a$

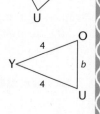

2. Given: $\triangle LUV$ with sides 10, 7, and *x*. What is the range of possible values for *x*?

3. Given: $\triangle YOU$ is an isosceles triangle with legs of length 4. What is the range of possible values for its base, *b*?

. . . .

[*] To brush up on finding one variable in terms of another, check out pp. 78–88 in *Hot X: Algebra Exposed.*

4. In the CUTE diagram, what is the length of $\overline{UE}$, and how do we know? Given this, what is the possible range of values for y?

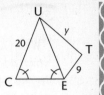

5. Given $\triangle BOY$ with sides $10a$, $7a$, and x. What is the range of possible values for x (in terms of a)?

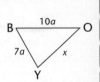

(Answers on p. 398)

Straight Lines and Triangles: 180° Soul Mates

Ever notice how two people can seem totally different on the outside but end up being super close? It could be a case of opposites attracting, but maybe they have something in common that creates a deep bond: perhaps a similar childhood, an unusual life philosophy, or even the same number of degrees in their angles. Wait. Well, that last one works for geometry, anyway . . .

Take, for example, a straight line and a triangle; they have way more in common than meets the eye. We know that a straight line—the "straight" angle—has 180° in it. (That's just because somebody decided that a full spin is 360°,* and this represents half of that. So the degrees add up to 180° just by definition.)

$360°$ $180°$ ←♡→

$a° + b° + c° = 180°$

....

* Why was 360° chosen for a full spin, and not, say 1000°? Well, 360 is divisible by tons of numbers, like 2, 3, 4, 5, 6, 10, and 12 to name a few, which makes it an easier number to work with!

But here's where it gets interesting: If we add up the angles in ANY triangle, we'll always get 180°! Yep, whether the triangle is obtuse, acute, scalene, isosceles, whatever—the sum of the angles will always equal 180°, no matter what.

 If the angles of a triangle are added together, the sum will equal 180°.*

For example, if we know that two of the angles in a triangle are 110° and 50°, then without even seeing the triangle, we know for sure that its third angle must be 20°, because 110° + 50° + **20°** = 180°.

And right now, we have the power to figure out the angles in an equilateral triangle! We know from p. 90 that all equilateral triangles have congruent angles; yep, all three angles are the same. Now we also know they must add up to 180°.

Well that means that $x° + x° + x° = 180°$, right? In other words: $3x° = 180° \rightarrow$ ***x*° = 60°.**

Notice that none of this had anything to do with the *size* of the equilateral triangle; the triangle could be as small as a piece of glitter or as big as the galaxy. If it's equilateral, then all of its angles must be 60°! Sounds like a theorem to me:

 The three angles in every equilateral triangle each measure 60°.

Let's practice all this new triangle info. (If you need hints, check out the Takeaway Tips on p. 106; it's a nice summary of what we've learned so far.)

. . . .

* To see the proof of this theorem, see GirlsGetCurves.com, but FYI, we'll be using stuff from Chapters 13 and 14, so don't check it out till you've read those!

Doing the Math

Find the desired values based on the Givens and diagrams. I'll do the first one for you.

1. In the WINK diagram to the right, Given: $\overrightarrow{KI}$ is the angle bisector of $\angle WKN$, $\triangle KIN$ is an isosceles triangle with base $\overline{KN}$, and $\angle W$ is $\frac{3}{2}$ the size of $\angle N$. What is $m\angle KIN$?

Working out the solution: It's tempting to say that $\angle KIN$ is a right angle, but there's no little square in the corner, so we'd better not assume anything. Let's start by labeling the three congruent angles with $n°$. (Go ahead and write them in your book!) Hmm, we can't figure out $m\angle KIN$ just using the triangle $\triangle KIN$, but if we use the biggest triangle, $\triangle WKN$, we can! In this big triangle, the angles are $\angle N$ (which we've labeled $n°$), $\angle WKN$ (which, because of the angle bisector, we know must be twice that size: $2n°$), and the third angle, $\angle W$, which we've been told measures $\frac{3}{2}$ the size of $\angle N$; in other words: $\frac{3}{2}n°$. Now for the big triangle, we can set up this equation: $n° + 2n° + \frac{3}{2}n° = 180°$, and solve for $n°$! This'll be easier if we multiply both sides by 2, and we get: $2n° + 4n° + 3n° = 360°$ $\rightarrow 9n° = 360° \rightarrow n° = 40°$. But this isn't the answer yet: We need $m\angle KIN$, which is part of the triangle $\triangle KIN$. Now we know the other two angles in $\triangle KIN$ are $40°$ and $40°$, right? And looking at $\triangle KIN$, since $40° + 40° + 100° = 180°$, the missing angle, $\angle KIN$, must measure $100°$. Phew!

Answer: $m\angle KIN = 100°$

2. In $\triangle SHE$, $m\angle S = (2x − 5)°$, $m\angle H = 3x°$, and $m\angle E = (x + 5)°$. Is $\triangle SHE$ acute, obtuse, or right? (Hint: First solve for x.)

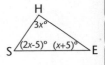

3. In the PICNK diagram, △PIC is an isosceles triangle with base $\overline{PC}$, and ∠CIK = 42°. What are the measurements for the three interior angles of △PIC? (*Hint: First label ∠P and ∠C with n°, and think about supplementary angles!*)

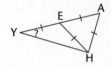

4. In the IMCOY diagram, △COY is an equilateral triangle, and ∠I = 25°. What is the measure of ∠M? (*Hint: vertical angles!*)

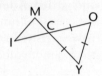

5. In the diagram YEAH, △AEH is equilateral, and △YEH is isosceles. What is m∠Y? (*Hint: Fill in all the angles we know, and use supplementary angles!*)

(Answers on p. 398)

Takeaway Tips

 Scalene, isosceles, and equilateral are three types of triangles. *Scalene:* all three sides and angles different. *Isosceles:* two angles congruent, two sides congruent. *Equilateral:* three sides congruent, each angle equals 60°.

 Every triangle is either acute, right, or obtuse, and the biggest angle will always be opposite the biggest side.

 Triangle Inequality: The sum of the lengths of any two sides of a triangle is always greater than the length of the third side.

Angles in a triangle: The sum of the interior angles of a triangle always equals 180°.

Making Exercise Fun

Exercise is so good for the mind, body, and spirit!* But when it becomes a chore, we're less likely to stick with it, right? The key to is to keep our workouts fun and varied. After all, working up a sweat doesn't have to be boring. Joining a sports team at school is a great way to make exercise fun and social. Here are some more fun ideas for staying motivated and fit.

Get Involved: Sign up for charity sporting events or walks/runs to benefit research leading to cures for cancer or other diseases. It's a great way to make new friends, and when you have a worthy goal in mind, exercise becomes even more gratifying.

Make It an Adventure: Plan an activity that you've always wanted to try like horseback riding, skiing, or surfing. Get serious about it, and physically prepare yourself by lap swimming before water sports or doing cardio to increase your stamina. Having something to look forward to is always a big motivator.

Make It a Love Match: Why not involve a guy you are interested in? Crash your boyfriend's (or guy friend's) pick-up football game—you can bond with him and get in shape at the same time. Plus, being one of the boys every once in a while can be fun.

Team Spirit: Channel your physical and mental energy by trying yoga. Breathing exercises and meditation are coupled with stretching and strength training. I even made a yoga/meditation DVD called *Daily Dose of Dharma* because I like yoga so much!

---

* See p. 282 for more on this!

How Do YOU Like to Stay Fit?

"*I* love running. Not only is it a good workout, but it also makes me really happy. I've recently begun doing yoga to help reduce stress and become a little more fit."
—Jade, 17

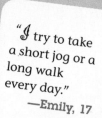

"*I* try to take a short jog or a long walk every day."
—Emily, 17

"*I* love swimming. My mind just goes somewhere else, and afterwards my muscles all feel great."
—Janna, 16

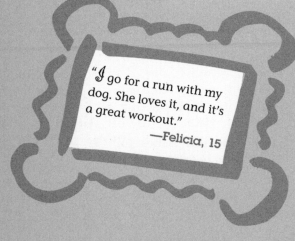

"*I* go for a run with my dog. She loves it, and it's a great workout."
—Felicia, 15

"*T*o stay healthy, I exercise by rollerskating almost every day. Find something you love to do that you wouldn't even consider exercising!"
—Megan, 14

Chapter 7

Romantic Walks on the Beach

Congruence Transformations

Have you ever taken a romantic walk on the beach—either with or without someone else? The gentle breeze blowing in your hair, the soft waves lapping at your feet (or maybe that's drool from your dog).

I like to look back at the long trail of footprints left in the sand, don't you? Those same footprints are going to help us with transformations!

You've probably heard the word *transformation* in a non-math context: "She cut her hair, changed her wardrobe—she completely *transformed* her look!" A **transformation** in geometry is where we take a shape and change its "look" somehow: its location, orientation, size, shape, you name it. We perform geometric transformations every time we retie a ribbon or spin the dial on an iPod.

You already know what congruent angles and congruent segments are; they are equal in measurement, right? Well, **congruent shapes** are equal in *all* their measurements. And when we do a *transformation* that results in an image that's congruent to the original shape, it's called a **congruence transformation**.

What Are They Called?

A **congruence transformation*** is the result of moving all the points of a geometric image in a "rigid" way, so that we end up with a *congruent image*. Here are some examples of congruent transformations:

. . . .

* Another name for *congruence transformation* is "isometry."

Translation: *Glide!* This is the easiest one to picture; all the points just glide over in the same direction: up, down, diagonal, whatever. Put your hand on the table and slide it forward (without rotating it). Ta-da! You've just done a translation. If only translating English to French (or sign language!) were so easy . . .

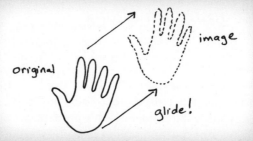

Reflection: *Flip!** This is when we reflect a shape across a line, just like in a mirror. For example, if you're standing still with your feet next to each other, they are a **reflection** of each other. Here's a neat factoid: The **line of reflection** is the *perpendicular bisector* of every segment connecting corresponding points. Just think about that for a second . . .

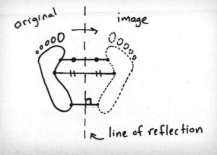

* In two dimensions, reflecting a shape across a line is the same as "flipping" it over that line. But in the 3-D world, flipping is different from reflecting. Hold your plate of dinner in front of the mirror. You're seeing its *reflected* image. Now take the mirror away, and *flip* the plate over to where it appeared to be in the mirror. And . . . now your dinner's on the floor.

Rotation: Um, *rotate*! This is when we rotate a shape with respect to a particular point. Did you ever decorate your bicycle wheels with LED lights or fun shapes? Spinning the wheel moves the shape, and that's all we're talking about here. The center of the wheel is called the **point of rotation**, and the amount that the wheel rotates is called the **angle of rotation**.

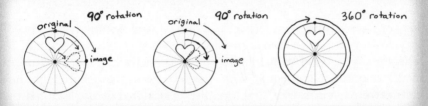

The point of rotation—the center of the wheel—can be inside, outside, or on the shape. (In the first diagram, it's outside the shape; in the second diagram, it's on the shape.) And if we rotate a shape by 360°, no matter *where* the point of rotation was, we'll always get back to where we started: The image will coincide exactly with the original shape!

Any time we're dealing with a rotation, I recommend imagining that the shape is on a bicycle wheel, with the center of the wheel as the point of rotation. It really helps.

QUICK NOTE   Strange preposition* rule: You'll notice that when we're talking about rotations, we say a shape rotates "about" a point, instead of "around" a point. It's actually the correct grammar; it just sounds weird at first.

. . . .

* Recall that a *preposition* is a part of speech, like a noun or a verb. Common prepositions are: *above, about, across, around, at, by, of, with.* You get the idea, but do an internet search to see the whole list!

QUICK NOTE   When we transform a shape, the result is called the "image." When we look in a mirror, we're seeing our image—same thing!

Types of Symmetry

Tons of shapes have symmetry—even letters! And symmetry has a very cozy relationship with transformations. Check it out: If a shape can be reflected across a line in such a way that the resulting image is identical to the original shape (same exact position, too), then that shape has **reflectional symmetry,** and the line we reflected across is called the **line of symmetry**.

Reflectional Symmetry

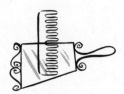

One way to think of reflectional symmetry is with a good 'ol mirror. Can we put a mirror somewhere that simultaneously blocks and also "re-creates" that half of the image? If so, it's got reflectional symmetry. Take a comb and a handheld mirror and see for yourself!

Another type of symmetry is called **rotational symmetry**. Just imagine we put a shape at the center of a bicycle wheel. Can we rotate the wheel (*less* than 360°, of course!) and end up with the same exact picture? If we can, then the shape has rotational symmetry.

Rotational Symmetry

Imagine the above shapes rotated the indicated amount; do you see how they'd match up perfectly again? By the way, if we rotate the wheel exactly halfway (180°) and get the same picture, then it's called **point symmetry**, or origin symmetry, but I like to think of it as upside-down symmetry, because those shapes look the same when you rotate them upside down.

Watch Out!

Don't confuse upside-down (point) symmetry with reflectional symmetry! Rotating something 180° is different from reflecting it across its line of symmetry. For example, the letter **E** has reflectional symmetry across a horizontal line running through its middle. But if you turn this book upside-down, you'll see that it looks wrong, because we rotated it, and <u>it doesn't have point symmetry</u>. On the other hand, the letter **N** looks the same upside down. (Turn this book upside down to see what I'm talking about.) But if we reflect it over a line, we'll end up with a shape that looks like a backwards **N**. Hold this book up to a mirror to see what I'm talking about! Of course, some shapes (and letters) have both kinds of symmetry and look the same upside down and backwards.

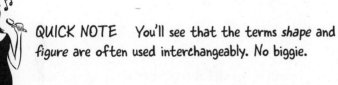

QUICK NOTE   You'll see that the terms shape and figure are often used interchangeably. No biggie.

Visualizing and working with transformations is great exercise for the brain, especially for creative spatial reasoning, which makes us better at everything from solving a jigsaw puzzle to packing a suitcase or organizing a dorm room. (It'll even help us with congruent triangles later on.) Let's practice!

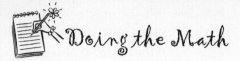
Answer the questions below on transformations. I'll do the first one for you.

1. Describe the rotational symmetry of a flower with 5 regular petals about its center point, by listing the ways we can rotate this flower (in degrees) so that it matches up with the original shape. Only give positive answers less than or equal to 360°.

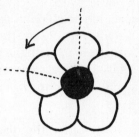

<u>Working out the solution:</u> Let's put this flower at the center of a bicycle wheel! Hmm . . . if a flower has 5 petals, we only have to rotate it $\frac{1}{5}$ of the way around before it matches up with the original, right? Well, what's $\frac{1}{5}$ of 360°? That's $\frac{1}{5} \times 360° = \frac{360°}{5} = 72°$. Great! But that's not all . . . we could also rotate it another 72° and get another match, right? That would be: 72° + 72° = **144°**. In fact, each time we spin the wheel 72° more, we'll get a match. So continuing to add 72° until we get to 360°, we get:

Answer: 72°, 144°, 216°, 288°, 360°

2. Draw the congruent shape resulting from reflecting the G over the indicated line: G ¦

3. Draw the congruent shape resulting from rotating the letter L counterclockwise 90° about its center point.

4. Name 5 capital letters that have **reflectional symmetry across a vertical line** (example: A).

5. Name 5 capital letters that have **reflectional symmetry across a horizontal line** (example: B).

6. Name 5 capital letters that have **rotational symmetry** (example: N). *(Hint: Imagine putting the letter on the center of a bicycle wheel.)*

7. Name 2 capital letters that have **rotational symmetry but *not* reflectional symmetry**.

8. Some letters look the same when facing a mirror. What kind of symmetry is this? Be specific: Name one of the bolded types of symmetry from the previous four questions.

9. Describe the rotational symmetry of a flower with 4 regular petals about its center point by listing the ways we can rotate this flower (in degrees) so that it matches up with the original shape. Only give positive answers less than or equal to 360°. *(Hint: See #1.)*

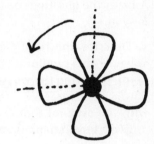

10. Describe the rotational symmetry of a flower with 6 regular petals about its center point by listing the ways we can rotate this flower (in degrees) so that it matches up with the original shape. Only give positive answers less than or equal to 360°.

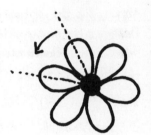

(Answers on p. 398)

Palindromes—Symmetry with Words!

Do you know about palindromes? This is a different kind of symmetry, and I'm including this because, well, I think they're pretty awesome. They can be a word (*kayak, noon, mom, dad, racecar*) or an entire phrase that reads the same forwards and backwards as long as you ignore punctuation and spaces. For example:

Did Hannah see bees? Hannah did.

Ma has a ham.

Don't nod.

Was it a cat I saw?

No lemon, no melon.

Yo, banana boy!

Can you make up your own? Start by looking at words around you and seeing what they spell backwards. Then see if you can make part of a word out of them, and go from there. When you do, email 'em to me at share@danicamckellar.com.

If only your name were Liam, then I'd say: "Liam, email!"

That Beach Walk: Composition of Transformations

Have you ever seen the word *compose*? It just means "to put together." We can compose an email out of words, compose a song out of notes, or compose a transformation out of . . . other transformations. This is called a **composition of transformations**.

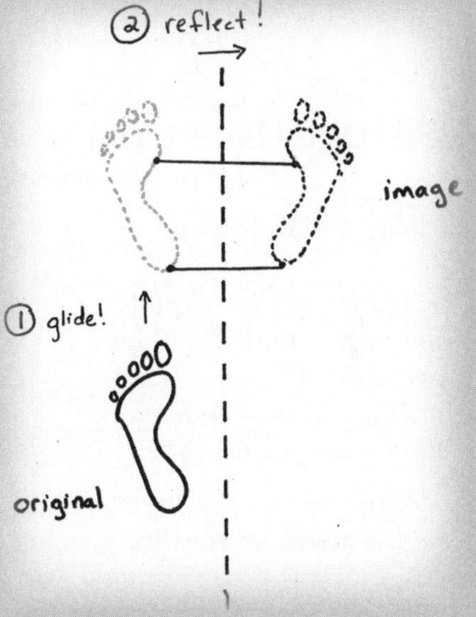

For example, we could take a shape and rotate it, and then reflect (flip) it over a line. Or we could rotate something and then flip it, and then glide it over, too! My favorite composition of transformations is footprints on the beach. They're called **glide reflections**. To get from one footprint to the next, we translate (glide) up and then reflect over the center line: "glide, reflect."

(diagram labels: ② reflect! · image · ① glide! · original)

So the next time you're making footprints, think to yourself, "Glide, reflect! Glide, reflect!" Or, um, don't. You're on the beach. Why are you doing math?*

Oh, and if you're feeling like a rock star, check out GirlsGetCurves .com to see how the Commutative Property works (or, um, doesn't) with compositions of transformations. Make sure to have your phone handy for some serious flipping action. I've also posted some other bonus stuff like how to do transformations on the coordinate plane.

---

* I know, I know, you can't help it. Me neither.

In this chapter, we only talked about rigid transformations, because those are the kinds that lead to congruent figures. In Chapter 17, we'll see some transformations that *don't* result in congruent figures. And they have nothing to do with plastic surgery. Okay, maybe a little.

Takeaway Tips

 Congruence transformations are functions that act upon a shape and change it in a way that creates an image that is congruent to the original shape.

 Translations, **reflections**, **rotations**, and **glide reflections** (footprints!) are all congruence transformations. The images will *always* be congruent to the original shapes.

 In reflections, the **line of reflection** is the *perpendicular bisector* of every segment connecting corresponding points.

 If there is a line of reflection that results in the image coinciding exactly with the original shape, then the shape has **reflectional symmetry** across that line.

 For rotations, imagine that the **point of rotation** is the center of a bicycle wheel that the shape rotates on.

 If a shape's rotation (of less than 360°) will result in an image coinciding exactly with the original shape, then the shape has **rotational symmetry** about that point of rotation.

Chapter 8

Cheerleading Tryouts

Congruent Triangles

$\mathcal{H}$ave you ever thought about trying out for the cheerleading squad or dance team? My schedule didn't allow me to (I was on TV at the time), but my sister did, and of course I watched her as often as I could. In addition to it being a great workout, there was also quite a bit of letter yelling, like "Gimmie an S!" As it turns out, that sort of thing comes in very handy when proving that two triangles are congruent.

As we saw in Chapter 7, we can use certain transformations (translations, rotations, and reflections) to change a shape into an image that is **congruent** to the original shape. Now, let's say we are given two triangles that *look* congruent. Even if we know that a few sides or angles are congruent, how do we know if the entire triangles are congruent? Was a transformation even involved? I know this would really bother you,* so now we're going to learn how to actually *prove* when triangles are congruent.

. . . And cheerleading is going to help us. (More on that in a moment.)

. . . .

* It's been keeping you up at night, right?

If all pairs of corresponding parts (all angles and sides) are congruent in two triangles, then they are **congruent triangles**.

Here, △CAR is congruent to △KYE.

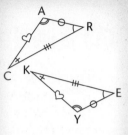

△CAR ≅ △KYE

If we imagine rotating and gliding the triangle △KEY until K is on top of C, and E is on top of R, and if we *then* imagine reflecting △KEY over the line $\overline{CR}$,* we can see how the triangles would overlap and the congruent parts would match up perfectly.

Yup, we have ourselves a pair of triangles that could *totally* impersonate each other in a jewelry heist, so we can put a moustache between them: △CAR ≅ △KYE. (See p. 27 for more on jewelry heists and moustache disguises.)

Watch Out!

In the example above, as tempting as it may be to write △CAR ≅ △KEY, we mustn't do it. It's simply not a true statement, because it matches up the wrong parts! We have to always list the three letters (vertices) of the triangle so that corresponding points match up exactly. Here are some ways to express congruency that would all be *correct*: △CRA ≅ △KEY, △ARC ≅ △YEK, △ACR ≅ △YKE, △RAC ≅ △EYK, △RCA ≅ △EKY.

Let's say we wanted to prove that two triangles are congruent. Doesn't it seem like it would take forever to deal with all those sides and angles? Good news—we don't have to! As it turns out, if we can just establish that a *few* pairs of corresponding things are congruent in two triangles, then the *entire triangles* have to be congruent. See the nifty shortcuts below.

....

* Or $\overline{KE}$; they'd be the same at that time!

Shortcut Alert

Here are four shortcuts to proving that two triangles are congruent. Check out the kitty scratches and other markings on the diagrams to see what each shortcut means.

 Side-Side-Side Rule (SSS): If the three corresponding sides of two triangles are congruent, then the two (entire) triangles are congruent. Why does this work? Take three straws of different lengths (or pencils or nail files or whatever), make a triangle by touching their ends together, and trace it on a piece of paper. Notice that no matter how you position them, if you create another triangle by touching their ends together, it will be congruent to the last one you did.*

 Side-Angle-Side Rule (SAS): For any two triangles, if two pairs of corresponding sides and the angle in between† are congruent, then the two (entire) triangles are congruent. Why does this work? Take two sticks and put their ends together to form an angle. Well, there's only one triangle we could make. We'd connect the two open "ends" and that's it! So side-angle-side is enough information to determine everything about the triangle.

 Angle-Side-Angle Rule (ASA): For any two triangles, if two pairs of corresponding angles and the side in between are congruent, then the two (entire) triangles are congruent. Why does this work? This time, let's get a pair of sunglasses that are 6 inches wide. Say we open the right side to 40° and the left side to 70°. Notice that the two sides intersect to create a triangle, and there's only one triangle that

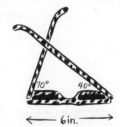

---

* However, if we do this with *four* straws of different lengths, we can easily make *non*-congruent quadrilaterals. Try it!

† Some books use the term *included angle* instead of the angle "in between." Same goes for sides.

the sunglasses can make! So if you had another pair of sunglasses that were also 6 inches wide, and if you made the same 40° and 70° angles, the resulting triangle would *have* to be congruent to the first. The full lengths of the other sides (the parts we're moving) don't matter; they intersect, and the parts *inside* the triangle will be forced to be congruent!

 Angle-Angle-Side Rule (AAS or SAA): For any two triangles, if two pairs of corresponding angles and a side *not* in between are congruent, then the two (entire) triangles are congruent. Why should this be true? It's harder to visualize this one, but once we get a little more geometry under our belts, it'll be easier to understand why it works.*

Watch Out!

Order matters! For example, SAS means the congruent angle must be the angle *between* the two congruent sides, just like in the name.

Something fantastic about these shortcuts is that when we use these Rules as the Reasons in two-column proofs, teachers almost always allow us to just write the little 3-letter abbreviations. We don't have to worry about writing out the entire "if . . . then" statements—just "SSS" or whatever. Nice, huh?

QUICK NOTE   In fact, it's time to start looking for other ways to abbreviate the things we write in the Reasons column. The "if . . . then" logic will always be in there, but it just takes too long to write everything out, especially as the proofs get

* Check out GirlsGetCurves.com for the explanation. You'll want to read Chapter 17 (on similar triangles) first!

longer. You can always write it out if you want to make sure the logic is airtight, but you'll get better at "seeing" the logic without having to see the "if . . . then" stuff all linked together. For example, we can just name the Rule that we're using, like "Definition of midpoint," instead of "If a point divides a segment into two congruent segments, then it's the midpoint." However, sometimes I'll write both, just to help make the logic clearer as you're reading.

Watch Out!

Just because we're shortening what we write doesn't mean we're relaxing the logic itself! Make sure each Statement in the first column is justified with a Reason that relies on Statements before it, and you'll be good to go.

Like, duh . . . The Reflexive Property

There's one more tool we'll need before we can move forward: **The Reflexive Property**. This says that a segment is congruent to itself (I'm not kidding).

It seems pointless at first, but it does serve a purpose: When we use any of the shortcuts from pp. 121–2, you'll notice that we need <u>three</u> pairs of congruent sides (S) and/or angles (A) each time, right? Well, sometimes the two triangles actually share a side, like on p. 124. And when we want to use that as a congruent "S" pair for one of these shortcuts, the Reflexive Property gives us something to write down in the Reasons column of our proof, other than "duh, it's obvious." That's . . . pretty much it.

What's It Called?

The Reflexive Property: Every segment (or angle)* is congruent to itself.

Obvious, right?

For example, if we were proving $\triangle OBS \cong \triangle VBS$, we might write $\overline{BS} \cong \overline{BS}$ in the Statement column and use that as an "S" for one of our shortcuts. We'd write "The Reflexive Property" in the Reasons column.

This will all make more sense in the context of a proof!

Notice that in the OBVS diagram, assuming $\triangle OBS \cong \triangle VBS$, we could *reflect* $\triangle OBS$ across the line $\overline{BS}$, and it would match up exactly with $\triangle VBS$. Maybe that's where they got the name "Reflexive" Property . . . ?

QUICK NOTE   Notice the kitty scratch mark on the shared segment $\overline{BS}$. It's optional, but adding that can help us to "see" the two congruent triangles better, like in #1 below.

Let's try some of this stuff, and practice our cheerleading at the same time . . .

Doing the Math

Answer these questions about congruent triangles. I'll do the first one for you.

1. In STAR, based on the diagram markings, $\triangle STR \cong$? Which shortcut proves it?

. . . .

* We'll use the Reflexive Property on *angles* when we prove similarity in Chapter 17.

<u>Working out the solution</u>: Let's tap a finger (or pencil) on the letters S-T-R, in that order. Now we can see that its reflection, in corresponding order, must be A-T-R, right? So we could write: $\triangle STR \cong \triangle ATR$. The kitty scratches tell us $\overline{ST} \cong \overline{AT}$; that gives us an "S"! And the two arcs with tick marks tell us $\angle STR \cong \angle ATR$; that gives us an "A"! Hmm, we need one more part, right? Well, wait a minute. There's a third "congruent pair," because the Reflexive Property lets us say $\overline{TR} \cong \overline{TR}$, so there's another "S." Since the congruent angles are <u>between</u> the pairs of congruent sides, we have SAS. Done!

Answer: $\triangle STR \cong \triangle ATR$ by SAS

2. For the STAR diagram, finish this statement: "$\triangle TSR \cong$?"

3. For the STAR diagram, if we were told that $\angle SRT \cong \angle ART$, then which shortcut(s) (besides SAS) could we use to prove that $\triangle STR \cong \triangle ATR$?

Use the LDY-BUG diagram to answer questions #4–7.

4. Which point in $\triangle BUG$ corresponds to the point Y in $\triangle LDY$?

5. Which point in $\triangle BUG$ corresponds to the point L in $\triangle LDY$?

6. Which segment in $\triangle BUG$ corresponds to the $\overline{YL}$ in $\triangle LDY$?

7. Which angle in $\triangle BUG$ corresponds to $\angle YLD$ in $\triangle LDY$? *(Use 3 letters in the angle name, in the correct order.)*

Use the markings on the diagrams A–D below to answer questions #8–12. Possible answers are: A, B, C, D, None, or *more* than one letter.

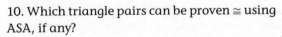
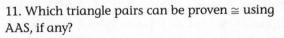
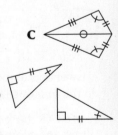

8. Which triangle pairs can be proven $\cong$ using SSS, if any?

9. Which triangle pairs can be proven $\cong$ using SAS, if any?

10. Which triangle pairs can be proven $\cong$ using ASA, if any?

11. Which triangle pairs can be proven $\cong$ using AAS, if any?

12. Which triangle pairs cannot be proven $\cong$, if any?

(Answers on p. 399)

Now that we've warmed up (it's very important to stretch before cheerleading), we're ready for triangle proofs!

Step By Step

Proving two triangles are congruent:

Step 1. Okay, first make sure you see *which* two triangles we want to prove are congruent, and pay attention to the order of the letters. Imagine reflecting/rotating/gliding one triangle onto the other to help "see" the congruence. There might be overlapping triangles, and with all those letters, it can be tricky to sort it out!

Step 2. Look at our Givens: Do we already have any sides or angles that we know for sure are ≅? Fill in what we know, using angle arcs and kitty scratch marks as needed (including on any *shared sides* for the Reflexive Property). Think about our shortcuts: SSS, SAS, ASA, AAS (and later, HL).* Which one is likely to be most helpful?

Step 3. We need *three* S's and/or A's to be ≅, so if we don't already have them, it's time to get 'em! What other info do we have? Key words like *midpoint* can give us ≅ segments, and *angle bisectors* or vertical angles can give us ≅ angles. Givens like *perpendicular bisector* or *isosceles triangle* can tell us info about ≅ sides <u>and</u> angles. And remember, when points or rays *bisect* or *trisect*, we end up with ≅ stuff, too! Remember the Reflexive Property if we're using a shared side!

Step 4. If the three S's and/or A's match up with one of our shortcuts, it's time to write out the proof. In triangle congruence proofs, the last Reason will usually be one of our shortcuts: SSS, SAS, ASA, AAS (and later, HL), followed by parentheses filled with the three Statement numbers corresponding to the three S's and/or A's that allowed us to use the shortcut. Phew, done!

QUICK NOTE   Whether doing homework or while reading the examples in this book or your textbook, I highly recommend redrawing the diagrams, even if just on a piece of scratch paper, and filling in kitty scratches and arcs for congruent stuff as you go. If it helps, try redrawing the triangles so they're

. . . .

* We'll learn the HL shortcut for right triangles on p. 143.

facing the same direction. It takes like two seconds, and it'll seriously keep your brain focused. Plus, it's good drawing practice! (You can even put your paper over the book and "trace" the diagrams to make them neater.)

When reading through these proofs, just follow along for now and get used to the logic; don't worry if you wouldn't know how to do these. The whole point of this chapter is to get you comfortable with triangle proofs. If you already knew how to do them, then I suppose you could have skipped this chapter, couldn't you?

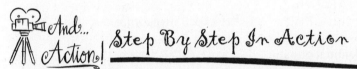

Given: In the IRELAX diagram, $\overline{RI} \cong \overline{XI}$, $\overline{RE} \cong \overline{XA}$, and $\overline{EL} \cong \overline{AL}$. Prove: $\triangle RIL \cong \triangle XIL$.

Step 1. It seems pretty clear that these two triangles should be congruent; just reflect one triangle across the center line, $\overline{IL}$, and we get the other triangle!

Step 2. The kitty scratch marks are already on the diagram, including the side $\overline{RI} \cong \overline{XI}$, which is fantastic. (Gimmie an "S"!) That's the only S or A we are simply "given" in the Givens. But notice that all the info we're given is about segment lengths (nothing about angles), so it sure looks like we'll be going after the SSS shortcut, doesn't it?

Step 3. We'd sure like to say that $\overline{RL} \cong \overline{XL}$ for a second "S," wouldn't we? Well, with the help of the Addition Property (see p. 71), we totally can! And finally, although it wasn't marked, we know the Reflexive Property lets us claim that shared side, $\overline{IL}$, as a third "S." Nice!

Step 4. Time to write out the proof with airtight logic, leading us to our SSS ending.

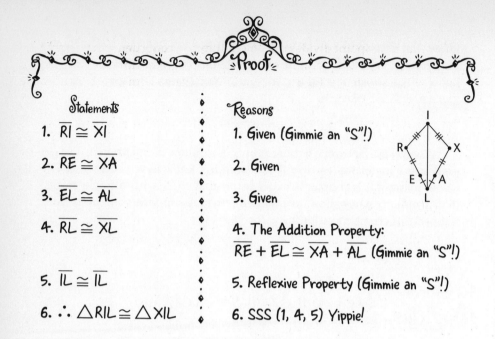

Proof

Statements	Reasons
1. $\overline{RI} \cong \overline{XI}$	1. Given (Gimmie an "S"!)
2. $\overline{RE} \cong \overline{XA}$	2. Given
3. $\overline{EL} \cong \overline{AL}$	3. Given
4. $\overline{RL} \cong \overline{XL}$	4. The Addition Property: $\overline{RE} + \overline{EL} \cong \overline{XA} + \overline{AL}$ (Gimmie an "S"!)
5. $\overline{IL} \cong \overline{IL}$	5. Reflexive Property (Gimmie an "S"!)
6. $\therefore \triangle RIL \cong \triangle XIL$	6. SSS (1, 4, 5) Yippie!

And there you have it: A real, live congruent-triangle proof!

Take Two: Another Example!

Given: In the CAT-COW* diagram, C is the midpoint of $\overline{TW}$, and $\angle A \cong \angle O$. Prove: $\triangle CAT \cong \triangle COW$.

Step 1. Which triangles are we proving congruent? Trace with a pencil from C to A to T, and then from C to O to W. We could imagine putting the $\triangle CAT$ triangle on a bicycle wheel, with the point C at the center of the wheel. Then if we spun the wheel 180°, $\triangle CAT$ would be right on top of $\triangle COW$, perfectly lined up with all the congruent parts, wouldn't it? That helps us to see that, for example, the point W matches up with the point T. Great! We're clear on which triangles they are and how they should be congruent.

Step 2. In the Givens, we're told that $\angle A \cong \angle O$. Gimmie an "A"! We're also given that C is the midpoint of $\overline{TW}$. Well, we

. . . .

* "Cat-cow" is the name of a yoga pose that I love, where you get on all fours and alternately arch and round your back. Stretching feels so good, doesn't it?

know that a midpoint divides a segment into two congruent parts, right? That means $\overline{TC} \cong \overline{CW}$. Gimmie an "S"! Great progress! Hmm, it's hard to know which shortcut we'll use yet, but probably one with an S and an A, like AAS, ASA, or SAS.

Step 3. It seems we've gotten all the useful info out of our Givens that we can . . . and we're still missing an S or an A. But wait, look at the diagram—we have vertical angles, which are *always* congruent (see pp. 28–9 for a refresher on chopstick math). That means $\angle ACT \cong \angle OCW$. Gimmie an "A"! We have two A's and an S. Should we use AAS or ASA? Well, are the congruent sides *in between* the two angles in each triangle? Nope! So we can't use ASA. Looking at $\triangle CAT$ and starting with that congruent side, it goes, "side-angle-angle," or SAA, which is the same as AAS.

Steps 4 and 5. Time to put this down in a proof:

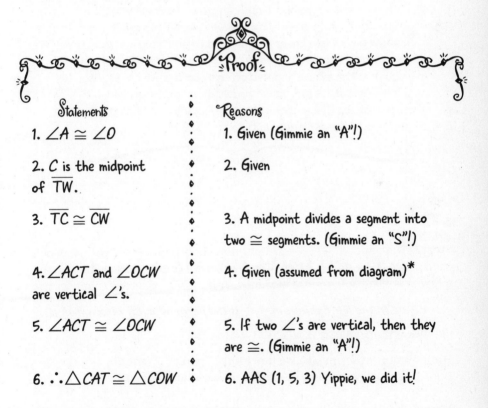

Proof

Statements	Reasons
1. $\angle A \cong \angle O$	1. Given (Gimmie an "A"!)
2. C is the midpoint of $\overline{TW}$.	2. Given
3. $\overline{TC} \cong \overline{CW}$	3. A midpoint divides a segment into two $\cong$ segments. (Gimmie an "S"!)
4. $\angle ACT$ and $\angle OCW$ are vertical $\angle$'s.	4. Given (assumed from diagram)*
5. $\angle ACT \cong \angle OCW$	5. If two $\angle$'s are vertical, then they are $\cong$. (Gimmie an "A"!)
6. $\therefore \triangle CAT \cong \triangle COW$	6. AAS (1, 5, 3) Yippie, we did it!

* Any time we can *assume something from a diagram*, like the presence of vertical $\angle$'s or straight $\angle$'s (see p. 39 for a full list), we don't need to write it in a proof. So that entire step (Statement and Reason) is *optional*. Sometimes I include it to make proofs easier to follow, but as you get more comfortable, feel free to leave it off (like we will on p. 139 and p. 232).

Watch Out!

Just because we can find three S's and/or A's doesn't mean we can actually use a shortcut. For example, if we find two S's and an A, we can use a shortcut only if the congruent angle is *between* the two sides; then we'd use SAS. See p. 133 for more on how to keep all these S's and A's straight. It's not so bad, I promise!

"There was this guy I had a huge crush on. He knew that I was smart, and that my friend (let's call her Jenni) acted dumb to try and get guys. One day he told me that when I hung out with Jenni, he could tell I was acting dumb, too, and that I wasn't my spunky, smart self. He said he'd always had a crush on me, and not her (and she is sooo much prettier than I am). I realized that no matter what guys think, I should be myself and I'll attract the life I want. I'll never play dumb again!" Lila, 15

Take Three: Yet Another Example

Given: In the NICE diagram, $\overline{NC}$ is the perpendicular bisector of $\overline{IE}$.

Prove that $\triangle NIC \cong \triangle NEC$.

Hmm, it almost seems like we don't have enough information, doesn't it? Let's follow the steps and see what happens!

Step 1. Using a pencil to tap on each of the points of $\triangle NIC$ and $\triangle NEC$ in the order they're written, it becomes clear which triangles we're proving congruent. We can just imagine reflecting $\triangle NIC$ across the segment $\overline{NC}$, and we get $\triangle NEC$, after all!

Steps 2 and 3. Hmm, we're not handed any congruent pairs of sides or angles in the Givens, but we've got a shared side, $\overline{NC}$, which means we'll use the Reflexive Property. Gimmie an "S"! Also, the definition of *perpendicular bisector* tells us that C is the midpoint of $\overline{IE}$, and that means $\overline{IC} \cong \overline{CE}$. Gimmie another "S"! Great!

Do we know anything about those last sides? Not really. Do we know anything about the angles *between* the two S's we have? Yep! Because $\overline{NC}$ is $\perp$ to $\overline{IE}$, two right angles are formed—it's also in the definition of *perpendicular bisector* (see p. 37). And two right angles must be equal to each other, right? That means $\angle NCI \cong \angle NCE$. Gimmie an "A"! And they're *in between* the S's we already have, so we'll use SAS. Ta-da! Time to put this into a proof.

Step 4. Hmm, in actually writing out the proof, how do we handle the part about the two right angles being congruent? We'll do it in two steps: "If two segments are $\perp$, then they create right $\angle$'s." And then, "If two $\angle$'s are right $\angle$'s, then they are $\cong$ to each other." Let's do it!

Proof

Statements	Reasons
1. $\overline{NC}$ is the $\perp$ bisector of $\overline{IE}$.	1. Given
2. C is the midpoint of $\overline{IE}$.	2. If a seg is $\perp$ bisector of a seg, then it passes through that seg's midpoint.
3. $\overline{IC} \cong \overline{CE}$	3. A midpoint divides a seg into two $\cong$ seg's. (Gimmie an "S"!)
4. $\overline{NC} \cong \overline{NC}$	4. Reflexive Property (Gimmie an "S"!)
5. $\angle NCI$ & $\angle NCE$ are right $\angle$'s.	5. If two seg's are $\perp$ (as happens with a $\perp$ bisector), then they create right angles.
6. $\angle NCI \cong \angle NCE$	6. If two $\angle$'s are right $\angle$'s, then they are $\cong$ to each other. (Gimmie an "A"!)
7. $\therefore \triangle NIC \cong \triangle NEC$	7. SAS (3, 6, 4) Yippee!

What's the Deal?

The Statements/Reasons 5 and 6 may have seemed a little confusing. It seems obvious that all right angles are congruent to each other, but are we really allowed to just *say* that in a proof? Is that a theorem or a property or something? I mean, if we haven't officially seen it,* what are we allowed to do?

Some of this is teacher-dependent, and some will have to do with which theorems and properties you've seen in your particular textbook up until that point. Most books and teachers are very strict about the "Reasons" column being filled with only theorems and properties (Rules) that have been covered in the class up until that moment. So what you're "allowed" to use as Reasons changes, depending on where you are in the book.

But the true point of these proofs isn't to be word-perfect on some definition. The true point of doing proofs is to master the art of clear, airtight logic. So if you are taking a test and simply can't remember if you've covered an obvious thing like "all right angles are congruent," then my advice is this: Use definitions as much as possible, be clear in your thinking, and if you can, write this clear thinking in "if . . . then" form, like we did above: "If two ∠'s are right ∠'s, then they are ≅ to each other."

You're doing awesome. Have I mentioned that lately?

QUICK NOTE   Notice in the NICE proof on pp. 130–1 that we use the two "halves" of the (converse of the) definition of *perpendicular bisector* in two different steps. In line 2 we used the bisector part, and in line 5 we used the perpendicular part.

---

* Here's the theorem, officially: **All right ∠'s are ≅.**

Minding Our A's and S's . . .

Yeah, we've got a lot of Rules to remember: SSS, SAS, ASA, and AAS (same as SAA). I like to think of it this way. All possible combinations of A's and S's will prove triangles congruent—*except:*

ASS: This is, um, a bad word. It just doesn't work, okay?*

SSA: This is "ass backwards"—equivalent to the above, so of course it doesn't work either.

AAA: All equilateral triangles have three 60° angles, right? So they all have "AAA." But a triangle the size of a gem on your finger and one the size of a pyramid in Egypt are certainly *not* congruent. In fact, by showing a counterexample, we've just *proven* that AAA is not enough to prove congruency. Ta-da!

All the other 3-letter combinations of A's and S's prove triangles congruent!

. . . .

* See the What's the Deal on p. 140 for why it doesn't (typically) work. However, we can use the A-S-S shortcut in one case: for right triangles. But they call it "HL" instead. We'll see more about this on p. 143.

Here are some tips to make finding congruent segments easier. Warning: You must read these <u>slowly</u> and follow along with the diagram on the side, or they are guaranteed to seem confusing. Heck, they're confusing to me if I read them too quickly!

Some Ways to Get Congruent Segments— Gimmie an "S"!

Key Items to Look for in the Givens	What It's Really Telling Us	How We Could Mark the Diagram
Midpoint or (segment) bisector: divides a segment into two ≅ segments. (p. 36)	Congruent segments! If we're told that F is the midpoint (or that $\overline{IF}$ bisects $\overline{NO}$), then we know $\overline{NF} \cong \overline{FO}$.	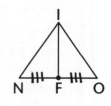
Median (We'll do this on p. 155.)	Congruent segments! If we're told that $\overline{AL}$ is a median of $\triangle TAG$, then we know $\overline{TL} \cong \overline{LG}$.	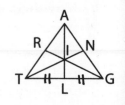
Trisecting points divide a segment into three ≅ segments. (p. 38)	Congruent segments! If we're told L and I trisect $\overline{PT}$, then we know $\overline{PL} \cong \overline{LI} \cong \overline{IT}$.	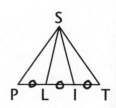

Key Items to Look for in the Givens:	What It's Really Telling Us	How We Could Mark the Diagram
Isosceles triangle: This tells us we have two ≅ segments. (p. 89)	**Congruent segments!** If we're told that the skinny △*LSI* is an isosceles triangle with base $\overline{LI}$ then we know that $\overline{SL} \cong \overline{SI}$.	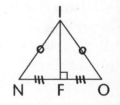
Perpendicular bisector: hits a segment at its midpoint (p. 37), and every point on the perpendicular bisector is equidistant from the segment's endpoints. (We'll do equidistance on p. 158.)	**Congruent segments!** If we're told that $\overline{IF}$ is the ⊥ bisector of $\overline{NO}$, then we know $\overline{NF} \cong \overline{FO}$. AND, because *I* must be equidistant from *N* and *O*, we also know $\overline{IN} \cong \overline{IO}$.	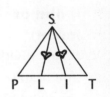

Here are some tips to make finding congruent angles easier. Warning: You must read these slowly and follow along with the diagram on the side, or they are guaranteed to seem confusing. Heck, they're confusing to me if I read them too quickly!

Some Ways to Get Congruent Angles— Gimmie an "A"!

Key Items to Look for in the Givens:	What It's Really Telling Us	How We Could Mark the Diagram
Vertical Angles are always congruent. (p. 29)	Congruent Angles! Just by looking at the diagram, we know that $\angle 1 \cong \angle 2$. And we've got an "A"!	
Angle bisectors (or angle trisectors) divide an $\angle$ into two (or three) $\cong$ $\angle$'s. (p. 38)	Congruent Angles! If we're told $\overrightarrow{SL}$ bisects $\angle PSI$, then $\angle 3 \cong \angle 4$! If we're told $\overrightarrow{SL}$ & $\overrightarrow{SI}$ trisect $\angle PST$, then $\angle 3 \cong \angle 4 \cong \angle 5$.	
Isosceles triangles: This tells us we have two $\cong$ $\angle$'s.	Congruent Angles! If we're told that $\triangle PST$ is isosceles with base $\overline{PT}$, then $\angle 7 \cong \angle 8$.	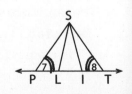

Key Items to Look for in the Givens:	What It's Really Telling Us	How We Could Mark the Diagram
Supplementary angles: ∠'s that are supp. to ≅ ∠'s are ≅. (p. 81)	**Congruent Angles!** Here, if we're told that ∠6 ≅ ∠9, then since we know ∠6 is supp. to ∠7 and ∠8 is supp. to ∠9 (because of the straight lines), we know ∠7 ≅ ∠8.	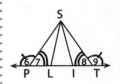
Complementary angles: ∠'s that are comp. to ≅ ∠'s are themselves ≅. (p. 81)	**Congruent Angles!** If we're told that ∠2 ≅ ∠3, then since the diagram tells us ∠1 is comp. to ∠2 and ∠4 is comp. to ∠3, we know ∠1 ≅ ∠4.	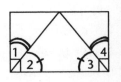
Perpendicular bisector: creates right angles (All right angles are ≅.)	**Congruent Angles!** If we're told that $\overline{IF}$ is the ⊥ bisector of $\overline{NO}$, then we know ∠IFN ≅ ∠IFO.	

And don't forget these Rules, which can also help us get those S's and A's:

Transitive Property (p. 52)	Substitution Property (p. 81)	Addition Property (p. 71)
Subtraction Property (pp. 72–3)	Multiplication Property (p. 79)	Division Property (p. 80)

"When I figure out how to do a proof, it makes my brain feel like it's turning 'on,' like something snaps into place." Sierra, 14

Doing the Math

Use the diagrams and the givens to write two-column proofs. I'll do the first one for you!

1. See the PARTY diagram below.

Given: $\angle 1 \cong \angle 2$ and $\angle APR \cong \angle YPT$.

Prove: $\triangle APR \cong \triangle YPT$.

Working out the solution: Okay, so our Givens tell us the outside angles are congruent and the skinny angles up top are congruent, and we're supposed to prove that the skinny triangles are congruent. So we want to collect congruent stuff for the skinny triangles. The Givens tell us that $\angle APR \cong \angle YPT$, which is an "A" right there, yay! Next, since we know those outside angles are congruent ($\angle 1 \cong \angle 2$), we can show that the supplementary angles *inside* the skinny triangles must also be congruent:* $\angle PAR \cong \angle PYT$. Stepping back and looking at the "big" picture—

---

* See p. 81 to review this Supplementary Angle Rule.

this means that the big triangle, △APY, must be isosceles. And that means its legs, $\overline{PA}$ and $\overline{PY}$, must be congruent. That's an "S"! And now we have ASA. Let's do it!

Statements | Reasons

Statements	Reasons
1. ∠APR ≅ ∠YPT	1. Given (Gimmie an "A"!)
2. ∠1 ≅ ∠2	2. Given
3. ∠PAR is supp. to ∠1.	3. If two ∠'s form a straight angle,* then they are supp.
4. ∠PYT is supp. to ∠2.	4. If two ∠'s form a straight angle, then they are supp.
5. ∠PAR ≅ ∠PYT	5. Angles supp. to ≅ angles are ≅. (Gimmie an "A"!)
6. △APY is an isosceles △.	6. If two ∠'s of a △ are ≅, then it's isosceles.
7. $\overline{PA}$ ≅ $\overline{PY}$	7. If a △ is isosceles, its legs are ≅.† (Gimmie an "S"!)
8. △APR ≅ △YPT	8. ASA (1, 7, 5) Yippie!

2. See the SONG diagram to the right.

Given: $\overline{SO}$ ≅ $\overline{NO}$, $\overline{SG}$ ≅ $\overline{NG}$. Prove: △SOG ≅ △NOG.

....

* See, we could have added a step before this, whose Statement would be, "∠PAR and ∠1 form a straight angle," but the Reason would have been "Assumed from diagram," so we can skip it! See the footnote on p. 129 for the beginning of this discussion.

† On p. 148, we'll see how to combine Steps 6 and 7 with the nifty "If angles, then sides" shortcut. You'll love it.

3. See the TABLES diagram to the right.

Given: $\angle E \cong \angle S$, $\angle T \cong \angle L$, and A & B trisect $\overline{TL}$.
Prove: $\triangle TAE \cong \triangle LBS$.

4. See the MUSIC diagram to the right.

Given: $\angle M \cong \angle S$, $\triangle IUC$ is an isosceles triangle with base $\overline{IC}$. Prove: $\triangle MUI \cong \triangle SUC$.

5. See the CHAIRS diagram to the right.

Given: $\overrightarrow{AI}$ bisects $\angle CAS$, $\overline{CH} \cong \overline{SR}$, and $\overline{HA} \cong \overline{RA}$.

Prove: $\triangle CIA \cong \triangle SIA$.

(Proof solutions at GirlsGetCurves.com)

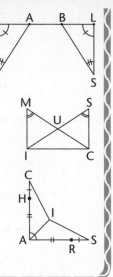

What's the Deal?

Why doesn't SSA (ASS) work? And why *does* it work for right triangles? The illustration below shows the case of SSA for two non-right triangles.

As you can see, two triangles could be made from the same SSA information, depending on which way that side swings.

But* why should SSA be enough for two right triangles? Well, as we'll see in Chapter 11, if we know the lengths of *any* two sides of a right triangle, the Pythagorean Theorem tells us what the third side has to be. That means if two right triangles have two pairs of congruent sides, then the third pair of sides has to be congruent, too! So for right triangles, if we have SS, we automatically have SSS, which means the two right triangles are indeed congruent.

. . . .
* Pun intended.

Takeaway Tips

 There are lots of ways to prove that two triangles are congruent, by just collecting a few pairs of congruent sides and/or angles. In fact, *all* the three-letter combinations of S's and A's work, *except* AAA and the, um, "bad words."

 If you feel lost during a proof, pay attention to the definitions of words in the Givens. One of their definitions could be the first Rule you use in the Reasons column (after "Given," of course!) See p. 385–6 for a summary of proof strategies.

We should always mark the congruent stuff we know with kitty scratches and/or arcs—even shared sides (to be used with the Reflexive Property). It really helps us "see" what we're doing.

TESTIMONIAL

Elisenda Grigsby

(Boston, MA)
<u>Before</u>: Insecure about math ability
<u>Now</u>: Mathematician and filmmaker!

High school was a tough time for me; I remember wondering what I had to do to get all of the "cool" kids to like and accept me, and at the same time, I began to feel very insecure in math class. I was easily intimidated when others (usually loud, show-off guys) seemed "quicker" or "brighter" than me in math class.

Whenever someone else effortlessly answered a question that I struggled

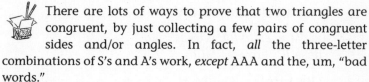

"I was easily intimidated when others seemed 'quicker' or 'brighter' ..."

with, I would beat myself up about it, thinking, "What's the use of working so hard to learn something that is so easy for someone else? I'm no good at this."

Eventually, though, I realized that math is never something that you just know or don't know. A new math concept can be challenging—but it feels so good when you finally *get* it! And it just doesn't matter whether others seem "quicker" or "brighter" than you. First of all, they probably aren't. And second of all, even if they are, who cares? Everyone has the ability to understand mathematical concepts, as long as they are willing to invest the time and effort to understand what is going on. When I finally realized this, I felt much better. And I sure don't feel intimidated anymore...

These days, my life is great. I have tons of friends, and I'm a professional mathematician at Boston College in an area of math called "topology." I work with 3- and 4-dimensional shapes, and it requires a lot of visualizing, just like in transformations and 3-D geometry. My main focus is the study of *knots*: Imagine taking a piece of string, manipulating it in some way, and then fusing the two ends together to form a closed loop. That's what a topologist means by a "knot," and a topologist considers two knots to be the "same" if one can be rearranged to look like the other one *without* cutting the string. Suppose you wanted to know whether a necklace in your jewelry box could be untangled, but you weren't allowed to touch it. Topologists have invented mathematical tools that tell you how to figure this out! Not only that, but a lot of these mathematical tools give us insight into "seeing" in four dimensions.

I also produce short films for a Cambridge-based nonprofit called "Girls' Angle," a math club for girls. Through films, an online magazine, and in-person get-togethers, Girls' Angle exposes girls like you to a bunch of awesome female role models and mentors.

It's amazing when I think of how far I've come from my days as an insecure teenager. I'm so grateful that I believed in myself enough to keep trying even when math seemed intimidating. Remember, math isn't for "other people"—it's for you!

Chapter 9

The Many Uses of Lipstick

CPCTC and More Proofs

Some beauty products are super-versatile—take lipstick, for example. Sure, we can put it on our lips, but depending on the color, we could mix it with a little moisturizer and use it as a cream blush, as an eyeshadow, or as a bronzer on our shoulders for a rosy, sun-kissed look. (Read p. 147 before you try this.) It's like the Swiss army knife of makeup!

As it turns out, congruent triangles are the "lipstick" of geometry. Proving that two triangles are congruent has seriously got to be the most useful tool for proofs, like, ever. And just like how lipstick needs moisturizer for its other uses, congruent triangles need an extra ingredient for their many uses, too. This ingredient is called CPCTC.*

Before we get into that, we're going to see one more way to prove that two triangles are congruent. Remember how ASS (or SSA) *doesn't* prove that two triangles are congruent? Well, if the triangles are right triangles, it *does* work! They gave it a different name, though. Big surprise, right?

Shortcut Alert

The Hypotenuse-Leg Theorem (HL)

If the hypotenuse† and leg of one right triangle are congruent to the hypotenuse and leg of another right triangle, then the two triangles must be congruent!

* If you ever see this "ingredient" on a package of food, run. Very far.

† Reminder: The **hypotenuse** is always the side opposite the right angle.

Yep, if the "A" is a right angle, then A-S-S is enough information to prove that the two triangles are congruent. The H-L makes me think it's saying hello, but with attitude, like, "**Hel-Lo**, the ASS works for me, okay?"

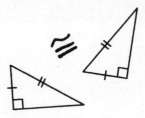

We'll use the HL Theorem in a proof on p. 146. I know you're excited to see HL in action, but a little patience goes a long way, okay?

Proving that two triangles are congruent is, you know, fabulous on its own. But mixed with this next nifty ingredient, we can use it as a tool to prove tons of other stuff, too. So when you're sitting there thinking, "How the heck am I supposed to prove such-and-such?" you'll remember the lipstick and the moisturizer, and you'll see the solution appear before your very eyes . . .

More Cheerleading: CPCTC

Gimmie a "C," "P," "C-T-C!"

Um, yeah, this acronym thing is getting a little out of hand, I totally agree.* But hey, cheerleaders are always shouting letters, right? If they can do it, so can we! So what does this long list of letters stand for?

 CPCTC: <u>C</u>orresponding <u>P</u>arts of <u>C</u>ongruent <u>T</u>riangles are Congruent

It's a mouthful, but all it really means is that if two triangles are congruent, then *all* their corresponding parts are congruent. (They could

. . . .

* Acronyms, like SAS or CPCTC, are combinations of letters that abbreviate a longer phrase, like USA: United States of America, or SCUBA: Self-Contained Underwater Breathing Apparatus!

have called this ACPCTC, to include the word *all,* but I guess they figured that was plenty of letters already. I think we can all agree with that.) If we rewrote this Rule into "if . . . then" form, we'd see that it's just the converse of the definition of *congruent triangles* from p. 120.*

Look, it's got a really boring name, but I like CPCTC, and I think you will, too. Not only is it a really short thing to write in the Reasons column of a proof, but it also lets us be sneaky. For example, in SLY-ONE, let's say we want to prove $\overline{SL} \cong \overline{ON}$. Looking at the markings on the diagram, we have SAS for the two triangles: $\triangle SLY \cong \triangle ONE$. And that means $\overline{SL}$ and $\overline{ON}$ *have* to be congruent, because they are corresponding parts of congruent triangles! Pretty sneaky . . .

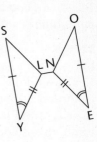

Here's the Step By Step!

Step By Step

Using CPCTC . . . to Prove Congruent Pairs of Angles or Segments:

Step 1. Which two segments (or angles) do we need to prove are congruent? And which two congruent-looking triangles have those as corresponding parts? (We might have to consider a few different pairs of congruent triangles before we find one that works.)

Step 2. To prove that the two triangles are congruent, we can use the Step By Step on p. 126. We'll be collecting S's and A's for SSS, SAS, ASA, AAS, or HL.

Step 3. The original segments (or angles) we wanted to prove congruent are now "corresponding parts of congruent triangles," which means they also must be congruent. We name CPCTC as our last Reason, and we're done!

* "If triangles are congruent, then all of their corresponding parts are congruent."

 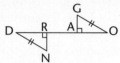

Given: In the DRAGON diagram, R & A trisect $\overline{DO}$, $\overline{DN} \cong \overline{OG}$, and ∠DRN & ∠OAG are right angles. Do a paragraph proof to show that: ∠D ≅ ∠O.

Step 1. If ∠D and ∠O just happen to be corresponding parts on two congruent △'s, then we can use CPCTC to do it! Hmm, can we figure out which two △'s are likely to be congruent, where ∠D corresponds to ∠O? Sure! There's only one pair of △'s here, and it sure looks like △DRN ≅ △OAG. All we'd have to do is rotate this diagram (perhaps centered on a bicycle wheel?), and we can see how one △ becomes the image of the other. So our new goal is to prove that these two △'s are congruent.

Step 2. Let's start gathering A's and S's, and figure out which shortcut we might use. Hmm, since ∠DRN and ∠OAG are right angles, we have an "A."* And since $\overline{DN} \cong \overline{OG}$, that gives us an "S." We're also told that R and A trisect the segment $\overline{DO}$, which by definition means that $\overline{DR} \cong \overline{RA} \cong \overline{AO}$. And that's great news, because the statement $\overline{DR} \cong \overline{AO}$ means we have another "S"! But wait: If we look closely at the order, this gives us, um, ASS. Not good. But wait again! Because these are <u>right</u> triangles, we can actually use HL to prove △DRN ≅ △OAG. Phew!

Step 3. Now comes the easy part. The two △'s are ≅, so that means all their corresponding parts must be ≅. And since ∠D corresponds to ∠O, CPCTC tells us that ∠D ≅ ∠O. Ta-da!

Watch Out!

It's easy to feel lost during proofs. All those letters can make anyone's head spin! It makes ALL the difference to keep referring to the diagram and tapping out the letters with a pencil as you read each segment, angle, or triangle. Do this every time, and you'll do great!

. . . .

* Oddly enough, when using HL in a two-column proof (we'll do one on pp. 151–2), we *don't* need to write a separate line stating that "If two ∠'s are right ∠'s, then they are ≅" (see p. 132), because once we state that we *have* the right ∠'s, we only need to "collect" H and L. If we were using something like SAS, and if the two right angles were the "A," then we *would* need to include that extra step so we could formally "collect" the "A" needed for SAS. Gimmie an "A"!

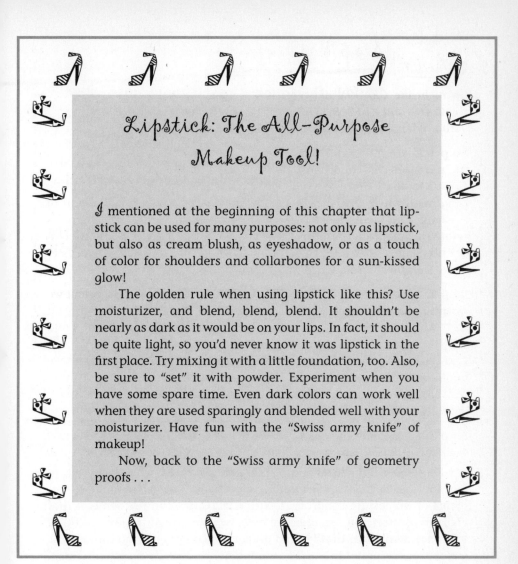

Lipstick: The All-Purpose Makeup Tool!

I mentioned at the beginning of this chapter that lipstick can be used for many purposes: not only as lipstick, but also as cream blush, as eyeshadow, or as a touch of color for shoulders and collarbones for a sun-kissed glow!

The golden rule when using lipstick like this? Use moisturizer, and blend, blend, blend. It shouldn't be nearly as dark as it would be on your lips. In fact, it should be quite light, so you'd never know it was lipstick in the first place. Try mixing it with a little foundation, too. Also, be sure to "set" it with powder. Experiment when you have some spare time. Even dark colors can work well when they are used sparingly and blended well with your moisturizer. Have fun with the "Swiss army knife" of makeup!

Now, back to the "Swiss army knife" of geometry proofs . . .

"SOS" Help

Don't worry, "SOS" doesn't stand for another triangle congruence theorem. This is the SOS in isosceles! Yep, sometimes when wanting to prove that two triangles are congruent, we can get some pretty serious help from our friend, the isosceles triangle. On p. 90, we learned that an isosceles triangle has two very good definitions: one based on its sides, and one based on its angles, right? Now we'll combine those definitions into a very nice shortcut.

Shortcut Alert

"If sides, then angles" and **"If angles, then sides"**

If a triangle has two congruent sides, then we automatically know it has two congruent angles, and vice versa:

 If a triangle has two congruent sides, then the angles opposite those sides are congruent. (If sides, then angles.)

 If a triangle has two congruent angles, then the sides opposite those angles are also congruent. (If angles, then sides.)

I like to see these two theorems as pictures:

If sides, then angles: If △, then △△

If angles, then sides: If △△, then △

We'll use some of this new isosceles stuff in the example on p. 149.

Remember our charts on pp. 134–7 that list a bunch of keywords to look for (like *midpoint* and *angle bisector*) that actually *tell* us that two segments or angles are congruent? With CPCTC, we can use some of that chart in reverse! You see, once we have two congruent triangles, CPCTC tells us that all pairs of sides and angles are congruent, and that means we can easily prove things like midpoints and angle bisectors. This'll make much more sense with some examples.

Our next proof requires a bit more strategizing. I can feel your brain getting stronger . . .

Take Two: Another Example

Given: In the STICKY diagram, $\overline{SC} \cong \overline{IY}$, $\triangle CKY$ is isosceles with base $\overline{CY}$, $\angle TCY \cong \angle TYC$.
Prove: T is the midpoint of $\overline{SI}$.

Step 1. Um, what? How are we supposed to prove something is a midpoint? Okay, first we breathe—deeply. Working backwards, if we look at the definition of *midpoint*, we can realize that all we have to do is prove that the point is in between two congruent segments: $\overline{ST} \cong \overline{TI}$. Sound good? Continuing to work backwards, how could we prove $\overline{ST} \cong \overline{TI}$? Well, if they are corresponding parts on congruent triangles, then we could use CPCTC, right? Gosh, $\triangle STC$ and $\triangle ITY$, sure look congruent! Great; so here's our strategy (in the forward direction): We'll try to prove $\triangle STC \cong \triangle ITY$, then use CPCTC to prove $\overline{ST} \cong \overline{TI}$, and then use the definition of *midpoint* to prove T is the midpoint. Let's see how it goes.

Step 2. Time to prove $\triangle STC \cong \triangle ITY$. We're given $\overline{SC} \cong \overline{IY}$, which is awesome. Gimmie an "S"! What else do we know? We're given that the bottom *little triangle* $\triangle CKY$ is isosceles, so that means $\overline{CK} \cong \overline{YK}$ and $\angle YCK \cong \angle CYK$, right? Hmm, neither of those are parts of $\triangle STC$ and $\triangle ITY$, but we'll mark 'em anyway. The last Given is $\angle TCY \cong \angle TYC$, which also isn't a part of the two triangles. But notice that this means center triangle $\triangle CTY$ is isosceles, which means its sides are congruent: $\overline{TC} \cong \overline{TY}$.
(We'll use a Rule from p. 148 to combine those into one step.) Gimmie an "S"! Marking that on the diagram, too, now we can ask ourselves: What else would we need to prove $\triangle STC \cong \triangle ITY$? With these two S's, for SSS we'd either need the remaining sides to be congruent, $\overline{ST} \cong \overline{TI}$ (but that's essentially what the problem is asking us to prove!), or for SAS we'd need the

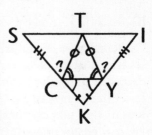

angles *in between* the known sides to be congruent: $\angle SCT$ and $\angle IYT$.

Let's get creative: If we use the Addition Property for angles, we can prove $\angle TCK \cong \angle TYK$, right? And those two angles are supplementary to $\angle SCT$ and $\angle IYT$. So the Supplementary Angles Rule from p. 81 will give us $\angle SCT \cong \angle IYT$: Gimmie an "A"! And SAS proves that $\triangle STC \cong \triangle ITY$. Phew!

Step 3. Now that we have $\triangle STC \cong \triangle ITY$, we know all their corresponding parts must be congruent, which means $\overline{ST} \cong \overline{TI}$, and of course CPCTC is what we write in the Reasons column. Finally, the definition of *midpoint* says that if a point (T) divides a segment ($\overline{SI}$) into two congruent segments, it's the midpoint. And that's what we wanted to prove!

Proof

Statements	Reasons
1. $\overline{SC} \cong \overline{IY}$	1. Given (Gimmie an "S"!)
2. $\angle TCY \cong \angle TYC$	2. Given
3. $\overline{TC} \cong \overline{TY}$	3. If angles, then sides. (Gimmie an "S"!)*
4. $\triangle CKY$ is isosceles with base $\overline{CY}$.	4. Given
5. $\angle YCK \cong \angle CYK$	5. If a $\triangle$ is isosceles, then its base $\angle$'s are $\cong$.
6. $\angle TCK \cong \angle TYK$	6. Addition Property (adding 2 and 5)†
7. $\angle TCK$ is supp. to $\angle SCT$.	7. If two $\angle$'s create a straight $\angle$, then they are supp.
8. $\angle TYK$ is supp. to $\angle IYT$.	8. Same as above
9. $\angle SCT \cong \angle IYT$	9. If two $\angle$'s are supp. to two $\cong$ $\angle$'s, then they are $\cong$. (Gimmie an "A"!)
10. $\triangle STC \cong \triangle ITY$	10. SAS (1, 9, 3) Phew! Almost there!
11. $\overline{ST} \cong \overline{TI}$	11. CPCTC
12. $\therefore T$ is the midpoint of $\overline{SI}$.	12. Definition of midpoint (If a point divides a seg into two $\cong$ seg's, then it's the midpoint.)

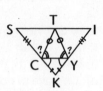

Pant, pant! Great job following that monster proof! Don't worry if you need to read it again to "get" it—that's totally normal!

Notice that we don't necessarily use the Givens in the order they are given. Personally, after I get my strategy down, I like to use the "easiest" Givens first and get an S or an A under my belt as soon as possible.

....

* See if your teacher will let you do the little drawing for your "Reason." Otherwise, you can just write, "If angles, then sides."

† Did we need to include the "(adding 2 and 5)" to this line? No, but it's a great idea to add these kinds of details so you can understand your own proof later, and so you can get partial credit in case you make a mistake somewhere!

Watch Out!

Sometimes when dealing with CPCTC, we'll start off trying to prove one pair of triangles congruent, only to find that we were barking up the wrong tree. It's really helpful to step back and first look for *all* possible congruent triangle pairs that might be useful. We'll do an example like this in exercise 1 below.

QUICK NOTE   If you ever feel panicky, especially when being asked to prove something (perhaps something that doesn't even involve congruence, like midpoints), take a breath, and then look up the *definition* of the thing you're supposed to prove. There's a good chance that its definition involves congruence, and then congruent triangles and CPCTC can probably get you there.

"*I like triangles the most, because there are so many clues that help you find the answer you're looking for.*" Jade, 17

Doing the Math

Complete the following with two-column proofs. I'll do the first one for you.

1. Given: In the KITES diagram, $\overline{KS} \cong \overline{TS}$ and $\angle IKS$ & $\angle ITS$ are right angles. Prove $\overrightarrow{IS}$ is the angle bisector of $\angle KIT$.

Working out the solution: Not going to panic! Let's check out the definition of *angle bisector* on p. 37. Great! So, to prove $\overrightarrow{IS}$ is the angle bisector of $\angle KIT$, we need to first prove that the two $\angle$'s on either side of it ($\angle KIE$ and $\angle TIE$) are $\cong$. Our strategy will

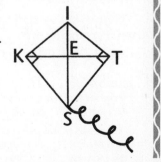

probably be to prove a pair of ≅ △'s, and then ∠KIE & ∠TIE will be corresponding ∠'s, right? Let's see . . . could we prove the two top △'s congruent, △KIE & △TIE? Looking at the bottom half, since △KST is isosceles (after all, we're given $\overline{KS} \cong \overline{TS}$), its two base ∠'s must be ≅, and we could use the Subtraction Property (pp. 72–3) to prove ∠IKE ≅ ∠ITE (an "A"!). Then "if angles, then sides" gives us $\overline{KI} \cong \overline{TI}$ (an "S"!) and the Reflexive Property ($\overline{IE} \cong \overline{IE}$) would give us another "S." But this is ASS, which doesn't work, and now we're stuck. Gosh, maybe we can't prove those top △s are ≅ after all . . . But wait! Let's step back and notice the two big right △'s: △IKS & △ITS. This is WAY easier! They share a hypotenuse, $\overline{IS}$, and we're given $\overline{KS} \cong \overline{TS}$. This means we can use HL (p. 143) to prove △IKS ≅ △ITS! Then CPCTC tells us ∠KIE ≅ ∠TIE. And finally, the definition of angle bisector tells us that $\overrightarrow{IS}$ is indeed the angle bisector of ∠KIT. Phew, done! Here's the proof:

Statements | Reasons

Statements	Reasons
1. ∠IKS & ∠ITS are right angles.	1. Given
2. $\overline{KS} \cong \overline{TS}$	2. Given (Gimmie an "L"!)
3. $\overline{IS} \cong \overline{IS}$	3. Reflexive Property (Gimmie an "H"!)
4. △IKS ≅ △ITS	4. HL (3, 2)
5. ∠KIE ≅ ∠TIE	5. CPCTC
6. ∴ $\overrightarrow{IS}$ is the angle bisector of ∠KIT.	6. Definition of ∠ bisector: If a ray divides an ∠ into two ≅ ∠'s, then it's the ∠ bisector. Ta-da!

2. Write out the DRAGON proof as a two-column proof, using the Givens and diagram from p. 146. *(First, try it without peeking at our strategy!)*

3. For PEARL, Given: ∠*EPR* ≅ ∠*LPA*, and △*EPL* is an isosceles triangle with base $\overline{EL}$. Prove: ∠*EAP* ≅ ∠*LRP*.

4. For LIPSTK, Given: $\overline{IP}$ ≅ $\overline{SP}$; ∠*I* ≅ ∠*S*; ∠*ILK* & ∠*STK* are right angles; $\overline{LK}$ ≅ $\overline{TK}$. Prove: *P* is the midpoint of $\overline{LT}$. *(Hint: What does $\overline{LK}$ ≅ $\overline{TK}$ tell us about angles in the diagram?)*

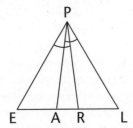 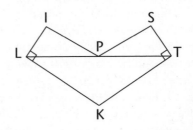

(Proof solutions at GirlsGetCurves.com)

Even if you feel frustrated sometimes, stick with it! It's sinking in, slowly but surely. It just takes some time to get used to it. If you keep reading through the proofs, the logic will start to make more and more sense. You're doing awesome!

"PROOFS" IN REAL LIFE: BEAUTY FOOD!

By nutritionist Jen Drohan

Did you know that there is something that can make your skin glow, your body toned, your brain sharp, and your energy soar through the roof? No, it's not the latest "miracle pill." It's food!

But most food companies pretend their products are healthier than they really are, so we have to be smart and savvy. (See p. 95 for some tricky advertising phrases!) As in geometry, just because something *seems* to be a certain way, doesn't mean it is! Here's how to "prove" that the food you are eating is the real deal:

1. **Know what's inside.** Words like *natural*, *fresh*, and *healthy* are all over food packaging. But those buzz words don't necessarily mean *anything*! Organic is better than not-organic, and always <u>read the ingredients</u> to know what is in your food. A good rule of thumb is this: If you can't pronounce it, send it packing. Hydro-syntho-maldextro-who-zee-whats??? No, thanks!

2. **Eat close to nature.** Food in its original form holds on to all the wonderful properties that make it so healthy. Eat foods that look the way they did when they fell off the tree or were pulled out of the ground, like apples, carrots, bananas, or broccoli.

3. **Listen to your body.** Let's face it; some foods make us feel light & energized, and others make us feel sluggish & restless. Learn to listen to your body's messages. It may turn out to be the only "proof" that you need.

For some great after-school snack ideas, check out p. 82!

Mountains, and More on Triangles*

Let's take a break and do a little mountain climbing. The fresh air, the birds chirping, the triangles . . . okay, maybe it's not really a break.

. . . .

* Not to be confused with "Moron Triangles." First joke I ever learned: *A big moron and a little moron were at the top of a mountain, and the big guy fell off. Why didn't the other one fall off? Answer: Because he was a little more on* . . . Hey, I was like five, okay?

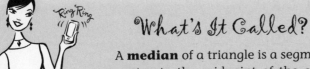

What's It Called?

A **median** of a triangle is a segment drawn from a vertex to the midpoint of the opposite side. In other words, a median bisects the opposite side into two congruent segments. No biggie.

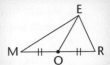

If a segment connects a vertex of a triangle to the midpoint of the opposite side, then it's a median.

For example, $\overline{EO}$ is a median of $\triangle MER$, so O is the midpoint of $\overline{MR}$, and $\overline{MO} \cong \overline{OR}$. When you see the word *median*, think: "Midpoint! Congruent segments, yay!"

Have you ever flown on an airplane? At some point the captain will usually say something like, "Today we'll be flying at an **altitude** of 30,000 feet," meaning that if we drew a perpendicular line from the plane to the ground (sea level), it would measure 30,000 feet. And if we went mountain climbing and decided to shimmy all the way down the rope to the ground, our rope would create a perpendicular line to the ground and would look just like the altitude of a triangle. Let's not do that, okay?

What's It Called?

An **altitude** of a triangle is the perpendicular segment drawn from a vertex to the opposite side; in other words, it creates a right angle. Sometimes an altitude will end up on the *outside* of its triangle, and we'll have to extend the opposite side in order to see the right angle (see the next page). If the "opposite" side lies flat on the ground, then imagine a little mountain hiker dangling on a rope from the vertex and dropping straight down. (Gravity creates a right angle between the rope and the ground.) The rope is our altitude, and that's how you know where to draw it!

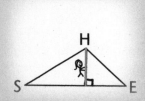 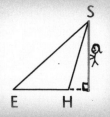 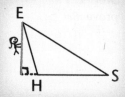

Above are the three altitudes of the *same* triangle, △*SHE*. (Can you see how the triangle rotates clockwise for each one?) And below is a picture of all three altitudes on the same triangle, sans the mountain climber. I think she was tired.

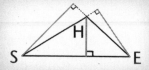

The three altitudes of △SHE

We can always draw a median and an altitude from every vertex of a triangle, so that means that every triangle has three medians and three altitudes!

Next, we're going to use the steps from p. 145 to prove that in an isosceles triangle, the altitude going from the vertex to the base is *also* a median . . . who knew?

And... Action! Step By Step In Action

In the BOTH diagram: Given that △BOT is an isosceles triangle with base BT and OH is an altitude, prove that OH is a median, in a paragraph proof.

Let's use the steps on p. 145.

Step 1. Working backwards, let's notice that if we can prove $\overline{BH} \cong \overline{HT}$, then we'd know that H is the midpoint of $\overline{BT}$. Then we could prove $\overline{OH}$ is a median in the very next step, just by using the definition of *median* in the Reasons column! So now that our goal is to prove $\overline{BH} \cong \overline{HT}$, how can we do that? It sure looks like we have two congruent triangles, doesn't it? If we do, then we can use CPCTC to say that $\overline{BH} \cong \overline{HT}$. Okay, now our goal is to prove $\triangle BOH \cong \triangle TOH$, right?

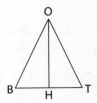

Step 2. Since $\triangle BOT$ is isosceles with base $\overline{BT}$, that means $\overline{BO} \cong \overline{TO}$ and $\angle B \cong \angle T$. Along with the Reflexive Property, we can get SSA, which doesn't quite cut it; we need something else. But hey, we haven't used the fact that $\overline{OH}$ is an altitude: That gives us two right angles, and we can either use HL or SAA to get us $\triangle BOH \cong \triangle TOH$; both will work.

Step 3. Now that we have $\triangle BOH \cong \triangle TOH$, we can use CPCTC to get $\overline{BH} \cong \overline{HT}$, and that means H is the midpoint of $\overline{BT}$, right? And then the definition of *median* tells us that $\overline{OH}$ is, in fact, a median.

Lesson learned: The next time we see the word *altitude* in the Givens, we should draw in the right angle marker right away!

Remember perpendicular bisectors? Sometimes we can skip over the whole congruent triangle/CPCTC thing by using these "equidistant"* Rules. Sounds like a . . . shortcut alert!

. . . .

* **Equidistant** just means "the same distance."

The Equidistant Rules

 If a line is the perpendicular bisector of a segment, then every point on the line is equidistant (the same distance) to the segment's endpoints.

This can be a great tool for finding pairs of congruent sides. For example, in CLIMB, if we were told that $\overleftrightarrow{LB}$ is the $\perp$ bisector of $\overline{CM}$, then we'd automatically know that L, I, and B are each equidistant to the points M and C, meaning that: $\overline{LC} \cong \overline{LM}$, $\overline{CI} \cong \overline{MI}$, and $\overline{CB} \cong \overline{MB}$. Take a moment to see how, along with the Reflexive Property, this gives us three pairs of congruent triangles with SSS. They are $\triangle LCI \cong \triangle LMI$, $\triangle CIB \cong \triangle MIB$, and $\triangle LCB \cong \triangle LMB$. How about that!

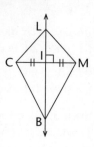

Conversely:

 If two points are each equidistant to the endpoints of a segment, then those two points determine the segment's perpendicular bisector.*

This is great for getting a right angle where we didn't have one before and also for collecting even more congruent sides. For example, in MOUNT, if all we were told is that $\overline{OM} \cong \overline{ON}$ (meaning that O is the same distance to M and N) and $\overline{TM} \cong \overline{TN}$ (meaning that T is the same distance to M and N), then we'd automatically know that O and T are on the $\perp$ bisector! And that means $\overleftrightarrow{OT}$ is the $\perp$ bisector, because two points are enough to determine (define) the entire line.* So we could totally draw in the right angle marker at the U!

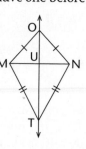

Here's an example of how this shortcut works:

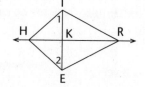

Given: $\overline{RI} \cong \overline{RE}$, $\angle 1 \cong \angle 2$. Prove: K is the midpoint of IE.

Normally, we'd be thinking—hmm, looks like congruent triangles (we'd actually have to prove two sets of $\cong \triangle$'s!) and CPCTC. It'd be a pretty long proof . . . But check this out:

. . . .

* See p. 389 in the Appendix for how two points "determine" a line.

Statements	Reasons
1. $\angle 1 \cong \angle 2$	1. Given
2. $\overline{HI} \cong \overline{HE}$	2. If angles, then sides.
3. $\overline{RI} \cong \overline{RE}$	3. Given
4. $\overleftrightarrow{HR} \perp$ bis. $\overline{IE}$*	4. If two points (H & R, from Steps 2 & 3) are each equidistant to the endpoints of a seg ($\overline{IE}$), then those two points determine the seg's $\perp$ bisector ($\overleftrightarrow{HR}$).
5. $\therefore$ K is the midpoint of $\overline{IE}$.	5. Definition of $\perp$ bisector (p. 37).

Ta-da! Nice shortcut, eh?

Also notice that in the NICE problem on pp. 130–1, we could have used this new equidistant stuff to prove $\triangle NIC \cong \triangle NEC$ with SSS, instead of SAS: Replacing Steps 5 and 6, we could have said "N is equidistant to I and E" with the reason "If a line is a $\perp$ bis of a seg, then its points are equidistant to the seg's endpoints." And our next statement would have been "$\overline{NI} \cong \overline{NE}$," and our reason could have been "Definition of equidistant." In fact, there are often many ways to do one proof . . .

QUICK NOTE   If the triangle is a right triangle, then its legs (the non–hypotenuse sides) will actually <u>be</u> two of the altitudes. After all, each leg <u>is</u> the segment going from one vertex and creating a right angle with the opposite side, isn't it?

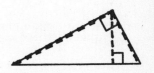

. . . .

* "$\overleftrightarrow{HR} \perp$ bis $\overline{IE}$" means "$\overleftrightarrow{HR}$ is the $\perp$ bisector of $\overline{IE}$." It's a common abbreviation in most textbooks. I guess it literally stands for "$\overleftrightarrow{HR}$ perpendicularly bisects $\overline{IE}$" or something. There's some creative grammar for ya.

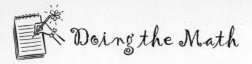

Doing the Math

Do these problems involving medians, altitudes, and equidistance. I'll do the first one for you.

1. In SPLIT, Given: L and I trisect $\overline{PT}$. Prove: $\overline{SL}$ is a median of $\triangle PSI$.

<u>Working out the solution:</u> It seems like that triangle doesn't have a median marked, but wait—there are big, medium, and skinny overlapping triangles here. If we pay attention, we'll see that $\triangle PSI$ is the medium-sized triangle on the left. In order to prove that $\overline{SL}$ is a median, we need to prove that L is the midpoint of $\overline{PI}$; in other words, that $\overline{PL} \cong \overline{LI}$, right? And hey, the definition of *trisect* means: $\overline{PL} \cong \overline{LI} \cong \overline{IT}$. Here's the proof!

Statements | Reasons

Statements	Reasons
1. L and I trisect $\overline{PT}$.	1. Given
2. $\overline{PL} \cong \overline{LI} \cong \overline{IT}$	2. Definition of trisect
3. L is the midpoint of $\overline{PI}$.	3. If a point divides a seg into 2 $\cong$ seg's, it's the midpoint.
4. $\therefore$ $\overline{SL}$ is a median of $\triangle PSI$.	4. Definition of median: If a seg connects a vertex (S) to the midpoint of the opposite side ($\overline{PI}$), then it's a median.

2. Name the three medians appearing in $\triangle RIK$, $\triangle DRN$, and $\triangle BAE$.

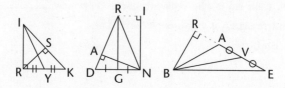

3. Name the five altitudes appearing in $\triangle RIK$, $\triangle DRN$, and $\triangle BAE$.

4. In △WHY, is $\overline{HY}$ an altitude? If so, why?

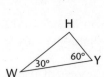

5. In MEDIAN, Given: $\overline{DN}$ is a median in the isosceles triangle △MDA with base $\overline{MA}$, and ∠ENM ≅ ∠INA. Prove: $\overline{EN} ≅ \overline{IN}$.

6. In BOTH (see above), Given: △BOT is an isosceles triangle with base $\overline{BT}$, and $\overline{OH}$ is an altitude. Prove: $\overline{OH}$ is a median. *(Hint: Try doing it without peeking, but see p. 157 for the logic if you need it! And if you want your answer to match my online solution, use HL instead of SAA.)*

7. In WAYS, Given: △WAS is an isosceles triangle with base $\overline{WS}$, and $\overline{AY}$ is a median. Prove: $\overline{AY}$ is an altitude. *(Hint: Use the equidistant stuff from p. 158 and the definition of altitude.)*

8. In TRAIL, Given: ∠1 ≅ ∠2 and ∠3 ≅ ∠4. Prove: $\overline{TA} ≅ \overline{IA}$. *(Hint: First prove $\overleftrightarrow{RL}$ is a ⊥ bis; see pp. 158–9.)*

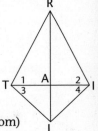

(Answers to #2–4 on p. 399; proofs at GirlsGetCurves.com)

By the way, altitudes are good for more than just mountain climbing; they'll help us find the area of triangles, too, which we'll do on p. 347.

Takeaway Tips

 HL: In two right triangles, if the hypotenuses and one pair of legs are ≅, then the triangles themselves are ≅.

 CPCTC stands for "Corresponding Parts of Congruent Triangles are Congruent." Once we prove two triangles are congruent, this theorem tells us that *all* their corresponding parts are congruent, and we can use this to prove tons of stuff!

 Isosceles triangles have two ≅ sides *and* two ≅ angles, and remember: "If angles, then sides" and "if sides, then angles." If we have one, we know we have the other!

 A **median** of a triangle is the segment drawn from a vertex to the midpoint of the opposite side, and an **altitude** of a triangle is the ⊥ segment drawn from a vertex to the (sometimes extended) opposite side.

 Using equidistance can make proofs much shorter. Keep an eye out for two sets of ≅ segments like in MOUNT on p. 158. That signals that we have a ⊥ bisector!

 Always tap out each letter on angles, segments, and triangles during proofs, and fill stuff in on the diagram as you go. It makes it much easier to "see" what's going on.

Danica's Diary
WHAT WE *REGULARLY* DO!

We all have ideas in our own heads about who we are. But who are we, *really*? Picture this: A popular girl talks all the time about how she's a writer and how she's writing a novel, blah blah blah. But when she goes home at night, she spends most of her time on social networking sites, and somehow she barely has time to finish her homework, let alone work on the "novel." Then there's the shy girl who goes home and writes poetry almost every day, silently filling the pages of her worn-out journal. *Which one is the writer?* Or consider the girl whose mom owns a yoga studio. She's always wearing yoga clothes and obviously thinks of herself as a "yoga person." But how often does she actually do yoga?

I've always thought of myself as the kind of person who is responsible, on time, etc. And yet I went through a period where I was always at least a few minutes late, wherever I was going. I remember my mom warning me, "Someday when you have a job, if you are always on time, then the *one day* you're 30 minutes late, your boss will be worried about you and relieved when you walk in the door. But if you're always 5-10 minutes late, then the *one day* you're 30 minutes late is the day you'll be fired."* It's not about being 30 minutes late that one day; it's how we define ourselves all of the other days. See what I mean? It's what we *regularly* do that defines us.

If we regularly do our homework on time, then the one day we stay up late to help a friend who really needs to talk isn't going to ruin our grades. If we regularly think nice thoughts about ourselves, then one self-inflicted insult isn't going to damage our self-esteem much. And if we regularly exercise and eat well, then a day spent on the couch with an ice-cream sundae isn't going to kill us. But if we keep ice cream

....

* Today, I'm almost never late. I've made it a habit to arrive 15 minutes early to work or other important things, because if there's traffic, then there's still a darn good chance that I'll be on time. It works great!

in the freezer for a nightly snack...well that's a different story.

What do you *regularly* do?

I'll be honest: Aside from big things like "go to school" and "soccer practice," it can be difficult to notice our habits during a busy day, and *thoughts* are even harder to keep track of. But if we can't step back and see the big picture, then it's hard to make the positive changes that we crave. My trick? Keeping track in a daily log of whatever habit I'm trying to break—or create. For example, because I tend to be a stress case, I programmed my phone to remind me to "Relax and think happy thoughts" every day. I'm not kidding—it works. And to help me stay on task when I'm writing my books, every day I keep track of the number of pages I've written on my computer calendar. It keeps me honest about my progress!

Are you always feeling the pressure of time? Then keep track of *how* you spend your time. You might discover that you're "wasting" more time than you realize. Are you always wishing you had better skin? Then keep track of the food and beverages you consume. You might be surprised to see that you aren't drinking nearly enough water or eating enough fresh veggies, and maybe you're eating more junk food than you intend to. Do you think of yourself as a clean person, but you often forget to wash your face at night? Then start keeping track of that. It'll help motivate you to remember.

Tracking your habits or thoughts isn't as much work as it might seem, and you can learn a lot by jotting down notes for even just one week. Try it! Use an app on your phone or keep a sheet of paper to log your habits. Find out what kind of person you *really* are—and then you'll be in a much more powerful position to make the positive changes you've been wishing for.

Chapter 10

Your Sister's Crush

Proofs by Contradiction (AKA Indirect Proofs)

There's a time and a place for everything—including making assumptions! Like, even if you've never tried a peanut-butter-and-sour-cream cookie before, it's pretty safe to assume that if you eat too many, you'll end up with a tummy ache. In life, we tend to assume things that we *think* are true, and sometimes it can actually help us (like with cookie restraint). But in math, we can only assume things that we're pretty sure *aren't* true.

Huh?

Yep, and this is how we start my favorite type of logic: Proofs by contradiction. If we want to prove that something is true, we first "assume" it's *not* true. Then we create an airtight argument that leads us from the assumption to a conclusion that's clearly false, which we call a **contradiction**. And this tells us that our assumption <u>must</u> have been false to begin with. This type of assuming is kind of like *pretending*. I'll show you what I mean.

Let's say your sister tends to have strong feelings, and when it comes to guys, either she totally likes a guy, or she's not interested at all: It's always one or the other. You currently suspect that your sister is crazy-obsessed with some guy named Mike, but she won't admit it. One way to "prove" that she really does like him would be to pretend—or "assume"—that she *doesn't* like him. You could say something like, "Oh, cool. You're not interested at all. Well then, I guess you wouldn't mind if I set Mike up with a friend of mine."

Your sister would freak out, which would be a contradiction: After all, if she didn't like him, she wouldn't be freaking out, would she? Now you know that the assumption that she <u>didn't</u> like him *must* have been false to begin with. The only other option is that she <u>does</u> like him . . . And you'd have finished your proof.*

Here's an example in math.

. . . .

* Of course, this would be cruel . . . and I have a feeling you're not the kind of person who would use her powers of logic for evil.

Given: $\overline{LI} \cong \overline{RI}$, $\angle LIA \not\cong \angle RIA$. Prove: $\overline{LA} \not\cong \overline{AR}$.

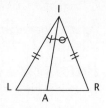

Proof: Look, either $\overline{LA} \cong \overline{AR}$ or $\overline{LA} \not\cong \overline{AR}$, right? It's gotta be one or the other. Our strategy will be to assume the opposite of what we're trying to prove (and then use that to reach an impossible conclusion).

So, we'll assume that $\overline{LA} \cong \overline{AR}$. We've been given $\overline{LI} \cong \overline{RI}$ (and, of course, $\overline{IA} \cong \overline{IA}$), so, using our assumption $\overline{LA} \cong \overline{AR}$, we could use SSS to prove that $\triangle LIA \cong \triangle RIA$, right? And then by CPCTC, we could conclude that $\angle LIA \cong \angle RIA$. But wait! This contradicts one of the Givens, which said $\angle LIA \not\cong \angle RIA$. So this means our initial assumption, $\overline{LA} \cong \overline{AR}$, <u>must</u> have been false. And $\overline{LA} \not\cong \overline{AR}$ is the only other option!

$$\therefore \; \overline{LA} \not\cong \overline{AR}$$

QUICK NOTE   If we're being asked to prove that something isn't true, then a proof by contradiction is probably the easiest way to do it.

Step By Step

Proofs by contradiction, AKA indirect proofs:

Step 1. Look at what we're being asked to prove, and make sure that there are only two possibilities for it: Either the thing is true, or it isn't.

Step 2. Being that there are only two possibilities, assume the *opposite* of what we're supposed to prove.

Step 3. Now we ask ourselves, "What if this assumption were true? What else would *have* to be true?" And we use the assumption to create an airtight argument (proof) leading to a conclusion that ends up being totally false, because it contradicts a fact from earlier in the proof (usually a Given). This tells us the assumption must have been false! (FYI—often, the assumption helps us "prove" two triangles are congruent, and the CPCTC helps us get to a contradiction.)

Step 4. Because we've shown that our assumption was false to begin with, this means the only other possibility—our desired conclusion—has been *proven true*. Done!

And... Action! Step By Step In Action

Refer to the diagram △GUY. Given: $\overline{GU} \not\cong \overline{YU}$.
Prove: $\angle G \not\cong \angle Y$.

Step 1. Look, either $\angle G \cong \angle Y$ or $\angle G \not\cong \angle Y$, right? It's gotta be one or the other!

Step 2. Since we're being asked to prove $\angle G \not\cong \angle Y$, we'll start this indirect proof by "assuming" that $\angle G \cong \angle Y$ and see what kind of trouble* it gets us into.

Steps 3 and 4. Well, what if $\angle G \cong \angle Y$? Then we'd know that △*GUY* is an isosceles triangle, right? But wait! Isosceles triangles have congruent legs (see p. 148 for the "If angles, then sides" Rule), which would mean that $\overline{GU} \cong \overline{YU}$, and that contradicts our Given. So that means our assumption, $\angle G \cong \angle Y$, *must* have been false to begin with, which means it *must* be true that $\angle G \not\cong \angle Y$. Done!

Proof

Statements	Reasons
1. $\overline{GU} \not\cong \overline{YU}$	1. Given
2. Assume $\angle G \cong \angle Y$.	2. Assumption leading to a possible contradiction
3. $\overline{GU} \cong \overline{YU}$	3. If angles, then sides. Contradicts #1!
4. ∴ $\angle G \not\cong \angle Y$	4. Our assumption ($\angle G \cong \angle Y$) must have been false, because Statements #1 and #3 contradict each other.

. . . .

* The good kind of trouble.

QUICK NOTE   Every teacher is different. You might be asked to do these indirect proofs in paragraph form or in two-column format. And now we've done an example of each!

Watch Out!

The only reason proofs by contradiction work is because the "assumption" and our desired conclusion (the thing the problem wants us to prove) are the <u>only two possible outcomes</u>. Either two segments are congruent, or they aren't congruent. Either a ray bisects an angle, or it doesn't bisect that angle. *That way, when we've disproved one, the other one is automatically proven.* In the sister example we did on p. 165, it was important that she had strong feelings and always either really liked a guy or didn't like him at all. That way, when we proved that "not liking him at all" was false, then it had to be true that she totally liked him. If a possibility was that she might sort of like him, then we wouldn't have proven she was obsessed.

Here's an example of the kinds of indirect proofs you'll do in pre-calculus and beyond. It might look very different, but it uses the *same exact* logic we've just learned.

Take Two: Another Example

Prove that there are an infinite number of odd numbers.

Steps 1 and 2. Not panicking, let's first realize that there are either a *finite* number or an *infinite* number of odd numbers, right? It's gotta be one or the other. So let's assume the opposite of what we're supposed to prove: We'll assume that there are a finite number of odd numbers.

Steps 3 and 4. Well, if the number of odd numbers is finite, there would have to be a *highest* positive odd number. Think about that for second—it would have to be true. Let's call that highest odd number N, because it's a free country after all. But then $N + 2$ would also be odd, wouldn't it? And $N + 2 > N$, no matter what N equals, right? So we've just found an odd number *higher* than N, which contradicts our assumption that it was the highest one. That means our assumption must have been incorrect—*there can be no highest odd number*, because we can always find one greater. And because the only remaining option is that there are an *infinite* number of odd numbers, we've finished our proof!

∴ **There are an infinite number of odd numbers.**

Pretty nifty, eh? I'm so proud of you for reading that.

Doing the Math

Do these proofs using indirect reasoning. Use a paragraph proof or two-column proof—your choice! I'll do the first one for you.

1. Given: The triangle $\triangle RUN$, $RU = 7$, $RN = 9$.

Prove: $UN \neq 20$.

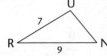

Working out the solution: Either $UN = 20$ or $UN \neq 20$, right? So assume that $UN = 20$, the opposite of what we want to prove. But if $UN = 20$, then this can't be a triangle! Remember the Triangle Inequality from p. 99? The sum of the lengths of any two sides of a triangle is always greater than the length of the third side; in other words, it would have to be true that: $7 + 9 > 20$. Oops! We've reached a contradiction, which means that our assumption, $UN = 20$, must have been false, and that means $UN \neq 20$. Nice.

∴ $UN \neq 20$

2. Refer to BIKE. Given: $\overline{BE} \cong \overline{KE}$, I is not the midpoint of $\overline{BK}$. Prove: $\overrightarrow{EI}$ does not bisect $\angle BEK$.

3. Prove that there are an infinite number of negative numbers. *(Hint: See pp. 168–9 for a similar example.)*

4. Refer to SWIM. Given: $\overrightarrow{SI}$ bisects $\angle WSM$, $\angle SIW$ is obtuse. Prove: $\overline{SW} \neq \overline{SM}$. *(Hint: First, "prove" two triangles are congruent, and then figure out how to contradict "$\angle SIW$ is obtuse" by proving something about $\angle SIM$ and $\angle SIW$. You can do this!)*

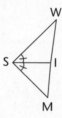

(Proof solutions at GirlsGetCurves.com)

"Logic helps in life, like for determining the most efficient route on road trips, getting a good deal on a purchase, budgeting, etc.—we need to be good at it." **Jacob, 15**

Takeaway Tips

 If a problem asks us to prove that something *isn't* true, then proofs by contradiction (AKA indirect proofs) are probably the best strategy to use.

 In a proof by contradiction, we first look at the thing we're supposed to prove, and we assume the opposite is true, which we then use—in an airtight proof—to arrive at a contradiction.

 The only reason proofs by contradiction work is because the "assumption" and our desired conclusion (the thing the problem wants us to prove) are the <u>only two possible outcomes</u>. That's why disproving the "assumption" actually *proves* the desired conclusion.

Spelling

You might think spelling isn't important in math class, but I have news for you: It's important *everywhere*. Good spelling makes us look smart and professional, and it's a great habit to get into. Here are some tips to help keep you from making some common spelling mistakes in geometry.

Word	How to remember the right spelling!
Para_ll_el	There are two L's in the middle—just like the parallel bars in gym class. (There's also some symmetry with "ARA" and "LEL.")
Perp_en_dicular	You might use a pen to draw perpendicular lines, right? The middle syllable is PEN!
Theorem	Did you know that REM means "rapid eye movement" and it's what happens when we're in a deep sleep? You know, from being totally bored (never by math class though . . . never, nope). Well, there was this guy, Theo, who was really bored in class and he always fell asleep: Theo-REM.
Postulate	Okay, so you and your brother took nearly the identical amazing picture at a concert. Your brother bragged that he'd get his posted online first, but you were first and said, "Post! U late! Me first!"

Word	How to remember the right spelling!
An<u>gle</u>	Often misspelled as "angel." Here's how to avoid this: Just before you write it, *say it out loud in your head*. *Angle* sounds like "ang–gul" and *Angel* sounds like "ain–gel"—gel, like hair gel. Do you want it to sound like "gel"? If not, then don't write it!
Bise<u>ctor</u>	I used to write "bisector" all the time, oops! The *bi* part was pretty easy . . . then I realized bisectors divide stuff into two equal <u>sect</u>ions, and I stopped making the mistake.
Isos<u>c</u>eles	The *C* is at the very Center of the word, and it's surrounded by SOS and ELE. Lots of symmetry, just like an isosceles triangle has . . .
Hypo<u>ten</u>use*	Hypo–ten–use: Just remember the number *ten* (we're doing math, after all) and the word *use* (because it's, um, useful)!

....

* We'll deal with hypotenuses in Chapter 11. Oh yes, we will.

Danica's Diary
INSECURITIES: WHAT GIVES?

We all have insecurities. These are negative thoughts we have about ourselves. They can be big and general, like "I'm not pretty enough," "I'm not smart," and "Nobody likes me." Or they can be specific ones, like "I hate my hair," or "I suck at math." But they can all be summarized as, "I'm not good enough," right? So where do the negative thoughts come from? Sometimes they are things other people have actually said to us (like an angry parent or sibling), and sometimes they come from our own imaginations, as a result of comparing ourselves to others. (See p. 78 for more on that.)

Wherever the negative thoughts came from, we need to deal with them so they don't hold us back. There are two different types of insecurities: helpful ones and not helpful ones. Helpful insecurities involve things about ourselves that we have the power to change. For example, "I'm no good at math," or "I'm too shy to make new friends" are both things you have the power to improve. Walk boldly through your shyness and introduce yourself to the new girl. Just put one foot in front of the other and do it. Yes, you might feel your cheeks flush or feel adrenaline pumping through your veins, but do it anyway. She's probably shyer than you are and will be grateful. And regardless, *walking through your fears* makes you one kick-ass young woman. As for the math insecurity, well, if you're reading this book, then you're already on the path to success, my friend.

On the other hand, some insecurities/negative thoughts involve things we *can't* change; for example, "That guy doesn't like me," or "I wish I didn't have freckles." Okay, sure, you can use makeup to cover those adorable freckles, but it's important to first *accept* them, *own* them, and then think of covering them with makeup as a fashion choice, like wearing a hat or scarf. It's a different look—it's not better or worse, just different.

In terms of the guy who doesn't like you, well, trying to change other people is a <u>huge</u> waste of energy. Time to let it go and move on, sister. We don't always get everything we (think we) want,* and it's really okay. Remember, the only person's behavior that we can change is our own.

There's a great poem called the Serenity Prayer that has really helped me:

SERENITY PRAYER

God,† grant me the serenity to accept the things
   I cannot change,
The courage to change the things I can,
And the wisdom to know the difference.

So when you're feeling insecure and wishing something were different, think about this poem and ask yourself, "Is this something I have the power to change or not?" If you can't change it (like when it's someone else's behavior), then let it go. Learning to accept things we can't change is an <u>amazing</u> life tool to practice and will help you in ways you can't even imagine. However, if the insecurity involves something you *do* have the power to change, then muster up that courage and do it—even if you just take one small step. We need to make sure our insecurities don't hold us back or keep us from joining that sports team or trying out for the school play, after all! If your insecurities are really getting you down, don't hesitate to reach out to a counselor or other trusted adult for help.

Be your own best friend and cheerleader, and *use* those insecurities to learn how to push through your fears and become the strong young woman you deserve to be. (Check out pp. 281-3 for some great ways to do this.) Remember, everyone has insecurities; it's how we *deal* with those insecurities that defines us.

. . . .

* See pp. 68–9 for more on "wanting to want" things.

† "God" has many definitions. I've personally never seen God as an old white guy; I like to think of God as *universal love,* which doesn't have a gender.

Chapter 11

Little Miss Know-It-All

Right Triangles and the Pythagorean Theorem

$\mathcal{N}$obody likes a know-it-all, but when you're right, you're right. And when you've got a right triangle and you know two sides, then you *do* know it all: Yep, if we know two sides of any right triangle, we can figure out the third side. Every time. And all thanks to a little-known Greek mathematician named Pythagoras.

Ring Ring What's It Called?

The Pythagorean Theorem is an equation describing the relationship between the three sides of any right triangle, where **a** and **b** are the two **legs** (the shorter sides) and **c** is the **hypotenuse** (the side opposite the right angle):

$$a^2 + b^2 = c^2$$

And this relationship is *always* true for *every* right triangle. So if we're given any two sides of a right triangle, we can figure out the third side—every time.

This little but powerful theorem is a perfect union of geometry and algebra, and it has been used for thousands of years. Finally, we get to join in. Oh, and speaking of algebra, here's a lightning-quick review of square roots!

Lightning-Quick Square Roots Review!*

The square root of a real number is the positive value (or zero), which when multiplied times itself gives you the original number. For example:

$$\sqrt{9} = 3 \quad \text{because} \quad 3^2 = 9$$

$$\sqrt{0.09} = 0.3 \quad \text{because} \quad (0.3)^2 = 0.09$$

$$\sqrt{64} = 8 \quad \text{because} \quad 8^2 = 64$$

$$\sqrt{\frac{1}{64}} = \frac{1}{8} \quad \text{because} \quad \left(\frac{1}{8}\right)^2 = \frac{1}{64}$$

For positive numbers, squares and square roots "undo" each other (they are inverse operations), so $\sqrt{3} \cdot \sqrt{3} = 3$. In other words, $(\sqrt{3})^2 = 3$. And as we can see above, $\sqrt{3^2} = 3$. In fact, this is true for all $a \geq 0$:

$$(\sqrt{a})^2 = a \quad \text{and} \quad \sqrt{a^2} = a$$

Also, an expression like $3\sqrt{5}$ *means* $3 \cdot \sqrt{5}$, so an exponent would distribute like this: $(3\sqrt{5})^2 = 3^2 \cdot (\sqrt{5})^2 = 9 \cdot 5 = \mathbf{45}.$[†]

We can multiply square root expressions with this rule:

$$\sqrt{a} \cdot \sqrt{b} = \sqrt{ab}$$

For example, $\sqrt{6} \cdot \sqrt{2} = \sqrt{6 \cdot 2} = \sqrt{12}$, which we'd then simplify by factoring out the (hiding) perfect square, 4, and we'd get: $\sqrt{12} = \sqrt{4 \cdot 3} = \mathbf{2\sqrt{3}}$.

We can add and subtract square root expressions if their square root parts are identical. For example, $3\sqrt{10} + 5\sqrt{10} = 8\sqrt{10}.$[‡]

....

* For a full review of squares and square roots (and how the Pythagorean Theorem can make your pool prettier), check out Chapter 19 in *Hot X: Algebra Exposed*.

† For distributing exponents (with Ms. Exponent herself!), see pp. 256–8 in *Kiss My Math*.

‡ Just pretend the $\sqrt{10}$ is x, and then it's easier to see: $3x + 5x = 8x$.

Let's see the Pythagorean Theorem in action. If we're given a triangle with legs measuring 2 and 4, then we can find the length of the hypotenuse by plugging $a = 2$ and $b = 4$ into the formula $a^2 + b^2 = c^2$, which gives us: $(2)^2 + (4)^2 = c^2 \rightarrow 4 + 16 = c^2 \rightarrow c^2 = 20$.

At this point, we take the square roots of both sides and get: $\sqrt{c^2} = \sqrt{20}$. Since we know c is positive, the left side becomes "c."* And we can simplify the right side like this: $\sqrt{20} = \sqrt{4 \cdot 5} = 2\sqrt{5}$. So our equation has become: $c = 2\sqrt{5}$. And ta-da! We've found the hypotenuse.

Usually when we do these problems, we end up with a square root in the final answer. But sometimes we'll end up with an answer like $\sqrt{289}$, which believe it or not, equals 17. Luckily, there's a shortcut to help us in situations like that . . . well, for certain *families*, anyway.

Shortcut Alert

"Pythagorean Triple" Triangle Families

There are some right triangles whose sides are *all* whole numbers—no square roots involved! And if we find one, we can skip the Pythagorean Theorem completely.

Four families according to their sides' ratios:

"Leg:Leg:Hypotenuse"

"3:4:5" "5:12:13" "7:24:25" "8:15:17"

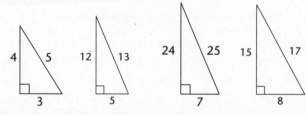

The reason they're called "families" is because each of the above triangles represents a whole bunch of triangles† with the same ratio! Consider this: If we multiplied all three lengths of a 3-4-5 triangle by any

....

* To totally understand what goes on when we take the square roots of variables, check out pp. 369–71 in *Hot X: Algebra Exposed*.

† An infinite number of triangles, actually.

positive number like 2, 100, or even $\frac{1}{101}$, we'd get another triangle in the 3:4:5 family—a right triangle with the *same exact shape* as the 3-4-5 triangle; it would just be bigger or smaller.* These new triangles would have lengths 6-8-10, 300-400-500, and $\frac{3}{101}$ - $\frac{4}{101}$ - $\frac{5}{101}$, respectively.

The reason these families are considered a *shortcut* is because if we're told that two legs of a right triangle are 5 and 12, then instead of using $a^2 + b^2 = c^2$, we could just say, "Hey, this is a 5-12-13 triangle!" and we'd automatically know that the hypotenuse is **13**. Or if we saw a triangle with legs 10 and 24, since we're clever, we'd say, "Hmm, this looks like it's in the 5:12:13 family where each triangle side was multiplied by 2, so the hypotenuse must be 2(13) = **26**!"

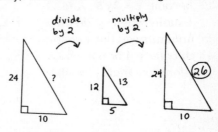

Sure, we could do it the long way by setting up the equation $10^2 + 24^2 = c^2$, squaring everything out and then taking the square root of both sides. But the numbers start getting big and annoying. I highly recommend learning these four families; it'll make you way faster on tests!

There are many such triplet "families" (an infinite number, in fact!) with *whole number ratios* between all of their sides, but these four involve the smallest numbers, so they're the most common ones to pop up on tests.

Step By Step

Using the Pythagorean Theorem to solve a right triangle:

Step 1. Identify the right triangle(s) involved, and make sure you know which side is the hypotenuse and which sides are the legs. Draw in the right-angle box if it's not already done.

Step 2. Could this triangle belong to one of our special Pythagorean triplet families from p. 178? If it's not obvious, then if the two known sides have a factor in common, see if reducing them (by factoring out the common factor)† reveals that the two sides are part of a triplet. If so, great! Skip Step 3, write down the missing triplet number, and then multiply the common factor back to it. That's the missing triangle side!

. . . .

* The 3:4:5 family is actually made up of all the triangles that are **similar** to the 3-4-5 triangle. For more on **similar triangles**, check out Chapter 17.

† See Chapter 2 in *Math Doesn't Suck* for factoring out common factors (like GCFs)!

Step 3. If there wasn't a triplet (or none found), then write down the Pythagorean Theorem, $a^2 + b^2 = c^2$, where c is the hypotenuse. Then rewrite it, filling in the parts we know. Solve for the missing variable by isolating it on one side of the equation, taking the square root of both sides (knowing the answer must be positive), and simplifying.

Step 4. Make sure you've found what the problem was asking for. Done!

 *And...* Step By Step In Action
Action!

In the MEOW diagram, Given: OW = 1, MO = 2, EW = 4, right angles as marked. Find ME.

Step 1. It looks like we have two right triangles, $\triangle MOW$ and $\triangle EMW$. We are given the two legs for $\triangle MOW$, so let's start by finding its hypotenuse and going from there.

Steps 2 and 3. Well, 1 and 2 sure don't seem to belong to any Pythagorean triplet family, so we'll go ahead and set up our equation: $a^2 + b^2 = c^2$. Plugging in $a = 1$ and $b = 2$, we get: $(1)^2 + (2)^2 = c^2 \rightarrow 5 = c^2 \rightarrow \sqrt{5} = \sqrt{c^2}$. Because c has to be positive, the right side of the equation just becomes c, and we get $\sqrt{5} = c$, or: $\boldsymbol{c = \sqrt{5}}$.

Step 4. So far, we've found the length of $\triangle MOW$'s hypotenuse; in other words, we've found MW. But the problem wants ME, so let's keep going! Looking carefully at $\triangle EMW$, we can fill in $MW = \sqrt{5}$ as a leg, and notice that 4 is the hypotenuse. So we can set up the formula again: $a^2 + b^2 = c^2$, this time plugging in $a = \sqrt{5}$ and $c = 4$: $(\sqrt{5})^2 + b^2 = 4^2 \rightarrow 5 + b^2 = 16 \rightarrow b^2 = 11 \rightarrow \sqrt{b^2} = \sqrt{11} \rightarrow$ (we know b is positive, so) $\rightarrow \boldsymbol{b = \sqrt{11}}$. And that's ME. Done!

Answer: $ME = \sqrt{11}$

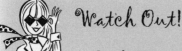

Watch Out!

Just because a triangle has two sides matching one of our "families" on p. 178 doesn't mean we automatically know the third side. First of all, it must be a right triangle! Second of all, the legs and hypotenuse must match up. Remember, the hypotenuse (the side opposite the right angle) is always the longest side.

The triangle below on the left isn't a member of the 3:4:5 family, because the hypotenuse is 4. That other leg *can't* be 5. We'd just have to use good 'ol Pythagoras to find the missing leg, which turns out to be $\sqrt{7}$. The same goes for the triangle next to it—it's not a member of the 5:12:13 family. In fact, the hypotenuse equals $\sqrt{313}$!

Not Members of the Families*

Must Be Members of the Families (even though the right angles aren't marked)

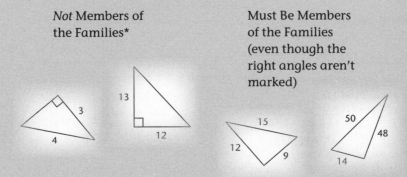

On the other hand, if *all three* sides match up with a "family," then <u>it must be a right triangle</u>. How do we know? Because SSS† tells us that three sides are enough to determine (define) a triangle. So, for example, any other triangle with the sides 3, 4, and 5 *must* be congruent to the 3-4-5 *right* triangle! And that means the original triangle must have been a right triangle, too. You, my dear, just got smarter. Yes, just by *reading* that. (Read this paragraph until you could explain it to someone else!)

---

* But don't worry, they have lots of great friends.
† See Chapter 8 for SSS triangle congruence . . . and lots of cheerleading.

Take Two: Another Example

Find the length of the altitude ⊥ to the base of the isosceles △ whose base measures 30 and whose legs measure 39 each.

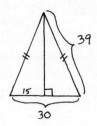

Step 1. Whoa, what? Let's draw this isosceles triangle the problem describes, including the altitude to its base!

Nice. On p. 157, we learned that the altitude to the base of an isosceles triangle is also a *median*, which means the altitude we drew must cut the base in half, giving one side of our new right triangle a length of 15. Great, <u>now we have a right triangle with **leg 15** and **hypotenuse 39**</u>, and we know our job is to find the other leg.

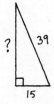

Steps 2 and 3. Do we have a Pythagorean triplet? Hmm, it's hard to tell. But if we factor out a 3 from the **15** and **39**, we get 5 for a leg and 13 for a hypotenuse, right? And yep, we have a member of the 5:12:13 family, multiplied by a factor of 3, which means the missing leg is $12 \times 3 = $ **36**.

Step 4. The problem asked for the length of the altitude, and that's exactly what we found. Done!

Answer: 36

If we hadn't noticed that the above right triangle was in the 5:12:13 family, finding the missing side could have gone a little something like this: We'd have substituted $a = 15$ and $c = 39$ into $a^2 + b^2 = c^2$ to get: $15^2 + b^2 = 39^2 \;\rightarrow\; 225 + b^2 = 1521 \;\rightarrow\; b^2 = 1296 \;\rightarrow\; b = \sqrt{1296} \;\rightarrow\; b = $ **36**.

If a calculator wasn't handy, we could do a factor tree for 1296 and either notice from its factors that it was a perfect square, or just simplify the radical by pulling out smaller perfect squares, one step at a time: ($\sqrt{1296} = \sqrt{4 \cdot 324} = 2\sqrt{324}$, and so on). I don't know about you, but it sure makes me appreciate the families shortcut!

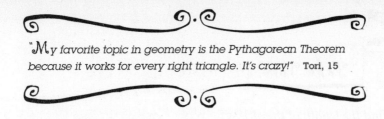

QUICK NOTE   A favorite type of right-triangle problem is finding missing sides, perimeters, or areas that require drawing in an altitude. We're always allowed to draw in an altitude from a vertex, making a right angle with the opposite side.* And that's what we'll do! See 1 below for a perimeter example. (We'll do tons of area problems in Chapter 20.)

What's It Called?

The **perimeter** of a two-dimensional shape is the total distance around it. For example, if a square's four sides each measure 5 inches, then its perimeter is 5 + 5 + 5 + 5 = **20 inches**. And the perimeter of the top triangle on p. 182 is 39 + 39 + 30 = **108**. Ever heard of a security guard "walking the perimeter"? That means she's walking along the edge of the entire property to make sure there are no intruders. She's walking along the *perimeter.*

---

* See p. 389 for the Perpendicular Line Postulate, which allows us to do this!

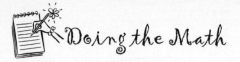

Doing the Math

Use the Pythagorean Theorem or the "families" from p. 178 to do these problems, and draw a picture if it helps. I'll do the first one for you.

1. Find the perimeter of the trapezoid above.

<u>Working out the solution</u>: Hmm, before we can find the perimeter, we need to find the missing side's length. So, here's the trick for doing these. We fill in an altitude to create a rectangle and a triangle.* And if we knew the length of the altitude, that would be the same as the missing trapezoid side, right? Well, notice that the little triangle's top side must equal 58−42 = <u>16</u>. (Read that until it makes sense!) Hmm, a leg of 16 and a hypotenuse that's 34—could it be a happy family member?

Factoring out their common factor, 2, we get 8 for a leg and 17 for the hypotenuse, so it seems we have an 8:15:17 triangle in disguise, with everything multiplied by 2! That means our missing triangle side is 15 x 2= **30**, which is the same length as our missing trapezoid side, right? And the perimeter is just all the outer sides added up: 30 + 58 + 34 + 42 = **164.** Done!

Answer: Perimeter = 164 inches

. . . .

* By drawing in an altitude, we've created a quadrilateral with three right angles. The fourth angle must also be 90°, and that means it's a rectangle, and that means opposite sides are congruent! (See Chapter 15 for more on this.)

For exercises #2–13, find the missing side:

2. Leg = 3, Hyp = 5   8. Leg = 15, Hyp = 25

3. Leg = 3, Leg = 5   9. Leg = 0.005, Leg = 0.012

4. Leg = 7, Hyp = 25   10. Leg = 45, Hyp = 51

5. Leg = 1, Leg = 2   11. Leg = 8x, Hyp = 17x

6. Leg = $5\sqrt{13}$, Hyp = $13\sqrt{13}$ 12. Leg = 8, Hyp = 12

7. Leg = $\frac{8}{71}$, Leg = $\frac{15}{71}$ 13. Leg = 45π, Hyp = 51π*

14. Find the missing side on this trapezoid and also its perimeter.

15. In the ANGEL diagram, find EL. (Hint: Do this one step at a time, first finding those internal lengths.)

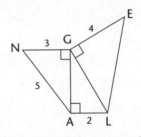

16. Find the missing side on this trapezoid and also its perimeter. (Hint: First draw in the missing altitude on the right. What must it equal? Then find the other leg of that new triangle with subtraction. And finally find the hypotenuse!)

17. A triangle has sides 21, 72, and 75. Is it a right triangle? Explain.

(Answers on p. 399)

. . . .

* Hint: Try pretending the π's aren't there. How would you do that problem?

Rationalizing the Denominator

Here's a nifty trick for "getting rid of" square root signs from the bottoms of fractions called **rationalizing the denominator** (most teachers don't like square root signs in denominators). So if we do a problem involving square roots and our answer is something like $\frac{6}{\sqrt{2}}$, where a square root is still on the *bottom* of a fraction, we'll want to apply this trick.

First of all, we're agreed that $\frac{\sqrt{2}}{\sqrt{2}} = \mathbf{1}$, right? It's a copycat fraction,* after all! Now, to "get rid of" the square root in an answer like $\frac{6}{\sqrt{2}}$, we can just multiply it by $\frac{\sqrt{2}}{\sqrt{2}}$; watch what happens:

$$\frac{6}{\sqrt{2}} \times \frac{\sqrt{2}}{\sqrt{2}} = \frac{6 \times \sqrt{2}}{\sqrt{2} \times \sqrt{2}} = \frac{6\sqrt{2}}{(\sqrt{2})^2}$$

(So far, so good, right? This is just fraction multiplication.) And because $(\sqrt{2})^2 = 2$, our answer becomes:

$$\frac{6\sqrt{2}}{(\sqrt{2})^2} = \frac{6\sqrt{2}}{2} = \frac{3\sqrt{2}}{1} = \mathbf{3\sqrt{2}}$$

Hey, so $\frac{6}{\sqrt{2}} = 3\sqrt{2}$. Crazy, huh?

This trick doesn't always get rid of the entire fraction, but it's sure nice when it does!

By the way, how did we know which copycat fraction to use in order to get the bottom of $\frac{6}{\sqrt{2}}$ to be rational? We needed a copycat fraction with the <u>same square root bottom</u>, $\sqrt{2}$. And if it's got $\sqrt{2}$ on the bottom, it better have $\sqrt{2}$ on the top, too, so that it still equals 1: $\frac{\sqrt{2}}{\sqrt{2}}$. Ta-da! We're so clever. (We'll do more of these later in the chapter.)

. . . .

* Copycat fractions always equal 1 because they have the same thing on top and bottom. See p. 64 in *Math Doesn't Suck* to meet these furry helpers!

More Special Triangles

Here are two *very* special families of triangles, and they're even more powerful than the other families we've seen. Why? Because they're famous for their angles, and if we know even <u>one</u> of their sides, we can figure out the rest.

Cutting Your Sandwich in Half: The 45°- 45°- 90° Triangle

What happens when we cut a square in half along a diagonal? You know, like if you're making a sandwich?* We *bisect* the 90° angle into two 45° angles! And if the sides of the square sandwich were 1 unit each, then we end up with two isosceles triangles (whose legs measure 1 unit), right? So we can plug $a = 1$ and $b = 1$ into the good 'ol Pythagorean Theorem and get: $a^2 + b^2 = c^2 \rightarrow 1^2 + 1^2 = c^2 \rightarrow c^2 = 2 \rightarrow c = \sqrt{2}$ for its diagonal. This is called a "45°- 45°- 90° triangle," and its sides are 1, 1, and $\sqrt{2}$.

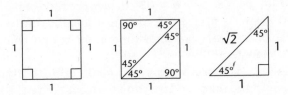

If the sandwich were bigger, then our two isosceles triangles would have different lengths, but their *ratio* would still be **1: 1: $\sqrt{2}$**. And that means that every single half-sandwich (45°- 45°- 90° triangle) has a hypotenuse that is $\sqrt{2}$ times one of its legs. So the sides can be expressed as n, n, and $n\sqrt{2}$. Check it out:

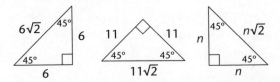

That means that if we know it's a 45°- 45°- 90° triangle, then all we need is <u>one</u> of the sides, and we can figure out the rest. If someone told us the leg of a 45°- 45°- 90° triangle was 6, we'd know the other leg was also 6, and the hypotenuse was $6\sqrt{2}$, right? All we did is multiply by $\sqrt{2}$. If

. . . .

* Mashed banana and almond butter (it's like peanut butter but with almonds) on whole-wheat bread. YUM.

instead we were told the *hypotenuse* was $9\sqrt{2}$, we'd know that the two legs were 9 each; we just divided by $\sqrt{2}$ that time!*

Half an Equilateral Triangle: The 30°- 60°- 90° Triangle

Now imagine if we made a sandwich out of bread shaped like equilateral triangles. We know that all three angles are 60°, right?

If we draw in an altitude to create a 90° angle, then we also *know* we've bisected the top angle into two 30° angles.†

Because we've created a right triangle, we can now figure out the <u>length</u> of that altitude. Let's say the equilateral triangle's sides were 2 units each, then the bottom of our right triangle will be 1,‡ and we can plug $a = 1$ and $c = 2$ into the Pythagorean Theorem to figure out the missing side: $a^2 + b^2 = c^2$ → $1^2 + b^2 = 2^2$ →

$1 + b^2 = 4$ → $b^2 = 3$ → $\boldsymbol{b = \sqrt{3}}$. In fact, the ratio of the sides of every 30°- 60°- 90° triangle is $\mathbf{1} : \sqrt{3} : \mathbf{2}$, and their lengths can be expressed as n, $n\sqrt{3}$, and $2n$. We just plug something in for n, and bam, we've got a 30°- 60°- 90° triangle!

And that means that if we know it's a 30°- 60°- 90° triangle, then all we need is *one* of the sides, and we can figure out the rest! For example, if someone told us the smaller leg of a 30°- 60°- 90° triangle were **7**, then we'd automatically know that the longer leg is $\boldsymbol{7\sqrt{3}}$ and the hypotenuse is 2(7) = **14**. (We'll do examples where we first get the longer leg or the hypotenuse on p. 190 and p. 192.)

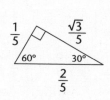

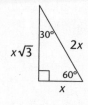

Here's the Step By Step!

. . . .

* The hypotenuse of a 45°-45°-90° triangle could also be a whole number, like 17. We'd still divide by $\sqrt{2}$ to find the legs, and we'd get $\dfrac{17}{\sqrt{2}}$, or, with a rationalized denominator: $\dfrac{17\sqrt{2}}{2}$.

† Why? Because if we know the other two angles are 90° and 60°, the third one had to have been bisected into 30° so they'll all add up to 180°. Pretty sneaky, huh?

‡ In every isosceles triangle, the altitude to the base *is* the median of the base (see p. 157), so that's how we know the altitude we drew also cuts the bottom side in half.

Step By Step

Solving 45°- 45°- 90° and 30°- 60°- 90° triangles:

Step 1. Make sure we're dealing with a 45°- 45°- 90° or a 30°- 60°- 90° triangle. It might be hiding! Remember, drawing an altitude in an equilateral triangle always creates two 30°- 60°- 90° triangles.

Step 2. If it's a 30°- 60°- 90° triangle, skip this step. If it's a 45°- 45°- 90° triangle, and if we have a leg's length, then we know the other leg is equal to it, and we just multiply by $\sqrt{2}$ to get the hypotenuse! If we're given the hypotenuse, then we divide by $\sqrt{2}$ to get the two legs' lengths.

Step 3. If it's a 30°- 60°- 90° triangle, and if we're given either the shorter leg or the hypotenuse, it's easy to get the other one, because one is twice as long as the other, so we either multiply or divide by 2—easy! And then we just multiply the shorter leg by $\sqrt{3}$ to get the longer leg. On the other hand, if we're first given the *longer* leg, we first divide it by $\sqrt{3}$ in order to get the shorter leg. Then, as usual, we just multiply the shorter leg by 2 to get the hypotenuse. Double check to make sure the triangle sides follow the $1 : \sqrt{3} : 2$ ratio.

Step 4. Simplify! If there's a square root in the denominator, then depending on how your teacher wants you to leave your answer, you might need to "rationalize the denominator": multiply the answer by a copycat fraction, usually either $\frac{\sqrt{2}}{\sqrt{2}}$ or $\frac{\sqrt{3}}{\sqrt{3}}$, and then simplify. (See p. 186.) Finally, make sure this is the final answer the problem was looking for. Done!

Watch Out!

It's easy to get confused with all the $\sqrt{2}$'s and $\sqrt{3}$'s floating around. Just remember: We've got a "$\sqrt{2}$" triangle and a "$\sqrt{3}$" triangle. How do we remember which is which? Well, the 45°-45°- 90° triangle is isosceles, which means <u>2</u> of its sides are the same, so it's the $\sqrt{2}$ triangle! And the "30°- 60°- 90°" triangle has <u>3</u> different sides, so it's the $\sqrt{3}$ triangle! And when we're finding missing lengths, are we supposed to multiply or divide by the square roots? Well, think about if we're trying to get a shorter or a longer length, and remember: *Multiplying* by $\sqrt{2}$ or $\sqrt{3}$ will make a number bigger, and *dividing* by $\sqrt{2}$ or $\sqrt{3}$ will make a number get smaller.

QUICK NOTE   When checking our work to verify the lengths we got for a 30°- 60°- 90° triangle, we should ask, "Can we get from the shorter leg to the longer leg by multiplying by $\sqrt{3}$? Can we get from the shorter leg to the hypotenuse by multiplying by 2?" If so, we did it right!

 *And...* *Step By Step In Action*

Find the lengths of the shorter leg and the hypotenuse on this triangle.

Steps 1 and 2. We already have a right triangle, in fact, a 30°- 60°- 90° triangle, so we know the ratio of the sides needs to be 1:$\sqrt{3}$:2. Great!

Steps 3 and 4. Looking at the diagram, we see that <u>the longer leg is 12,</u> so in order to find the shorter leg, we need to *divide* by $\sqrt{3}$, right? That's

$\frac{12}{\sqrt{3}}$. Now that we have the shorter leg, we can just multiply it by 2 to get

the hypotenuse: $\frac{12}{\sqrt{3}} \times 2 = \frac{24}{\sqrt{3}}$. Nice!

But both our answers have square roots in their denominators, so let's

rationalize them by multiplying them by the copycat fraction $\frac{\sqrt{3}}{\sqrt{3}}$. For

the leg, we get: $\frac{12}{\sqrt{3}} \cdot \frac{\sqrt{3}}{\sqrt{3}} = \frac{12 \cdot \sqrt{3}}{\sqrt{3} \cdot \sqrt{3}} = \frac{12\sqrt{3}}{3} = \mathbf{4\sqrt{3}}$. How about that—

there isn't even a fraction left! Similarly, we can multiply $\frac{24}{\sqrt{3}}$ by $\frac{\sqrt{3}}{\sqrt{3}}$ to

get $\mathbf{8\sqrt{3}}$.

Awesome. So our lengths are $4\sqrt{3}$, 12, $8\sqrt{3}$. Does this follow the "1:$\sqrt{3}$:2" ratio? Well, we can totally get *from* $4\sqrt{3}$ *to* $8\sqrt{3}$ by multiplying by 2, so that's good! Can we get *from* $4\sqrt{3}$ *to* 12 by multiplying by $\sqrt{3}$? Let's try: $4\sqrt{3} \times \sqrt{3} = 4\sqrt{3}^2 = 4 \cdot 3 = 12$. Yep, we did it right!

Answer: The shorter leg is $4\sqrt{3}$, and the hypotenuse is $8\sqrt{3}$.

QUICK NOTE   Once we know the shortest side of a 30°- 60°- 90° triangle, it's easy to find the rest! Notice that the shortest side on a 30°- 60°- 90° triangle will always be opposite the smallest angle—the 30° angle. (See pp. 89–90 for why.)

Take Two: Another Example

A regular hexagon is made up of six equilateral triangles. In this particular hexagon, each side measures 10 cm. What is the total height of the hexagon?

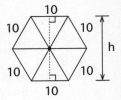

Hmm, first things first. Let's take a look at one of these little equilateral triangles. Since all sides are equal (by definition), that means each side measures 10 cm. And in order to find the total height of the hexagon, we need the height of a little triangle, and then we can just multiply it by 2, right?

Okay, so that's our new goal: Get the height of one of these little suckers!

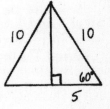

Step 1. Let's draw an altitude in the equilateral triangle, which creates two 30°- 60°- 90° triangles, and we'll just focus on one of them.

Steps 2 and 3. We know the hypotenuse is 10, and that means the shorter leg must be 5. Now that we have the shorter leg, we can find the longer leg: $5 \times \sqrt{3} = 5\sqrt{3}$.

Step 4. It's already simplified, but this isn't the final answer; we were asked to find the total height of the hexagon, which is equal to the height of *two* of these little triangles, right? And that's just $5\sqrt{3} \times 2 = 10\sqrt{3}$. Great!

Answer: $10\sqrt{3}$ cm

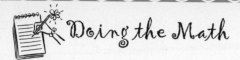 Doing the Math

Answer the questions below, <u>drawing a picture</u> if needed. I'll do the first one for you.

1. A square's diagonal is $5\sqrt{2}$ inches. What is its perimeter?

<u>Working out the solution:</u> Let's draw a picture!
If the diagonal of a square is $5\sqrt{2}$ inches, what is a side of the square? We can always find the leg of a 45°- 45°- 90° triangle by dividing the hypotenuse by $\sqrt{2}$, right? So that's: $5\sqrt{2} \div \sqrt{2} = \dfrac{5\sqrt{2}}{\sqrt{2}}$ (canceling a factor of $\sqrt{2}$ from top and bottom) = 5. And that's also the length of the side of the square. The perimeter is made up of 4 of these identical sides; in other words: 4 x 5 = **20.**

Answer: The perimeter of the square is 20 inches.

2. The short leg of a 30°- 60°- 90° triangle is 7. Find the other two sides.

3. The hypotenuse of a 30°- 60°- 90° triangle is 20. Find the other two sides.

4. The hypotenuse of a 30°- 60°- 90° triangle is 5. Find the other two sides.

5. The long leg of a 30°- 60°- 90° triangle is $3\sqrt{3}$. Find the other two sides.

6. The long leg of a 30°- 60°- 90° triangle is 18. Find the other two sides.

7. The diagonal of a square is 5 inches. What is its perimeter?

8. An equilateral triangle's perimeter is 6 feet. What is the length of one of its altitudes?

9. In a 30°- 60°- 90° triangle, the shortest leg is x. Find the missing sides in terms of x.

10. In the BACON diagram to the right, an equilateral triangle (with each side = 2) is inscribed in a rectangle. Find *BO*. *(Hint: BO is the same as the height of △OAN's altitude from A; draw it in!)* Is this rectangle a square?

11. In the CHIKN diagram to the right, an isosceles triangle is inscribed in a square. The square's sides are 2 inches each. Find *HK*. *(Hint: Don't assume anything about the angles!)* Is this isosceles triangle equilateral?

12. A regular hexagon is made up of 6 equilateral triangles. If the total height of the hexagon is $4\sqrt{3}$ inches, what is the length of one of its sides? What is the entire perimeter?

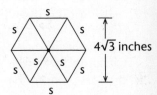

(Answers on p. 399)

Getting fluent at 45°- 45°- 90° and 30°- 60°- 90° triangles is an amazing tool that will make so many things in geometry easier (finding areas, for example!), and they're also fundamental concepts in trigonometry. You'll be like, "Trig? Piece o' cake!" Or maybe piece o' sandwich.

The Distance Formula

I know you're totally blown away by the power of the Pythagorean Theorem, right? Well here's one more reason to be impressed: It can find the distance between any two points on the coordinate plane. Yep! Take any two points, say (5, 2) and (8, 6), and draw a straight line between them.* Then draw lines to show the "rise" and the "run" like we did when first learning about slope.† How far to the right did we travel to get from the first point to the second point? That's the "run." In this case, it's 3 units, because we went from 5 to 8, and $8 - 5 = 3$. And the "rise" is how far up we traveled from 2 to 6, which is **4** units, of course!

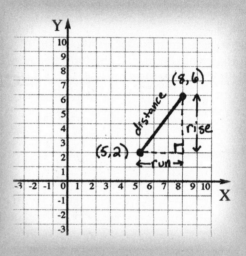

But instead of using the 3 and 4 to create a slope fraction (the slope of this segment is $\frac{rise}{run} = \frac{4}{3}$), we'll use these numbers as the *sides of the right triangle*, because as you can see, the hypotenuse is the distance between the two points! So using the Pythagorean Theorem, with the rise & run as the legs and the distance as the hypotenuse, $a^2 + b^2 = c^2$ becomes: $run^2 + rise^2 = Distance^2$, in other words:

$$Distance = \sqrt{run^2 + rise^2}$$

And in this case, we'd do:

$$Distance = \sqrt{3^2 + 4^2} = \sqrt{9 + 16} = \sqrt{25} = 5.$$

. . . .

* If we picked two points with the same *x*-values or *y*-values, like (1, 2) and (1, 5), then we couldn't draw a triangle, but the distance formula still works!

† For an awesome review of rise, run, and slope (including sausages and bubble-gum, but not, um, together), check out Chapter 18 in *Kiss My Math*.

Now let's say our two points are labeled as (x_1, y_1) and (x_2, y_2). Since the "run" is just the distance between the two x-coordinates $(x_1 - x_2)$, and the "rise" is the difference between the two y-coordinates $(y_1 - y_2)$, we can also write the distance formula like this:

$$\text{Distance} = \sqrt{(x_1 - x_2)^2 + (y_1 - y_2)^2}$$

The formula looks more complicated than it is—just learn the rise/run form if it's easier, but you don't really have to memorize it at all: You know how to figure it out on your own by doing a quick sketch on the coordinate plane (do it now!) and remembering Pythagoras. I know I mentioned this before, but you're a total badass.

Takeaway Tips

 For any right triangle with legs a & b and hypotenuse c, this is true: $a^2 + b^2 = c^2$, and we can use that to find the missing side in right triangles.

 The four triangle families (3:4:5, 5:12:13, 7:24:25, and 8:15:17) can be used as a shortcut for finding missing sides on right triangles.

 Every 45°- 45°- 90 triangle is half a square, and the ratio of legs to hypotenuse is $1:1:\sqrt{2}$.

 Every 30°- 60°- 90° triangle is half an equilateral triangle. The ratio of shorter leg to longer leg to hypotenuse is $1:\sqrt{3}:2$. Always find the shorter leg first; it makes the others easier to find!

 The Distance Formula is just the Pythagorean Theorem on the coordinate plane.

TESTIMONIAL

Amy Chen

(Orlando, FL)
<u>Before</u>: Geometry lover
<u>Now</u>: World-class professional pool player!

For as long as I can remember, I've always loved a good challenge. In high school, math fulfilled that craving, and because I'm an artistic, visual person, geometry was my favorite math class. But I didn't really think too much about how I would apply math in my life until I got to college. Freshman year, my guy friends started inviting me to play pool (AKA billiards). They would taunt me and tell me that I would never be able to beat them. They'd say it in a joking way, of course, but I knew it was true at the time, and it really bothered me. I decided to view it as a challenge in hopes to prove them wrong, so I practiced long hours to learn and conquer the game. It was fun! I discovered how to apply geometry, physics, and probability to get even better, which made the game more complex and stimulating all at the same time. And I admit, it was pretty satisfying to beat my guy friends and shut them up once and for all. It was during this time that I realized how much I love pool!

"I've always loved a good challenge."

I started entering competitions and soon became a two-time U.S. Amateur Women's Champion in Billiards. Today, I'm a full-fledged professional pool player. And believe it or not, math gives me the edge to win tournaments: I use geometry to calculate the angle of the object ball or cue ball when it bounces off the cushion. An understanding of physics is crucial for deciding the speed and spin that should be used. After playing for a while, we all start to understand that we're better at some shots than at others. This is where I use probability—to calculate my chances of making a shot or missing it, and either going for a shot or playing a safety. And it can make all the difference in my game.

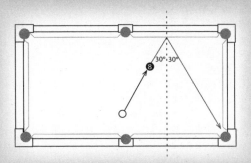

   One of the things I love most about playing professional pool is being able to travel all around the world to compete. I've been everywhere from Korea to the tropical Philippines! I love competition, and it's also really great meeting new and interesting people.
   I encourage you all to embrace the challenges in front of you. And if somebody tries to tell you that you "can't" do something because you're not smart or talented enough, well, work harder and prove 'em wrong! Challenges make us stronger, and they can even lead to an awesome career that you might never have otherwise imagined.

CONFESSIONAL: HOW DO YOU ATTRACT GUYS?

Can you be yourself around your crushes, or do you feel like you have to change yourself in order to attract them?

Look, when we get a crush on a guy, it can be like having blinders on. Suddenly he seems like the most perfect guy in the whole world, and we feel like we'd do anything to get him to like us back, right? If we think we might not be his "type," it can be tempting to try to change things about ourselves in order to attract the guy. But does that even work? Here's what you said!

"*Last* year I totally fell in love with a senior, but he was 'fast.' I wanted him to notice me, so I started dressing more provocatively to impress him and get his attention. But I wasn't being myself, and my friends started acting strange around me. I did start getting some guys' attention, but it wasn't the kind of attention that made me feel good about myself. The things they said just made me feel sort of slutty and like everything I'd done to earn the respect of my peers was going out the window. I stopped dressing like that, and I'm so glad. No guy is worth feeling bad about yourself!" —Marie, 16

"*There* was this guy, let's call him Brad, who I had a massive crush on. One day we were assigned as lab partners on a project, and I was so excited that I thought I was going to pass out! It was like my dreams came true. But then I realized: I'm a pretty good student, and he always hangs around the ditzy girls. I was desperate for him to like me, so I decided I would act ditzy. For the next week, I would giggle and twirl my hair, pretending not to know things. But it totally backfired. He actually said, 'Did you lose 20 IQ points?' I was mortified! What was I thinking?" —Angela, 17

"*Last* year, I was always trying to get a boyfriend. I was like obsessed with the idea or something. I tried everything. I'd flirt, smile at guys, but nothing worked. One day I just gave up and thought, 'I guess I'm going to be single for the rest of my life.' I think I started just being myself. And that's when I became friends with this guy who ended up becoming my boyfriend a few months later! Crazy, right?" —Melissa, 17

"*Never* change yourself for a guy. It can only end two ways. One: He figures out you've been putting on an act and you end up being embarrassed. Two: You're stuck not being true to yourself, and that feels horrible. Just be yourself. If the guy doesn't like you, that's his problem." —Caitlin, 16

"I once wanted to impress this guy, so I decided to get in better shape. I convinced my friend to go running with me every day, and I started eating healthier and drinking more water. It's funny; the guy ended up transferring out of our school, so the plan didn't get a chance to 'work,' but now my clothes fit better and my skin has cleared up. I feel great! I still run every day and eat better. Instead of impressing a guy, I'm impressing myself." —Olivia, 17

"Instead of trying to attract any one guy, just be yourself and see who is attracted to you. There's no rush to get a boyfriend, after all. The best relationships happen when we least expect it, and often with a guy we never considered before." —Jenny, 19

The goal isn't to get into a relationship with a particular guy at any expense. It simply won't result in happiness. The goal in romance is to eventually connect with a guy who makes you feel great about yourself and who likes you for all the same reasons that *you admire yourself.*

So . . . are you his "type"? Who cares? Stop the guessing game and focus on being *your* "type." By focusing on doing things that make you admire yourself (getting good grades, being a great friend, eating healthy, etc.), you'll continue to build confidence and self-esteem, which will naturally attract people to you for the rest of your life—including the kind of guy who will love you for *you!*

For more on dealing with obsessive crushes, see pp. 39–41 in Hot X: Algebra Exposed. *And if you've ever been tempted to dumb yourself down, check out pp. 50–2 in* Kiss My Math!

Chapter 12

School Dance Nightmare

Intro to Polygons . . . and Deriving Formulas

$\mathcal{A}$t your school, does everyone get a medal at sports, even the losing team? I could be wrong, but, um, doesn't that sort of defeat the purpose of having medals in the first place? It seems like schools are awfully worried about being ultra "fair" now.*

Can you imagine how embarrassing it would be if the principal extended this kind of thinking to school dances and made everybody dance with everybody else, just to be fair and not hurt anyone's feelings? Total nightmare. Can you even imagine how many dance pairings that would end up being? With polygons, we'll figure it out, convince the principal that it's a horrible idea, and yes, save the day. Nice.

What Are They Called?

Polygons, Diagonals

A **polygon** is a plane figure—no, not a *plain* figure, but an outline of a shape that lies flat on a 2-D plane. We'll deal with **convex polygons**, meaning that if we connect any two points on the polygon with a straight segment, the segment will always stay *inside* the polygon.

---

* I mean, we all know that life isn't *fair* most of the time; that's just the way the world is, and it's our job to be mature enough to deal with it and make the most of what we get. Know what I mean? It's part of what makes us strong and independent!

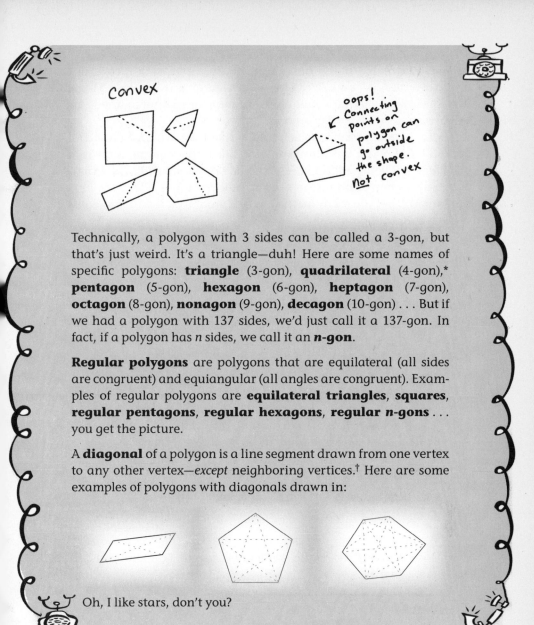

Convex

oops! Connecting points on polygon can go outside the shape. Not convex

Technically, a polygon with 3 sides can be called a 3-gon, but that's just weird. It's a triangle—duh! Here are some names of specific polygons: **triangle** (3-gon), **quadrilateral** (4-gon),* **pentagon** (5-gon), **hexagon** (6-gon), **heptagon** (7-gon), **octagon** (8-gon), **nonagon** (9-gon), **decagon** (10-gon) . . . But if we had a polygon with 137 sides, we'd just call it a 137-gon. In fact, if a polygon has *n* sides, we call it an **n-gon**.

Regular polygons are polygons that are equilateral (all sides are congruent) and equiangular (all angles are congruent). Examples of regular polygons are **equilateral triangles**, **squares**, **regular pentagons**, **regular hexagons**, **regular n-gons** . . . you get the picture.

A **diagonal** of a polygon is a line segment drawn from one vertex to any other vertex—*except* neighboring vertices.† Here are some examples of polygons with diagonals drawn in:

Oh, I like stars, don't you?

Vroom, Vroom: Deriving Formulas

Deriving is like driving, but with an extra *e*. And just like driving, *deriving* is a skill that gives you power and puts you in control. So what does it

* We'll meet many 4-gons (quadrilaterals) in Chapter 15.

† A line segment drawn between two neighboring vertices would just be one of the polygon's sides!

mean to derive a formula? It means to go from exploring specific examples to discovering how to write the general form, using *n* instead of numbers, in such a way that no matter what number you put in, the formula will always give the correct answer.

How can we possibly do something like this? Well, you're about to find out! You might not always totally follow the logic right away, and don't worry if you couldn't have come up with it yourself. It's *such* good exercise for your brain just to read these derivations. Plus, this kind of thinking will get you ready for the kinds of proofs you'll do in pre-calculus and beyond. Let's do it!

Hollywood Walk of Fame: Deriving the Diagonal Formula

Have you ever seen the Hollywood Walk of Fame? It celebrates the most influential "stars" on the stage and screen . . . by putting their names into brass stars embedded in the sidewalk for people to walk all over.* And you can bet that the machine that makes the stars for the Hollywood Walk of Fame uses a pentagon somewhere in its machinery, because the *diagonals* of a pentagon create a star! Let's find out more about diagonals.

You may have seen this formula in your book already: Diagonals for *n*-gon = $\frac{n(n-3)}{2}$. But where did it come from? And how the heck did somebody count diagonals on a shape with *n* sides? Pretty soon, you'll know exactly how, and you'll feel so much smarter, it'll be awesome. Let's do it!

When we want to count something involving *n*, it's always a great idea to first try it with actual numbers. It's easy to count the number of diagonals in a quadrilateral; there are two, right? As we can count on p. 201, in a pentagon (a star), there are five diagonals, and in a hexagon, there are *nine* diagonals. Hmm, what's the pattern?

Let's step back and think about an *n*-gon. They're hard to picture, because we don't know what *n* is, but we'll do our best to sketch (part of) one, and then start drawing diagonals from one vertex. My diagram looks more like a carnivorous

----

* I'll admit, it's pretty fun to see Marilyn Monroe's star.

Venus Flytrap* eating a question mark than I intended . . . hmm.

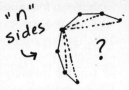

"n" sides →

Anyway, let's be a smarty-pants and notice that the number of diagonals we can draw from just *one* vertex will be $n - 3$. After all, from any vertex, we can draw a diagonal going to all *n* vertices *except* itself and its 2 neighbors—meaning we have to skip those 3, so there are **$n - 3$** places we can draw diagonals to!

So that's the number of diagonals from just *one* vertex. Now, how many *total* diagonals can we draw in an *n*-gon? Well, we can draw these $n - 3$ diagonals for <u>each</u> of the *n* vertices, right? That means for <u>each</u> of the *n* starting spots, we draw to $n - 3$ other spots. So that would be **$n(n - 3)$**. So far, so good? Let's test this for a 4-gon, and we get $4(4 - 3) = 4$ diagonals. But wait, we know there are only 2 diagonals in a square or rectangle, right? And for a pentagon, we'd get $5(5-3) = 10$ diagonals, which is also twice the correct answer, since we know it's 5. Hmm . . . our formula must not be quite right yet.

Here's what's going on: It's true that if we blindly drew diagonals from every vertex to every other (non-neighboring) vertex, then we'd indeed make $n(n - 3)$ pencil strokes. But notice that each diagonal would be drawn *twice* with our pencil—once starting at one end and then later from its other end.

Try it with a square or a pentagon *right now*. It'll take you two seconds. From each vertex, draw its two diagonals. If you keep going around the pentagon until you've used each vertex as a starting point, you'll end up drawing *two* overlapping stars.

So there are exactly half as many diagonals as there are pencil strokes, which means the number of actual diagonals is <u>half</u> of $(n)(n - 3)$, in other words:

Number of Diagonals in *n*-gon = $\dfrac{n(n - 3)}{2}$

Yay, we derived the general formula! And, armed with this nifty formula that we understand inside and out, we can figure out the number of diagonals in any size polygon. Got 20 sides? Then there are $\dfrac{20(17)}{2} = 10(17) = 170$ diagonals. Nice.

. . . .

* These are plants that <u>eat</u> flies and ants. I'm not kidding; look it up.

QUICK NOTE: **Solving Quadratic Equations**   If we're told the number of total diagonals in a polygon and we're asked to find the number of sides, we'll end up needing to solve a *quadratic equation*. For example, if we're told a polygon has a total of 27 diagonals, we could set up the equation: $27 = \dfrac{n(n-3)}{2}$, and solving for n, we'd first multiply both sides by 2 and get: $54 = n(n-3)$. Simplifying, we get: $54 = n^2 - 3n \rightarrow \underline{n^2 - 3n - 54 = 0}$. Then, factoring this,[*] we'd get $(n + 6)(n - 9) = 0$, which results in the two solutions, $n = -6$ or 9. We can't have a negative number of sides on a polygon, so the answer is **9**. Yep, nonagons have 27 diagonals. Who knew?

Mermaids and Seashells: Deriving the Angles Inside a Polygon

I admit, I've always loved mermaids, and I mean, who hasn't pretended to have a mermaid tail? You know, swimming with your legs together like a fin? Anyone? (uncomfortable beat) Er, something else I like to do is find shells on the beach and imagine what kind of creatures used them as their home or even as a hair decoration.[†] Drawing the diagonals in polygons often makes them look like seashells—clams, maybe? And they're going to help us understand the angles inside polygons.

We already know that the sum of the angles in *every* triangle is 180°, right? What about quadrilaterals? A square has four 90° angles, so the sum of any square's angles is 360°. But are all quadrilaterals the same? Let's take a look at any 'ol quadrilateral like QUAD here. Drawing in a diagonal from just one vertex, we can split it up into two triangles. Let's label the angles, just to make them easier to talk about:

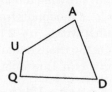 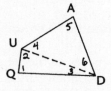

....

* To brush up on solving quadratic equations by factoring, and to see my answer to a jewelry dilemma, check out Chapter 22 in *Hot X: Algebra Exposed*.

† Like, um, a mermaid.

Well, we know that $\angle 1 + \angle 2 + \angle 3 = 180°$ and $\angle 4 + \angle 5 + \angle 6 = 180°$, right? It's just a couple of triangles. And that means $\angle 1 + \angle 2 + \angle 3 + \angle 4 + \angle 5 + \angle 6 = 360°$. With me so far? Using a little substitution, we can show that:

$$\overset{\angle Q + \angle U + \angle A + \angle D}{= \angle 1 + (\angle 2 + \angle 4) + \angle 5 + (\angle 3 + \angle 6)}$$

We already know that the second line equals 360°, so that means $\angle Q + \angle U + \angle A + \angle D = 360°$! And since QUAD wasn't "special" in any way, we've just proven that the sum of the interior angles in *every* quadrilateral is 360°. And all we needed to do was split it up into triangles. Nice.

As it turns out, we can use the same logic to find the sum of the interior angles for any *n*-gon, and yep, we can even come up with a formula for it. On p. 203, we talked about how in an *n*-gon, we can draw $n - 3$ diagonals from any single vertex. Looking at some examples below, if we take the number of diagonals drawn from one vertex and add 1, we get the <u>number of triangles created</u>. So when 3 diagonals are drawn, 4 triangles are created.

 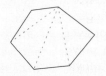

Pentagon: 5 sides, 2 diagonals from a single vertex, 3 triangles created!

Hexagon: 6 sides, 3 diagonals from a single vertex, 4 triangles created!

Heptagon: 7 sides, 4 diagonals from a single vertex, 5 triangles created!

And how many triangles are created by drawing diagonals from one vertex in an *n*-gon? Hmm, well, there are $n - 3$ diagonals from one vertex, right? So that's $(n - 3) + 1$ triangles, in other words, **$n - 2$** triangles. And each triangle contributes 180 degrees, right? So that's a total of $180(n - 2)°$ for the sum of all the interior angles in an *n*-gon. How about that!

Sum of Interior Angles in an *n*-gon = $180(n - 2)°$

So in any polygon with, say 20 sides, the sum of its interior angles has to be $180(\mathbf{20} - 2)° = (180)(18)° = 3240°$. Um, that's a lot. But it makes sense, because a polygon with 20 sides would have 20 really obtuse angles on its interior, right?

Oh hey, our new formula even works for a triangle, because $n = 3$ and then we get: $180(n - 2) \rightarrow 180(3 - 2) = 180°$. Can't you just feel your brain getting bigger? I'm so proud of you.

By the way, if the polygon is a *regular* polygon, then we can figure out the measure of *each* interior angle by dividing this whole thing by n, the total number of vertices (and angles!) in the polygon:

Each Interior Angle in a <u>Regular</u> n-gon = $\dfrac{180(n - 2)°}{n}$

So if we're asked, "What is the measure of each interior angle in a regular hexagon?" we could just insert 6 wherever we see n and we'd get:

$\dfrac{180(n - 2)°}{n} \rightarrow \dfrac{180(6 - 2)°}{6} \rightarrow 30(4)° = 120°.$*

In Case You Forget This Formula!

Say we're asked on a test, "What is the measure of each interior angle in a regular hexagon?" but we can't remember the formula. No problem! We'd quickly sketch a hexagon, and we'd be able to draw **3 diagonals** from a single vertex, right? And that would create **4 triangles** that we would see right there, which means there would be a total of $180° \times 4 = $ **720°** of interior angles in the entire hexagon. Great progress! Because it's a *regular* hexagon, those 720 degrees are evenly divided among 6 angles, and that means each angle would equal $\dfrac{720}{6} = $ **120°**. Ta-da!

QUICK NOTE: The Secret to Polygons   Here's the secret to understanding polygons: <u>By drawing in enough diagonals, polygons can always be split up into a bunch of triangles.</u> And we know a thing or two about triangles, don't we?

* Try using a protractor to measure an interior angle of the regular hexagon on p. 207 and see for yourself! (Check out p. 387 in the Appendix for how to use a protractor.)

Watch Out!

Don't be fooled into thinking that we can decrease the number of triangles in a polygon by simply drawing in fewer diagonals—because they wouldn't all be triangles. For example, if we only drew 2 diagonals from a vertex in a hexagon, we'd end up with 2 triangles . . . and a quadrilateral. Before we go using our knowledge about triangles, we better be sure to count the sides and make sure we've really got *triangles*.

Exterior Angles

So far, we've only dealt with interior angles, but polygons have *exterior* angles, too. We get them by extending the sides of the polygon.* Notice that each exterior angle is *supplementary* to the interior angle at the same vertex. Think about that for a second. For example, $\angle 1 + \angle 7 = 180°$. In this case, we have a regular hexagon, so we know that $\angle 1 = 120°$ (see pp. 205–6 for why). This means that the exterior angle $\angle 7$ equals 60°, because $120° + \mathbf{60°} = 180°$. Make sense? Now, if we add up all six of the exterior angles (just one exterior angle per vertex), we get: $6(60°) = \mathbf{360°}$. Amazingly, the sum of the exterior angles for every single polygon will always be 360°! See p. 213 for more. It's kind of mind-blowing.

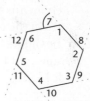

Sum of Exterior Angles for any *n*-gon = 360°

So what's the measure of just one exterior angle of a regular *n*-gon? It's just 360° divided by the total number of angles, *n*:

One Exterior Angle for a regular *n*-gon = $\dfrac{360}{n}$

Let's see how to keep all of these new formulas straight!

. . . .

* Notice that at each vertex, we could extend either side to create its exterior angle, but we chose one direction (counterclockwise) to keep things nicer looking.

Keeping the Polygon Formulas Straight!

# of diagonals from one vertex of an *n*-gon	**(n – 3)**	This is how many diagonals can be drawn from *one vertex only*, and it creates (*n* – 2) triangles, by the way!
Total # of diagonals in an *n*-gon from all vertices	$\dfrac{n(n-3)}{2}$	This is how many *total* individual diagonals can be drawn in an *n*-gon.
Sum of interior ∠'s in an *n*-gon	**180(n – 2)°**	There are (*n* – 2) triangles created from a single vertex's diagonals, and each of those contributes 180°!
Each interior ∠ in <u>regular</u> *n*-gon	$\dfrac{180(n-2)°}{n}$	This is just the total number of degrees in the regular *n*-gon, divided by *n*.
Sum of exterior ∠'s in an *n*-gon	**360°**	Yep, it doesn't matter how many sides there are; it's *always* 360°.
Each exterior ∠ in <u>regular</u> *n*-gon	$\dfrac{360°}{n}$	This is just the total number of exterior degrees in the *n*-gon, divided by *n*. Ta-da!

Reality Math:
School Dance Nightmare

Remember the principal from the beginning of the chapter who wanted to make everyone dance with everyone else, just to be "fair"? Let's see how many dance pairings there would actually be.

This is very similar to the diagonal stuff we did on p. 203, but we also need to include the neighboring vertices, because um, you have to dance with those two people, too (in this case, Zac and Bob).

Let's say there are *n* people at the dance, and everyone has to dance with everyone else—yes, girls too, because otherwise someone might complain about sexual-orientation discrimination, right? Let's show this on an *n*-gon with each person at a vertex. And first, let's just think about how many dances *you'll* have. Hmm, you'll dance with Max, Amy, Zac, Bob, Jen, Zoe, Ian . . . and everyone else on the *n*-gon. In fact, all *n* of

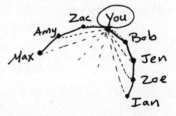

them—except yourself! So that's $(n - 1)$ dances for you. Since each of the *n* people have to do $(n - 1)$ dances, that's a total of $\boldsymbol{n(n - 1)}$ dances, right? But just like on p. 203, we've double counted, because, for example, you dancing with Max is the same as Max dancing with you. So the actual number of total dances becomes half of $\boldsymbol{n(n - 1)}$, in other words, $\dfrac{\boldsymbol{n(n - 1)}}{\boldsymbol{2}}$ dances.

So, let's see. Even if only 100 people showed up to the dance (maybe because they heard about the principal's plan), there would be $\frac{n(n-1)}{2} \rightarrow \frac{100(99)}{2} = 50(99) =$ **4,950** dances. Even if you only had boy-girl pairings (assuming there are about the same number of guys as girls), there would still be about $\frac{100(50)}{2} = 2,500$ dances.* Yeah, um . . .

Armed with this information, you politely approach your principal, explaining your solution, and you successfully save the entire school from public humiliation. Nicely done.

By the way, in probability (usually covered in Algebra II), this is also known as the "handshake" problem: If there are a bunch of people in a room who all want to meet each other and shake hands, how many handshakes will happen? And now YOU know how to solve it—with a polygon and its diagonals!

Doing the Math

Answer these questions about polygons, using the chart on p. 208 and sketching your own polygons when needed. I'll do the first one for you.

1. Each interior angle of a regular polygon is 150°. How many sides does it have?

<u>Working out the solution:</u> We know that in a regular *n*-gon, each interior angle measures $\frac{180(n-2)^\circ}{n}$. So in this case, that means $150° = \frac{180(n-2)^\circ}{n}$, right? And now we just need to solve for *n*, because that will tell us the number of sides on the *n*-gon, after all! Hmm. First, let's get rid of that fraction by multiplying both sides by *n*, and we get:  $n \cdot 150° = n \cdot \frac{180(n-2)^\circ}{n} \rightarrow$

....

* I didn't use the formula for this one; I multiplied 100×50 for the 100 people each having to dance with the (approx.) 50 people of the opposite sex, and then I divided by 2 for the double counting, and got $\frac{100(50)}{2}$.

$150n° = 180(n - 2)°$. Ah, much nicer. Distributing the 180 and then continuing to simplify, we get: $150n° = 180n° - 360° \rightarrow$ $-30n° = -360° \rightarrow n = 12$. To double check, we could plug **12** into the original formula, and we'd get: $\dfrac{180(n - 2)°}{n} \rightarrow$ $\dfrac{180(\mathbf{12} - 2)°}{\mathbf{12}} = \dfrac{1800°}{12} = 150°$. Yep, it worked!

Answer: The polygon has 12 sides.

2. Draw a picture of a *non*-convex pentagon, and show how it can still be divided into three triangles.

3. If a polygon has 10 sides, how many diagonals can be drawn from *one* vertex? How many triangles does that create? How many total diagonals can be drawn?

4. How many *total* diagonals can be drawn in a 1000-gon?

5. One interior angle of a regular polygon is 135°. How many sides does it have? What is the sum of its interior angles?

6. One exterior angle of a regular polygon is 12°. How many sides does it have? What is the sum of its interior angles?

7. What is the total number of degrees inside an 11-gon? What is the measure of a single angle in a regular 11-gon?

8. The sum of the interior angles of a polygon is 8640°. How many sides does it have?

9. What is the sum of the exterior angles of a 9876-gon?

10. What do we call a polygon with a total of 14 diagonals? How about with 20 diagonals? How about with 35 diagonals? *(Note: This involves quadratic equations. Brush up in Chapter 24 of* Hot X: Algebra Exposed *if you need to!)*

11. We know the measure of a single exterior angle in a regular n-gon is $\dfrac{360°}{n}$. We also know that a single interior angle is supplementary to this.

   a. Using these two facts, derive another expression for the measure of a single interior angle in a regular n-gon.

   b. Now combine your answer into a single fraction, using n as the common denominator, and factor out the GCF.* Does this answer look familiar?

....

* To brush up on factoring out the GCF (Greatest Common Factor), see Chapter 3 in *Hot X: Algebra Exposed.*

12. If you have 19 members of your extended family over to your house (plus you, that's 20 people total) and everybody hugs every other person exactly once, how many hugs happen? *(Hint: See the Reality Math on pp. 209–10.)*

(Answers on p. 400)

Keep It Classy: Technology Do's and Don'ts

Technology keeps us connected, entertained, and informed, but it can be abused if we're not careful. So here are a few tips on how to keep it—and *you*—safe and classy!

Do: Search online for extra help with a confusing math topic.

Don't: Search online for tips on how to cheat on your math test.

Do: Tweet about your winning softball game.

Don't: Tweet about your latest make-out session.

Do: Text friends to let them know you got home safe.

Don't: Text and drive.

Do: Chat with friends online.

Don't: Chat with strangers online.

Do: Correspond with old friends via email.

Don't: Break up via email.*

Do: Use Facebook to search for friends.

Don't: Use Facebook to bully or embarrass others.

Do: Check your phone for missed calls after school.

Don't: Check your phone during a date or class.

*Check out p. 140 in *Hot X: Algebra Exposed* for my 9th-grade break-up story.

What's the Deal?
Deriving the Sum of Exterior Angles Formula

Let's look at the exterior angles of our favorite polygon, the good 'ol triangle. Notice that for each vertex, the exterior angle and interior angle are supplementary, so for example: $\angle 1 + \angle 4 = 180°$, $\angle 2 + \angle 5 = 180°$, and $\angle 3 + \angle 6 = 180°$. And that means that the sum of all six angles is $180° + 180° + 180°$, right? In other words:

$$\angle 1 + \angle 2 + \angle 3 + \angle 4 + \angle 5 + \angle 6 = 3(180)°$$

Make sure you follow that before reading on.

We also know that $\angle 1 + \angle 2 + \angle 3 = 180°$, because it's a triangle, after all! Now we can substitute 180° for $\angle 1 + \angle 2 + \angle 3$ in the bolded equation above, and simplify: $180° + \angle 4 + \angle 5 + \angle 6 = 3(180)°$ → $\angle 4 + \angle 5 + \angle 6 = 540° - 180°$ → $\angle 4 + \angle 5 + \angle 6 = 360°$. And we've just shown that the sum of all three *exterior* angles of a triangle equals 360°. This can be done with any size polygon. Try it for a pentagon!

Are you ready for a real challenge, some seriously grown-up math? Check out GirlsGetCurves.com for the proof of the *n*-gon version . . . you'll love it!

Takeaway Tips

Polygons can always be split into triangles by drawing in enough diagonals from a single vertex.

A polygon with *n* sides is called an *n*-gon. The number of diagonals we can draw from a single vertex is $(n - 3)$; think about the seashells. And the number of total diagonals we can draw from all *n* vertices is $\dfrac{n(n-3)}{2}$. Remember the double star inside the pentagon!

The sum of the interior angles in an *n*-gon is always $180(n-2)°$; that's because the $n - 3$ diagonals from one vertex create $(n - 2)$ triangles, and each triangle contributes 180°. The measure of one interior angle of a *regular n*-gon is just the total divided by *n*, in other words: $\dfrac{180(n-2)°}{n}$.

The sum of the exterior angles in any *n*-gon is independent of the number of sides; it always equals 360°!

Chapter 13

Getting Around the Mall
Parallel Lines and Transversals

Do you ever secretly "pose" while riding up the escalator at the mall? You know, in case someone important is watching you? I . . . never do that. Nope, never. Anyway, before we go to the mall, let's take a moment to notice the fabulous prefix *trans-*, which often means "across." For example, <u>trans</u>lucent material (like flimsy silk or sheer gauze) allows light to travel *across* it. A <u>trans</u>atlantic flight means you're kickin' it at 30,000 feet in the air, traveling *across* the Atlantic Ocean.

And <u>trans</u>versals? They go *across* two coplanar lines.

What's It Called?

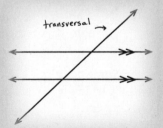

transversal →

A **transversal** is a line that intersects (goes *across*) two coplanar lines, hitting them at two distinct points. If the two coplanar lines are parallel, they'll often be marked with double arrows.

I think **transversals** look like skinny escalators, taking us from downstairs to upstairs, or vice versa. You know, *trans*porting us around the mall . . .

The Funhouse Mall—and Pairs of Angles

I wonder if anyone's ever built a mall with slanted floors. It might be kind of fun for the customers (and good exercise), but I guess not for the vendors when all their stuff falls off the shelves . . . Hmm, maybe this is a bad idea after all. Anyway, the indoor/outdoor mall diagram to the right shows a whole bunch of angles that are formed by a transversal, *t*, cutting across two lines, *u* and *d* (upstairs and downstairs). Right now, our little shopper is strolling by some lovely rooftop stores on a bright, sunny day. How nice.

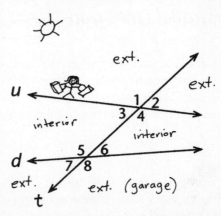

Isn't it the best when your favorite stores are easy to find? Like if you visited store 1 and then rode down the escalator and went to store 5, it would be super convenient because 5 is in the exact same "position" as 1, except on a lower floor. This is an example of the positioning of **corresponding angles**. Other pairs of corresponding angles are ∠2 & ∠6, ∠3 & ∠7, and ∠4 & ∠8. Do you see how each pair of angles is in the same position, except that one is "upstairs" and one is "down-

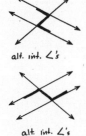

stairs"? Above, ∠3 & ∠6 are called **alternate interior angles** because they're on the *interior* of the mall, and on *alternate* sides of the escalator. Same goes for ∠4 & ∠5. These pairs always make an "N" or "Z" shape* (see on the left), like the "oh-no-he-didn't" Z-snap. Sometimes the N or Z will even be all stretched out.

On the other hand, ∠1 & ∠8 are called **alternate exterior angles** because they're on the *exterior*†—the roof and garage—and on *alternate* sides of the esca-

lator. Same goes for ∠2 & ∠7. These pairs always look like two V's trying to get as far away from each other as they possibly can.

. . . .

* Or a backwards "N" or "Z".

† Did you know that "INT" and "EXT" stand for "interior" and "exterior" in movie and TV scripts for setting the location of a scene? For example: EXT. ABANDONED GARAGE—NIGHT: With earbuds in place, Brittany listens to music as she walks to her car, oblivious to the heavy footsteps behind her . . . CUE: Scary music.

But wait, there's more! Can you guess which pairs of angles are the **same side interior angles**? Hmm, same side of the escalator, interior . . . that'd be ∠3 & ∠5 and also ∠4 & ∠6. And the **same side* exterior angle** pairs are "outside" (rooftop or garage) and on the *same side* of the escalator as each other, so that'd be ∠1 & ∠7 and also ∠2 & ∠8.

Normal Malls . . . and The Corresponding Angles Postulate

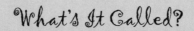

Most malls have level floors, meaning they are parallel to the ground and to each other, which is good news for more than just the vendors. You see, a special thing happens when an escalator (er, transversal) crosses two *parallel* lines—the corresponding angles are congruent!

What's It Called?

The Corresponding Angles Postulate: If two *parallel* lines are crossed by a transversal, then the corresponding angles are congruent. And this makes sense, because the exact same kind of intersection is happening at the top and bottom of the escalator. It's just happening on different floors!

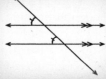

If ∥ lines, then corr. ∠'s are ≅.

As you'll see on p. 218, the abbreviated versions of these transversal Rules tend to be pretty darn short; they even leave out the word *transversal*! Feel free to write out the whole thing till you get used to it, but I have a feeling you'll be using these super-short Rules in no time.

- - - - - - - - - - - - - - -
* Some books say "consecutive" instead of "same side."

The image below on the left is labeled to show that each pair of **corresponding angles** is congruent. Take a look. But wait, we also know that all pairs of **vertical angles** are congruent to each other: The chopsticks strike again! That means all the ★'s have the same measure as the ♥'s, right? So we can just use ♥'s to mark both types of angles—I mean, why not? And that also means the ❀'s have the same measure as the ☺'s, so we can just use ☺'s to mark these angles, too. Relabeling our diagram more efficiently (in the center, below), now it seems we have just two different kinds of angles here.

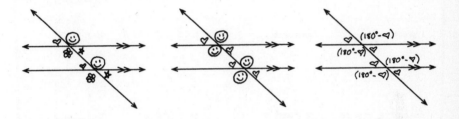

But we actually know even *more* information, because whenever two angles add up to a straight angle, then they are supplementary, right? And that means ☺ + ♥ = 180; in other words: ☺ = (180 – ♥), so we can fill in (180 – ♥) wherever we see ☺ (see above on the right) and make our escalator diagram even *more* chock full 'o information. Because every pair of angles is either congruent or supplementary, notice that if someone tells us what just *one* of the angles is, then we can fill in the rest of them. How about that? And all because the mall floors were parallel . . .

Here's a chart to summarize the relationships we've just discovered. Don't feel pressure to memorize this whole thing, but definitely use it to find the angles on the diagram below it and see these relationships for yourself. After using these rules for a while, you'll be able to create this chart without even looking—because once you find them on the diagram, you'll see that they just make sense.

When a Transversal Crosses
Two <u>Parallel</u> Lines

Angle Pairs	Relationship	An Example (See p. 219)	Rule
Vertical ∠'s	congruent!	$\angle 1 \cong \angle 4$	If ∠'s are vertical, then ∠'s are ≅.
Corresponding ∠'s	congruent!	$\angle 1 \cong \angle 5$	If ‖ lines,* then corr. ∠'s are ≅.
Alternate Interior ∠'s	congruent!	$\angle 3 \cong \angle 6$	If ‖ lines, then alt. int. ∠'s are ≅.
Alternate Exterior ∠'s	congruent!	$\angle 1 \cong \angle 8$	If ‖ lines, then alt. ext. ∠'s are ≅.
Same Side Interior ∠'s	supplementary!	∠3 & ∠5 are supp.	If ‖ lines, then same side int. ∠'s are supp.
Same Side Exterior ∠'s	supplementary!	∠1 & ∠7 are supp.	If ‖ lines, then same side ext. ∠'s are supp.

I only named one example per category; can you find more examples for each row?

. . . .

* Instead of "If ‖ lines," we could say "If lines are ‖." Abbreviations in proofs will vary a bit from book to book, so just double-check for your teacher's preferences. The most important thing is to make it clear which theorem you're using!

Notice that the last two rows of the table on p. 218 describe pairs of angles that are *supplementary*, not congruent. It's not too hard to remember because they're also the only angle pairs with "same side" in their names: *same side, supplementary*; I guess the S's stick together!

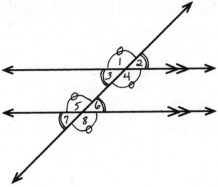

QUICK NOTE   The two parallel lines won't always be horizontal. They could both be tilted or even vertical! As long as they are parallel <u>to each other</u>, then our chart applies whenever there's a transversal cutting across them. Sometimes we'll just tilt our heads to "see" the mall. No biggie.

Who knew an escalator at the mall could be so handy for homework?

Doing the Math

Answer the questions based on the diagram below. $\overleftrightarrow{BE}$ is a transversal cutting across the <u>parallel</u> lines $\overleftrightarrow{AC}$ and m. I'll do the first one for you.

1. Name the 2 pairs of alternate interior angles.

<u>Working out the solution:</u> This diagram is oriented sideways; we can just tilt our heads to the left to help see the mall/escalator setup we're used to! We could trace some "Z" shapes to find alternate interior angle pairs. But even without that, first we could just pick an interior angle—how about ∠1? Then its *alternate interior angle* would be on the interior, on the alternate side, which doesn't have a number on it, but we can call it ∠CDE. (We also could have called it ∠EDC.) Similarly, the other pair is ∠2 & ∠ADE (which also could be called ∠EDA).

Answer: Alt. int. angle pairs are ∠1 & ∠CDE and ∠2 & ∠ADE.

2. Name the 2 pairs of alternate exterior angles.

3. Name the 2 pairs of same-side interior angles.

4. Name the 2 pairs of same-side exterior angles.

5. Name the 2 pairs of *numbered* vertical angles.

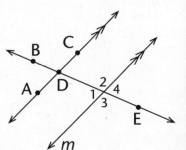

6. Name the corresponding angle to ∠CDE.

For the next two questions, let's assume that ∠BDA = 82°.

7. Name the *numbered* angles that measure 82°.

8. What is the measure of ∠BDC?

9. Given the diagram to the right, with m ∥ n, find x°. *(Hint: Draw a line that is parallel to both m and n, passing through the "crook" where x° is. First find y°, and use alt. int. ∠'s to fill in what you can.)*

(Answers on p. 400)

The Angle Game

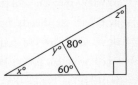

Now we're going to play a game—yes—the angle game! These problems show up on standardized tests all the time. We're given a bunch of angles, and we need to find the missing ones. Here's one using a triangle:

Hmm, let's start by filling in what we can. Because of supplementary angles, we know that $y° + 80° = 180°$, right? So that means $y° = 100°$. Great start! What else do we know? Looking at the little triangle above (and remembering that the angles in any triangle add up to 180°), we know that $x° + y° + 60° = 180°$; in other words, $x° + 100° + 60° = 180°$, which means $x° = 20°$. Fabulous!

Next, we can look at the great big right triangle and do the same trick: $x° + z° + 90° = 180°$; in other words, $20° + z° + 90° = 180°$, which means that $z° = 70°$. Nice.

Answer: $x° = 20°$, $y° = 100°$, and $z° = 70°$

These "angle game" problems often use parallel lines as well as triangles, so here are some steps to help navigate 'em.

Step By Step

Angle Game Problems:

Step 1. Start by giving the diagram a quick glance, looking for supplementary angles, vertical angles (chopsticks!), parallel lines/transversals (escalators at the mall!), and triangles. Remember, triangles' angles add up to 180°, and those triangles could be small or big, and could be overlapping or inside each other.

Step 2. What's the angle we're trying to find? Is it part of a triangle? Perhaps it's part of more than one triangle. Is it supplementary to something else? Come up with a "wish angle" (If only I knew THIS angle, then I could find THAT one . . .).

Step 3. Keep hunting and filling stuff in as you go (if you can't write in your book, then copy the diagram on a piece of paper), getting closer and closer to finding the "wish" angle and/or the angle the problem is asking for. That's it!

Watch Out!

Sometimes it's challenging to see the "escalator in the mall" within the diagram, because the mall is tilted, or worse—there's more than one transversal! They're just trying to confuse us, but we won't let that happen. If you're having trouble "seeing it," try extending lines, tilting your head, and paying attention to which escalator each angle is touching.

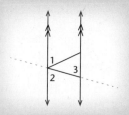

For example, in this diagram, ∠1 & ∠2 aren't touching the same transversal, so although we can tell that ∠2 ≅ ∠3, we can't figure out any relationship between ∠1 & ∠2 at all. When trying to use the theorems on p. 218, make sure the angle pairs are touching the same escalator, and you'll be golden.

And... Action! Step By Step In Action

Using the diagram given, find w°.

Wow—that angle marked with $w°$ seems so far away from the angles we know. How can we possibly find it? Have a little faith.

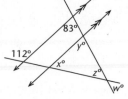

Step 1. We have a set of parallel lines! And um, two escalators. (Just tilt your head to the left to see 'em.) We also have two triangles—one big and one small. Cool.

Step 2. The angle we want is not part of a triangle, but because of chopsticks, $w° = z°$. So our wish is: "If only I knew $z°$ . . ."

Step 3. Hmm, let's tilt our head way to the left and see what we can figure out. By the "same side exterior angle" rule, we know that x is supplementary to 112°, which means that $x° + 112° = 180°$ → $x° = 68°$. And because of congruent corresponding angles (but using a different escalator), we can see that $y° = 83°$. Also, if we know two angles in a triangle, we can find the third, right? So we know that: $68° + 83° + z° = 180°$ → $151° + z° = 180°$ → $z° = 29°$. We have our wish! Because of chopsticks (vertical angles being congruent), this means $w° = 29°$. Done!

Answer: $w° = 29°$

"*I love when math is like a puzzle. Even when it's hard, it doesn't feel as much like 'homework.'*" Sonia, 15

Take Two: Another Example

Using the diagram given, find x°.

This looks like, um, most of a star or something. Well, let's see what we can do!

Step 1. There are no parallel lines marked, but there are vertical angles and supplementary angles, and many triangles.

Step 2. It looks like $x°$ is part of a big triangle that's lying on its side: $\angle T$ is one vertex of the big triangle, and so is $\angle R$. See it now? And since $\angle R = 20°$, that means $\angle T + x° + 20° = 180°$, right? So our wish is: "Gosh, if only we knew the measure of $\angle T$, we could find $x° \ldots$"

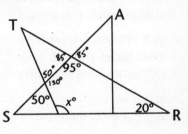

Step 3. Let's fill in what we can. Since there is an angle supplementary to 50°, we'll fill that in as 130° (because 50° + 130° = 180°). Notice that there are *two* angles that are supplementary to 95°. Let's fill in 85° for both of those (since 85° + 95° = 180° in each case). Also, we can fill in a congruent angle—the one vertical to 50°.

And lookie there, our "wish angle," $\angle T$, is part of that small triangle on the upper left! That tells us: $50° + 85° + \angle T = 180°$. This becomes: $135° + \angle T = 180° \rightarrow \angle T = 45°$. Nice progress! The big triangle tells us that: $\angle T + x° + \angle R = 180°$; in other words: $45° + x° + 20° = 180°$. And this becomes: $x° + 65° = 180° \rightarrow x° = 115°$. Ta-da!

Answer: $x° = 115°$**

. . . .

** We also could have found the answer by first finding $\angle S$. It's shorter; try it!

Be sure not to assume something just because it looks like it might be true. Diagrams can have lines that *look* parallel, angles that *look* like 90°, and more—but don't be fooled! (See more about this in Chapter 3.) Only use what you <u>know</u> to be true, even if at first it doesn't seem like enough information to find what you're looking for. Fill some stuff in, and before you know it, you'll have your answer.

Doing the Math

Based on the diagrams given, find what is asked for. I'll do the first one for you.

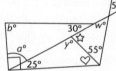

1. Given the diagram, find $b°$.

<u>Working out the solution</u>: So, $b°$ looks like it equals 90°, but we can't assume it. However, we do know **30°** is one angle in the triangle, right? Also, we know that $a°$ is complementary to 25°, because those two angles make up a right angle! So, $25° + a° = 90° \rightarrow a° = 65°$. Now that we have <u>two</u> angles in the triangle, we can find the third angle: $a° + b° + 30° = 180°$; in other words, $65° + b° + 30° = 180° \rightarrow 95° + b° = 180° \rightarrow b° = 85°$.

Answer: $b° = 85°$

2. In the above diagram, find ☆.

3. In the above diagram, find ♡. *(Hint: Use what you found in #2.)*

4. In the diagram on the right, find $x°$, and state the rule that makes it true! *(Hint: Extend the straight lines and tilt your head!)*

5. In the diagram on the right, find ☺ . *(Hint: Remember supplementary angles!)*

6. In the HOT diagram, find $\angle 1$.

7. In the HOT diagram, find $\angle HOT$. *(Hint: Find $\angle H$ first.)*

8. Given the SHOP diagram below. If $\overleftrightarrow{SH} \parallel \overleftrightarrow{OP}$:

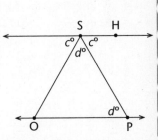

a. Write $\angle SOP$ in terms of $c°$. *(Hint: Extend $\overline{OS}$ into a line, and use your finger to cover up $\overline{SP}$. Can you see the mall and the "Z"?)*

b. Finish this equation, looking at those top angles: $c° + d° + c° = $ __.

c. Finish this equation, looking at the triangle: $\angle SOP + d° + d° = $ __.

d. Use all of this to find $c°$. Is $\triangle OSP$ equilateral?

(Answers on p. 400)

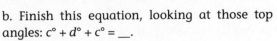

 Takeaway Tips

 When a transversal cuts across two *parallel* lines, it looks a lot like an escalator at the mall, and every pair of angles is either congruent or supplementary.

 Tilting our heads and extending lines can make the "escalator at the mall" easier to see.

When playing the angle game, we start by filling in whatever we can. Often supplementary and vertical angles can give us a great start! We should also look for hidden, overlapping triangles, because the angles in every triangle always add up to 180°.

Danica's Diary
NEVER TOO LATE: PROCRASTINATION BUSTER FOR BIG PROJECTS

We all know the pitfalls of procrastination. We have some enormous project to work on, whether it's a paper, studying for Friday's big test, or something else.* The project just seems so huge that we want to wait until we have a long chunk of time to dedicate to it, and we figure we should "take care of" a whole bunch of other little tasks first—you know, to clear the way for this big project, right? The problem is, those little things can often fill the day, especially because we might also get a funny text or decide to quickly look up something online . . . and the next thing we know, it's practically time for bed and we figure, "Oh, well, this evening's wasted anyway," and we don't even attempt to start whatever the project was. No!

In my experience, it's almost *never* too late! Even if I only have 10 minutes, I've found that starting a big project—even just outlining how I might start it—plants the seeds for how I'll continue, and then the next day, instead of the big project looming over me, I've already dipped my toe in the water and I can slide right into working on it. Getting into this habit will save you time and time again-and it's a great skill for later in life, no matter what career you end up in. Try it. It works!

(Check out pp. 275-6 in *Kiss My Math* for more time management ideas that really work, and pp. 226-8 in *Kiss My Math* for some of your procrastination horror stories!)

R· · · · · ·R· · · · · ·R

----

* For me, it might be writing chapters of my math books!

Chapter 14

Musical Theater Emergency

Proving Lines Are Parallel

You've been thinking about getting involved in musical theater, and although it's too late to audition for this year's big musical (it starts tonight!), you decided to stop by and watch some of the last-minute rehearsals. When you arrive, everyone's running around in a panic: The enormous, slanted platform used in the big dance number has just collapsed! Luckily, no one was hurt, but the crew is trying to build a new one as fast as they can, and everyone is pretty much freaking out.

They have the engineer on the phone, who says the most important thing is that the legs be parallel to each other. Otherwise, it could collapse again. The actors—understandably—refuse to dance on it unless it's *proven* to be safe.

Looking around, you see a hammer, some nails, and a ball of string. You just came from geometry class, so you also have your protractor in your bag. Can you *prove* that the legs are parallel and save the day? Yes! (In just a few pages, that is . . .)

Remember on p. 216 how we saw that if two lines are parallel, then their corresponding angles are congruent? As it turns out, the *converse* is also true. So, if any two corresponding angles on a transversal are congruent, then we know the lines it crosses must be parallel. We're going to prove this new theorem with an indirect proof—yep, a proof by contradiction, just like we talked about in Chapter 10.

Here's how we'll set it up: We'll draw our transversal diagram, marking (any) two corresponding angles congruent, and then we'll assume the two lines *aren't* parallel. When we reach a contradiction, we'll have proven they *must* have been parallel all along!

Given: The transversal $\overleftrightarrow{AP}$ crosses $\overleftrightarrow{PL}$ & $\overleftrightarrow{AY}$, and $\angle 1 \cong \angle 2$.
Prove: $\overleftrightarrow{PL} \parallel \overleftrightarrow{AY}$

Either $\overleftrightarrow{PL} \parallel \overleftrightarrow{AY}$ or $\overleftrightarrow{PL} \nparallel \overleftrightarrow{AY}$. Let's face it, it's gotta be one or the other.

Let's assume $\overleftrightarrow{PL} \nparallel \overleftrightarrow{AY}$.

Well, if we assume the lines *aren't* parallel, then they must cross somewhere, right? We'll call that point of intersection "Q." It might be a loooong way away, but if the lines aren't parallel, that point would *have* to exist. Let's redraw the diagram to show this proposed scenario (see below).

We're going to label an angle $\angle 3$ because it's a free country (and you'll see that it'll come in handy). Well lookie there, now it seems we have a triangle: $\triangle PAQ$.* The plot thickens . . .

Because of supplementary angles, we know that $\angle 1 + \angle 3 = \mathbf{180°}$. We are given that $\angle 1 \cong \angle 2$, so we can substitute $\angle 2$ for $\angle 1$ in the above bolded equation, and we get: $\angle 2 + \angle 3 = \mathbf{180°}$. So far, so good?

This diagram is impossible, and we're going to prove it!

Notice that the three angles of our new triangle are $\angle 2$, $\angle 3$, and $\angle LQY$. The angles in every triangle add up to 180°, and that means: $\angle 2 + \angle 3 + \angle LQY = 180°$, right? But wait: We already know from our second bolded equation that $\angle 2 + \angle 3 = 180°$, which means that $\angle LQY$ would have to measure zero degrees!

This is impossible for a triangle; we've totally reached a contradiction. And that means our assumption, $\overleftrightarrow{PL} \nparallel \overleftrightarrow{AY}$, was false. So it *must* be true that $\overleftrightarrow{PL} \parallel \overleftrightarrow{AY}$, because that was the only other possibility.

$$\therefore \overleftrightarrow{PL} \parallel \overleftrightarrow{AY}$$

Ta-da! And because there was nothing "special" about our diagram, we've actually proven a new theorem: the *converse* of the Corresponding Angles Postulate (see p. 229).

. . . .

* Of course, if the point of intersection is a loooong way away, then the triangle would be really long and skinny. But this page is only 6 inches wide, after all! So this picture might not be "drawn to scale," but that doesn't matter for the proof.

Don't worry if you wouldn't have known how to do that proof on your own. In fact, any time you see a proof and you're like, "Um, I would have had no idea how to do that . . . ," don't let your neck feel all prickly with panic! Instead, take a breath, read it slowly, and do your best to follow the logic of the proof. Remember, if you already knew this stuff, you wouldn't be taking geometry in the first place. Next, identify the part(s) of the proof that seemed the hardest, and realize that you now have at least one new trick up your sleeve!

And what's the new trick here? It's that if two lines aren't parallel, then they <u>must cross somewhere</u>, and that means we're allowed to draw in their intersection point. Depending on the diagram, sometimes a triangle will appear, which can be very helpful.*

Whether you realize it or not, this stuff is trickling in, and you are becoming one powerful proving machine, sister! Here's the theorem we just proved:

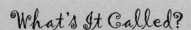

What's It Called?

Converse of the Corresponding Angles Postulate: If two lines cut by a transversal form a pair of corresponding angles that are *congruent*, then the two lines must be parallel. (This is the *converse* of the Corresponding Angles Postulate from p. 216.)

If corr. ∠'s are ≅, then lines are ‖.

As it turns out, *all* of the transversal/parallel line theorems from p. 218 have true converses. Check 'em out on the next page!

....

* In fact, any time we see a triangle, we know *for sure* that none of its sides are parallel, which can be handy in proofs by contradiction.

Rules That Can Prove Lines Are Parallel

If corr. ∠'s are ≅, then lines are ||.

If alt. int. ∠'s are ≅, then lines are ||.

If alt. ext. ∠'s are ≅, then lines are ||.

If same side int. ∠'s are supp., then lines are ||.

If same side ext. ∠'s are supp., then lines are ||.

For example, if we're told that ∠1 ≅ ∠2 in the diagram SHOW, then "**If alt. int. ∠'s are ≅, then lines are ||**," tells us that $\overline{SH} \parallel \overline{WO}$. Or if we were told that ∠2 & ∠3 are supplementary, then by "**If same side int. ∠'s are supp., then lines are ||**," we'd automatically know that $\overline{HO} \parallel \overline{SW}$. Pretty powerful stuff, huh?

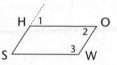

Let's prove: **If alt. int. ∠'s are ≅, then lines are ||.**

Given the diagram to the right: ∠1 ≅ ∠2.

Prove: m || n.

How are we going to do this? We might not be sure how to start, but let's think about this: On p. 228,

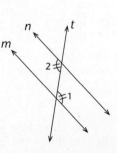

we proved the converse of the Corresponding Angles Postulate, which states "If corr. ∠'s are ≅, then lines are ||." So we could use that Rule to help prove that *m* || *n*. In fact that Rule would be a great *ending* to our proof, wouldn't it?

So how do we get there? Hmm. Thinking about the chain of logic, we can only use *that* Rule if we <u>first prove that we have a pair of congruent corresponding angles</u>, so that's our new mission! Let's do this two-column style and start by filling in the Givens and the end:

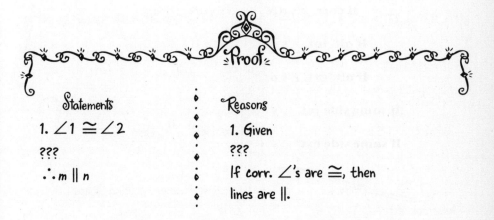

Proof

Statements	Reasons
1. ∠1 ≅ ∠2	1. Given
???	???
∴ *m* || *n*	If corr. ∠'s are ≅, then lines are ||.

Working backwards, we need to figure out *which* corresponding angles we can prove are congruent, so that we can use that Rule. Hmm, the first thing we'll need to do is to label a pair of corresponding angles so we can even talk about them in the first place! Well, what's the corresponding angle to ∠1? It would be opposite ∠2, and let's label it, oh, I don't know, how about ∠3?

So our greatest wish at this moment is to prove ∠1 ≅ ∠3, because then we'd have a pair of congruent corresponding angles, and then we could use the Rule we filled in above as our last Reason. With me so far?

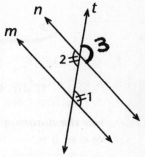

Remember the Vertical Angles (Chopsticks) Theorem from p. 29? It's about to be useful— again! Let's notice that because ∠2 & ∠3 are vertical angles, they must be congruent: ∠2 ≅ ∠3.

We were given the fact that ∠1 ≅ ∠2, and now we know that ∠2 ≅ ∠3, so guess what? Because of the Transitive Property (see p. 52), we now know that ∠1 ≅ ∠3. Nice.

(see p. 52)

Do we have a solid chain of logic? Let's fill in what we have so far and see if there are any holes.

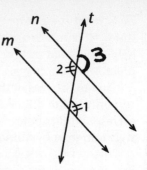

Proof

Statements	Reasons
1. ∠1 ≅ ∠2	1. Given
2. ∠2 ≅ ∠3	2. If vertical ∠'s, then ∠'s are ≅.
3. ∠1 ≅ ∠3 (Now we have corresponding angles congruent, yay!)	3. Transitive Property, using Steps 1 & 2: If ∠1 ≅ ∠2 and ∠2 ≅ ∠3, then ∠1 ≅ ∠3.
4. ∴ m ∥ n	4. If corr. ∠'s are ≅, then lines are ∥.

Looks good!

QUICK NOTE   Notice a strategy we used above that can come in really handy in proofs: If we don't know how to start, we can try to figure out how it will end! And then we can often work backwards from that point to create our chain link of logic.

It's time to solve that musical theater emergency from p. 227...

from p. 227

Reality Math: The Show Must Go On

Remember back on p. 227 how you walked into the panicked theater rehearsal? The crew has just finished rebuilding the slanted platform and is now desperate to prove to the actors that the platform won't collapse again. You overhear a particularly cute stagehand say to his crewmate, "I wish the engineer were here! How else can we prove that the legs are parallel?" You step up and boldly say, "I can help."

The room goes silent. All eyes are on you. "I'll need some nails, a hammer, and a long piece of string," you announce. The stagehand grabs them for you, and you quickly nail the string on the two front legs, which we'll call *a* and *b*.

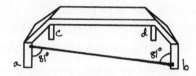

Notice that it doesn't matter where the nails go on the legs; *any* transversal will work. With your trusty protractor, you measure two alternate interior angles, getting 81° for each; sure enough, they're congruent, which means $a \parallel b$! Next, you nail the string between legs *b* and *c*, and those interior angles both measure 76°, which means that $b \parallel c$. You keep going until you've checked every pair of legs, and each pair of angles is indeed congruent.

You turn to the actors and say, "By the converse of the Alternate Interior Angles Theorem, these legs have all been proven parallel to each other, and the show can go on!" Cheers erupt, and yes, you have saved the day. You give a wink to the stagehand (who is too stunned to wink back) and silently thank yourself for paying attention in geometry class. The End.

Nice story, huh?

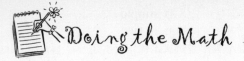

Doing the Math

Answer the questions based on the diagrams below. I'll do the first one for you.

1. In the FASHN diagram, if ♡° = 40°, can we prove that any line pairs are parallel? If so, which lines?

<u>Working out the solution:</u> In order for a pair of lines to be parallel, they certainly can't intersect each other, right? The only pair of lines that *doesn't* intersect are $\overleftrightarrow{AH}$ & $\overleftrightarrow{SN}$, so that's the only pair that *could* be parallel. Our question has now become: If ♡° = 40°, can we prove $\overleftrightarrow{AH} \parallel \overleftrightarrow{SN}$? In a "mall" diagram, these lines would be the two floors. Tilting our heads to the right, let's imagine if we extended those lines outward. Hmm, looks like we've got <u>two</u>

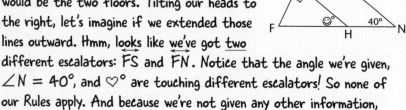

different escalators: $\overleftrightarrow{FS}$ and $\overleftrightarrow{FN}$. Notice that the angle we're given, ∠N = 40°, and ♡° are touching different escalators! So none of our Rules apply. And because we're not given any other information, we simply can't conclude anything about ANY of the other angles shown here.

Answer: Nope! We can't prove the lines are parallel.

2. In the FASHN diagram, if ☺° = 40°, can we prove that any line pairs are parallel? If so, which lines?

3. In the FASHN diagram, if ☆° = 40°, can we prove that any line pairs are parallel? If so, which lines? Can we prove anything else about this diagram?

4. In the LOVED diagram: Given: ∠1 ≅ ∠2. Which pair of lines is parallel? Which Rule guarantees that? *(Hint: Extend the lines and find the escalator!)*

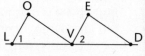

5. In the LOVED diagram, Given: ∠1 ≅ ∠2, and ∠O and ∠E are right angles. Can we figure out ∠OVE? If so, what is it? *(Hint: Use the result from 4.)*

6. In the SINGR diagram, Given: ∠3 is supplementary to ∠4. Prove: $\overleftrightarrow{SN} \parallel \overleftrightarrow{IG}$, using a two-column proof. *(Hint: This is a short proof!)*

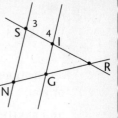

7. As shown in the SINGR diagram, the lines $\overleftrightarrow{SI}$ & $\overleftrightarrow{NG}$ intersect. Prove ∠ISN is not supplementary to ∠SNG with a proof by contradiction in paragraph form. *(Hint: Start by assuming they ARE supplementary.)*

8. Refer to the STORE diagram. Given: $\overline{ST} \cong \overline{RE}$. Prove: $\overline{SE} \not\cong \overline{TR}$. *(Hint: Draw $\overline{SR}$ and use a proof by contradiction like we did in Chapter 10. Look for congruent triangles, and then using CPCTC, show that two angles would have to be congruent, which would then mean that a particular pair of lines are parallel. But this is a contradiction because they intersect on the diagram. You can do this!)*

9. Prove this Rule: **If two lines are both ⊥ to a third line, then they must be ∥**. Draw a diagram with two horizontal lines, *m* and *n*, both ⊥ to a vertical (up-and-down) line, *l*. Use a Rule from this chapter to prove that *m* ∥ *n*.

(Answers to #2–5 on p. 400; proof solutions at GirlsGetCurves.com)

Takeaway Tips

   The converses of the Transversal Rules are all true, and we can use them to prove lines are parallel. Look for the mall escalator!

Working backwards can be a great strategy for challenging proofs.

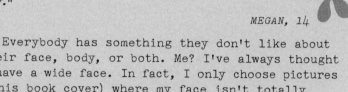

Danica's Diary
BODY IMAGE

"Every girl is beautiful in her own special way."

MEGAN, 14

Everybody has something they don't like about their face, body, or both. Me? I've always thought I have a wide face. In fact, I only choose pictures (like for this book cover) where my face isn't totally faced forward; it's got to be angled to one side or the other so my cheekbone is showing and not the widest part of my jaw. Hey, I'm an actress. I have to know these things about my face, right?

In terms of my body? When I was younger, I hated my wrists and arms. In the 6th and 7th grades, my wrists were a lot smaller than my classmates' wrists, and I thought mine looked too bony and freakish. I seriously used to go home and pray for my wrists to get thicker. And my arms? Don't get me started on my arms. I was born with a lot of hair, okay? And my arms were, well, pretty furry as a young child. I have dark hair, so it was really noticeable. As I grew from a pre-teen to a teenager, things got better. I guess because as my arms grew, the same number of hair follicles then spread to a wider surface area.* All I know is once I was in my twenties, it didn't bother me as much. (And then after college, when I got back into acting, I actually started bleaching my arm hair.)

I also wished that I had curly hair. Have you ever heard the expression, "The grass is always greener"? Most girls are obsessed with certain aspects of their appearance that they wish were different. As it turns out, most girls in my class with curly hair wished that they had straight hair like mine!

We're only human, and it's normal for us to compare ourselves to others in order to try to see if we're normal and okay. But when we compare ourselves to others, it almost always leads to unhappy feelings—it's just not productive. Plus, think about who we might be comparing ourselves to. Are the images even real, or are they manufactured with airbrushing, tons of makeup, and careful camera angles? Worse still, are they real models who are starving themselves?

. . . .

* See p. 363 for the definition of *surface area*.

Being surrounded by images of airbrushed women in magazine ads or billboards is enough to make any gal wish she looked different. The flawless skin, hair, etc., on the at-least-10-pounds-underweight models set an impossible, unhealthy standard.* But many girls look at these ads and become obsessed with their weight, even developing eating disorders in the process. (See pp. 339-41 for one 16-year-old's story.) And by the way, guys are way more attracted to the girl who's confident and truly comfortable with herself than to the girl who's underweight and obsessing about her appearance all the time. Ask anyone! So . . . how do we avoid becoming fixated on being super skinny?

Some people take things to the opposite extreme. You might have seen this on TV (in between the ads filled with skinny models) where talk show hosts proclaim that "big is beautiful" and that nobody should judge anyone for being overweight, because it's not their fault. There are even so-called fat activists. However, it doesn't take a rocket scientist to tell that being overweight is generally not healthy and that it can be a sign of someone not taking care of herself.

It's not always easy for us to be our best selves—it takes effort! But I'm not going to lie and tell you that looks don't matter. *Of course* they matter. When we walk into a job interview, our potential boss will size us up in a matter of seconds: Does she take care of herself? Does she keep herself healthy? Does she put some effort into her fitness and physical well-being? That boss will be judging you partially based on your appearance, because she knows that as an employee, you'll represent the

* See p. 17 for why advertisements *want* us to be dissatisfied with how we look!

> "*I* go home every day and I say to myself, 'Why are you like this? Why can't you get up and exercise, so you can wear what the other girls wear?' I blame it on the fact that I don't have anybody right by my side doing the exercises with me or encouraging me to do them . . . but I know that I should be able to motivate myself. I need to get up off my butt and exercise." —Brittnee, 16

> "*I* know that caring about how I look doesn't have to turn into something unhealthy—just the opposite! And I can make sure of this by eating fresh food and getting exercise. A healthy body image is so important if we want to fulfill our true potential." —Katie, 14

entire company when you come into contact with a customer or a business associate.

Plainly put, being at a healthy weight is a sign of self-respect. Should we care about how we look? Yes. Should we obsess over it? No way! The key is balance. Eat healthy, but splurge once in a while. Exercise, but make it fun (see p. 107 for ideas). Admire other girls, but find things you love about *your* body, too.

How Do YOU Keep a Healthy Self-Image and Take Care of Your Body?

> "*E*ating healthy and being very active is the right way to stay fit. Just doing those things makes me feel good about myself." —Chloe, 15

> "*I*t is my top priority to eat energy foods, healthy and preferably organic foods. Water is the only thing I drink (I refuse to drink soft drinks; the artificial stuff is even worse than the sugar), and I try to drink a gallon per day. And I do my best to avoid oily foods because that usually reflects on the skin. It works!" —Lia, 17

"*R*ather than comparing myself to other people at my school, I do something different. As I pass people in the hallways, I do my best to pick out one thing about that person that is beautiful or that I like about that person. And instead of getting jealous, I'll look for ideas: hairstyles, makeup ideas, clothing ideas, etc., so I can spice up my own style. I've found that doing this has made me a stronger, more confident, and more accepting person." —Bailey, 17

Try taking the quiz on pp. 46–9 to check the status of your body image!

TESTIMONIAL

Melanie Soto-Medina

(Long Island, NY)
Before: Had trouble focusing in school
Now: Inspects and designs bridges!

I always struggled in math, but things got worse for me in 7th grade when tragedy struck our family: My younger brother was killed by a speeding driver. In addition to that, my family moved from New York to Pennsylvania, so everything about my life was turned upside-down. I had to deal with terrible loss while adjusting to a new school. It was hard to focus on anything, and my mind would wander in class—whose wouldn't? But over the years, it got worse, not better, and my grades kept slipping.

"My mind would wander in class..."

My turning point happened in the 10th grade. My geometry teacher asked my mother to meet with us, and I remember them both staring at me in disappointment as they discussed my grades and my attitude. I was so ashamed and embarrassed that I resolved to change: to begin studying and actually applying myself. I quickly learned that if I studied, geometry was neither boring nor as difficult as I had imagined it to be.

Newly focused, I perceived class time passing much more quickly. I was determined to complete all homework assignments, and I also discovered that tests and quizzes were much easier when I'd done the homework first! I transitioned from a "D" in the first quarter to an "A" by the end of the year. I was on fire, and it felt great. As a senior, I studied AP calculus and physics, and in college, I earned a degree in architectural engineering. It was amazing to realize I'd had the power to succeed inside of me all along.

Today, I am a structural engineer. I get to inspect and design bridges! I love it, and every aspect of bridge engineering requires geometry: the design of shapes, consideration of symmetry, balance, you name it. But most importantly, *safety* depends on geometry.

When a bridge crosses over the ground or a body of water, it often does so at an angle. When the superstructure (everything above the bridge "legs") is not perpendicular to the substructure (the bridge "legs" and foundation), a skew angle is created, which can be anything from 1 degree to more than 50 degrees. If the sides of the bridge are parallel, then because of transversal lines, the corresponding skew angles will be congruent. The force carried by the bridge (people, cars, wind, snow, etc.) can be dangerously multiplied, depending on these angles used, so a deep understanding of those skew angles and transversal lines is crucially important, not only to the beauty of the bridge, but also to the safety of such a large superstructure!

Each time I drive over or under a bridge that I have been a part of the design team for, I feel a tremendous sense of accomplishment, and I know my brother would be proud of me. I still face challenges today, but it's my duty and privilege to maintain the progressive attitude I found that day in the 10th grade, when I decided to live up to my fullest potential.

Despite whatever significant personal and financial hurdles we may face, we can all push through and pursue our dreams, if we believe in ourselves and focus on success. I'm living proof of that!

Chapter 15

Designer Shoes and Handbags

Introduction to Quadrilaterals

People who are obsessed with designer clothes, shoes, and handbags tend to be snobby, don't you think? It's like they believe if it's not designer, it's "less than." I've never thought designer clothes are important. It's more about the way we put things together that can really make the outfit, you know what I mean?* Sometimes the handbags *themselves* are snobby . . . well, when they're shaped like trapezoids, anyway (more on that soon).

Quadrilaterals might seem like a whole new topic, and sure, there are some new definitions to learn, but the strategies almost all come from stuff we did with parallel lines and triangles! You see, *every single* quadrilateral can be divided into two triangles, just by drawing in the diagonal like we talked about on p. 205. And here's the golden rule of quadrilaterals: <u>If you don't know where to start, draw in a diagonal</u>. You'll be amazed at how much easier it gets.

But first things first: A **quadrilateral** is any four-sided polygon, and it's time to meet some specific types.

𝒲hat 𝒜re 𝒯hey 𝒞alled?

Here are some *categories* of **quadrilaterals**:

A **rectangle** is a quadrilateral whose angles each measure 90°.

---

* See p. 249 for how your *posture* can do wonders for any outfit.

A **square** is a rectangle whose sides are *all* congruent. A square is the equiangular, equilateral quadrilateral! Say that three times fast. Or seriously, don't.

A **parallelogram** is a quadrilateral in which *both* pairs of opposite sides are parallel. It's an easy name and definition to remember, because it's all about being, um, parallel. Imagine if a rectangle had hinges at its corners and somebody sat on top of it. The whole thing would start to flatten, and we'd end up with the "typical," slanty parallelogram!* Notice that once it was flattened, it would have the same perimeter, but its area would be less—maybe a lot less, depending on how much we flattened it. If we flattened it too much, we'd end up with zero area, wouldn't we?

A **rhombus** is a parallelogram whose sides are all congruent.† This is the "square version" of a parallelogram. It's like if a square had hinges at the corners and somebody sat on top of it. The sides would all still be equal to each other, but the whole thing would flatten a bit, and we'd get the "typical" rhombus!

A **trapezoid** is a quadrilateral with *exactly one* pair of parallel sides, called its *bases*. It's like a triangle whose top was sliced off— *parallel* to the bottom side.

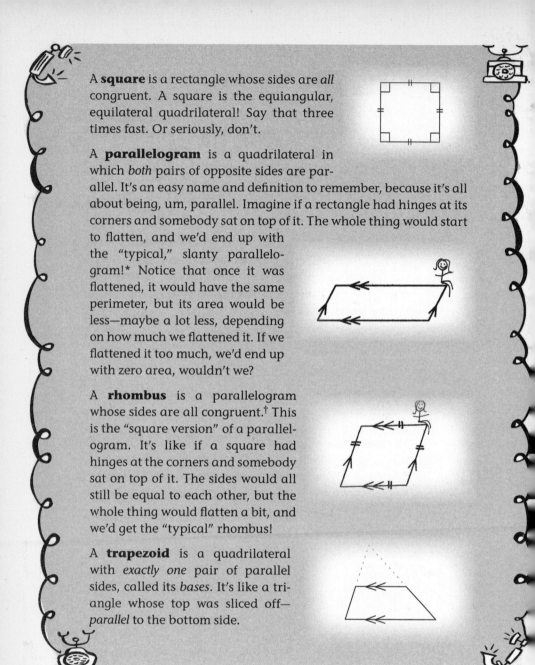

---

* Rectangles and squares are actually *special types* of parallelograms, so not all parallelograms *have* to be slanty. We'll see more about that on p. 245.

† As it turns out, if a quad has 4 ≅ sides, then its opposite sides are forced into being ||, which means it has to be a parallelogram, and therefore, a rhombus. (Use 4 toothpicks and try it on your own. Opposite sides have to be ||!) We'll *prove* this on p. 256.

An **isosceles trapezoid** is a trapezoid in which the (nonparallel) sides are *congruent*. An isosceles trapezoid looks like an isosceles triangle . . . if someone chopped off the top.

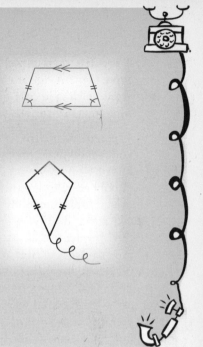

A **kite** is a quadrilateral in which two <u>disjoint</u> pairs of *consecutive* sides* are congruent. "Disjoint" just means that all four sides are distinct—no side gets "used" twice (see below for more on that). In kites, <u>opposite</u> sides are *not* usually congruent, and none of the sides are usually parallel to each other. And yep, they look like the kites that we fly in the air!

QUICK NOTE   In case you were wondering, here's an example of a quadrilateral with two pairs of consecutive congruent sides ($\overline{FA} \cong \overline{AK}$ and $\overline{AK} \cong \overline{KE}$) that *aren't* disjoint, so the shape *isn't* a kite. I mean, hey, it's two pairs, right? But they're not disjoint; the side $\overline{AK}$ gets used twice. Very sneaky. And now you know why "disjoint" is so important in the definition of *kite*. I mean, that thing does *not* look like it would fly.

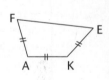

* **Consecutive sides** are sides that are next to each other. Some textbooks say "adjacent," which means the same thing as "consecutive," for <u>sides</u>. But *the two terms mean different things for <u>angles</u>*. See p. 388 in the Appendix for more.

The Categories of Quadrilaterals . . . and Shoes

Many shapes fall into more than one quadrilateral category, and it can be challenging to understand the hierarchy. To help get a feeling for how it works, we're going to take a look at . . . shoes.

Let's map out a (partial) diagram to describe the hierarchy of shoes. Each time we draw an arrow down, we're naming a subcategory. So, for example, high-heeled boots are a subcategory of boots; it's a more specific type of boot.

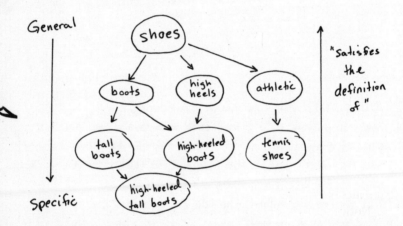

If we pick *any* specific shoe category and follow the chain directly up, the category we started with will satisfy the definition of any other category up along the way. Notice, for example, that we can't reach "athletic" by following the chain directly up from "high heels." But that makes sense, because nobody would say that high heels satisfy the definition of "athletic shoes"! Notice that "high-heeled, tall boots" satisfy the definitions of <u>four</u> categories: boots, high heels, tall boots, *and* high-heeled boots. I suppose some would say that high-heeled, tall boots are very, um, satisfying . . . ?

Now let's check out the quadrilateral categories:

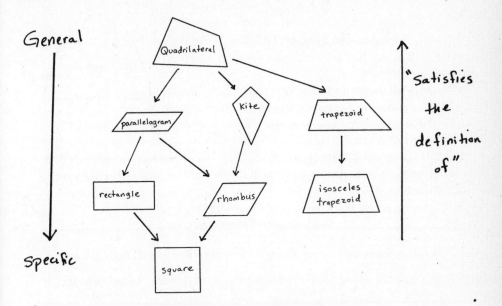

Yep, this works just like the shoes. And we don't have to know anything about quadrilaterals to tell—just by looking at this hierarchy—that <u>a rectangle satisfies the definition of parallelogram</u>. Think about that for a second. Also, we can tell that a rhombus satisfies the definitions of parallelogram *and* kite, and that a square satisfies the definitions of rhombus, rectangle, kite, and parallelogram! (Squares are the high-heeled, tall boots of quadrilaterals.)

Notice, however, that we *couldn't* say, "An isosceles trapezoid satisfies the definition of parallelogram." Even though the isosceles trapezoid looks lower on the tree, we actually can't get from it to the parallelogram by following only *up* the arrows. Isosceles trapezoids just aren't a subcategory of parallelograms.*

---

* By the way, don't worry about memorizing this hierarchy just yet. In Chapter 16, we'll see a story about a totally dysfunctional family that'll make learning it a snap!

"Always, Sometimes, Never" for Quadrilaterals: Trapezoids Are Snobs

Yes, trapezoids are total snobs. Maybe it's because they look like designer handbags, but they only talk to each other and no one else. More on that in a moment.

Because there are so many types of quadrilaterals, and tons of overlap between them, "always, sometimes, never" questions are really popular on tests. We'll see a sentence like *A rhombus is a kite,* and then we'll have to decide if the statement is always, sometimes, or never true.* We did some of this in Chapter 6, but here's a trick that works particularly well for hierarchies of shapes (and shoes).

Replace the word "is" with the words ***satisfies the definition of.***

Replace the word "are" with the words ***satisfy the definition of.***

For example, if someone said, "High heels are shoes," we could change it to "High heels *satisfy the definition of* shoes," and then say, yep, always! Let's go in reverse now: "Shoes are high heels" would become "Shoes *satisfy the definition of* high heels." Well, *sometimes.* Can we think of some type of shoe that satisfies the definition of high heels? Yes: high heels! And can we think of some type of shoe that *doesn't* satisfy the definition of high heels? Yep: flat boots. So the answer is *sometimes.*

It works the same way with quadrilaterals. The sentence "A square is a rectangle" would become "A square *satisfies the definition of* a rectangle." And yep, a square *always* satisfies the definition of a rectangle, because squares always have four 90° angles, right? So the answer is *always.* On the other hand, "A rectangle is a square" would become "A rectangle *satisfies the definition of* a square." We know it's not always true, because not all rectangles have four congruent sides. But can we think of *some* type of rectangle that satisfies the definition of a square? Sure: A square does! So the answer is *sometimes.*

. . . .

* The answer is "always"; see the second QUICK NOTE on p. 247 for more on this.

QUICK NOTE   Squares are super satisfying: When we're looking for a shape that could satisfy multiple categories, squares usually work! In fact, the only categories that squares *don't* satisfy are those snobby trapezoids and isosceles trapezoids. This means that for these quadrilateral *always/sometimes/never* questions, as long as no trapezoids are involved, the answer will be "always" or "sometimes." Crazy huh?

What's the Deal?

Why do trapezoids (and isosceles trapezoids) seem like such snobs? Because they're the only shapes that have *restrictions* on them: They only have exactly one pair of parallel lines; they're *not allowed* to have two sets of parallel lines. As soon as a shape has two sets of parallel lines, it just can't be a trapezoid.

QUICK NOTE   Looking at the chart on p. 245, we can see that rhombuses satisfy the definition of kites. It's easier to see why "a rhombus is a kite" if we draw a rhombus like

this: ◇ , instead of on its side like they're usually drawn.

Practicing Precision

By the way, are you really going to need to remember the exact difference between a kite and a rhombus later in life? Maybe not. But organizing our thoughts enough to be precise about definitions is a great brain sharpener. And who doesn't like a sharp brain?

Doing the Math

Unless otherwise stated, answer these questions with *always*, *sometimes*, or *never*, using the flowchart on p. 245 if it helps. I'll do the first one for you.

1. A kite is a parallelogram.

Working out the solution: Let's rewrite that: "A kite *satisfies the definition of* a parallelogram." Hmm, well, we know there are kites that *don't* satisfy the definition of ‖-ogram (the "typical" kite like on p. 243, for example, doesn't have two pairs of opposite ‖ sides). So the answer isn't *always*. But is it *sometimes* or *never*? Well, a square satisfies the definition of a kite (congruent, consecutive sides) and the definition of a ‖-ogram (‖ opposite sides).* So I guess it's possible for a kite to be a ‖-ogram—when it's a square! So the answer would be *sometimes*.

Answer: Sometimes

2. A trapezoid is a parallelogram.

3. A kite is a rectangle.

4. A square is a rhombus.

5. A trapezoid is an isosceles trapezoid.

6. A parallelogram is a rectangle.

7. An isosceles trapezoid is a trapezoid.

8. A rhombus is a kite.

9. A square is an isosceles trapezoid.

10. A rhombus is a square.

11. Name a shape that is a parallelogram but that isn't always a square or rhombus.

12. Name a shape that is both a kite and a rectangle.

13. Which shape is equilateral but not equiangular?

(Answers on p. 400)

. . . .

* A rhombus also satisfies the definitions of both kite and parallelogram!

Danica's Diary.
FASHION & POSTURE

Let's face it; we don't all have a gazillion dollars to spend on expensive new outfits. But I'm going to let you in on a little secret: We don't have to! Save up for these three items: an awesome belt, a great handbag, and some killer shoes. Those three things will dress up tons of otherwise "ordinary" outfits and make you feel fabulous. If you're going outside, sunglasses don't hurt, either. Oh, and I should mention another great best fashion accessory, which also happens to be totally free: *posture*.

Put on an outfit, any outfit. Now go look in the mirror and relax your shoulders so they slump forward. This might be how you normally look, sitting in class or even standing around.* Now, pull your shoulders back and down. Feel your shoulder blades coming together in the center of your back, but don't let your ribs or chest jut forward: Keep your rib cage *where it was* as your shoulders pull down and back. As you do that, feel your neck stretching longer, and firm up your tummy a little. But keep breathing! Now look in the mirror and notice how much better the outfit looks. It's like a whole new top, right? With practice, you can maintain this good posture and look extra-fabulous all the time (and it's actually better for your spine to stand like this, too).

Good Posture—the best fashion accessory since the smile!

Hello, Old Friends: Using Triangles to Prove Stuff with Quadrilaterals

Now that we've met all of the quadrilaterals, it's time to prove things about them. For example, if somebody says, "Hi, I have a polygon that has four sides with two pairs of parallel opposite sides" (this is the

* Check out pp. 305–6 in *Hot X: Algebra Exposed* to see why we girls get bad posture in the first place, and pp. 182–4 in *Math Doesn't Suck* for a totally embarrassing story about my posture as an actress on a TV show!

definition of *parallelogram*), we can actually *prove* that the opposite sides must also be congruent! In fact, we'll do that on pp. 251–3. And this proves that *every* parallelogram has congruent opposite sides, so it becomes a "property" of parallelograms ... and every other shape below it: the rhombus, rectangle, and square. From that point forward, any time we're told that we have a parallelogram,* we can use the fact that its opposite sides are congruent, which can be very useful during proofs. How can we prove things like this? Mostly with this trick for quadrilaterals:

Trick for Proofs with Quadrilaterals:
Draw in diagonals **and look for congruent △'s, isosceles △'s, and/or a transversal line (a mall escalator).**

We're total pros at proving stuff about triangles and transversals now, and it's such a nice feeling to find congruent triangles, isosceles triangles, and mall escalators inside of quadrilaterals. It's like, "Oh hello, old friends! Thank you for helping me with this weird quadrilateral problem."

QUICK NOTE   By the way, just like with triangles, you might see parallelograms with their own little symbol before the letters, too. So if the four vertices of a parallelogram are (in order) S, H, O, and P, then we could call it ▱SHOP.

Step By Step

Dealing with quadrilaterals:

Step 1. If it's unclear how to start, follow the above trick for proofs with quadrilaterals. Draw in one or both diagonal lines to make triangles! Have you created some (possibly) congruent triangles? How about a transversal situation (parallel mall floors and escalator)? What about an isosceles triangle?

Step 2. If we now have what seem like congruent triangles, it's time to prove that they're congruent. If we now have what might be a transversal

. . . .
* Or rhombus or rectangle or square.

diagram, try extending the lines to see the mall floors and escalator* and determine which pair(s) of congruent angles might be helpful. If we found an isosceles triangle, it's probably time to use either "if sides, then angles" or "if angles, then sides" from p. 148.

Step 3. Take what we've learned so far about the diagram and see how it can help with what we're supposed to find or prove. We might need CPCTC or one of the transversal rules from p. 218 or p. 230. You can do this!

Step By Step In Action

Given: ▱VEST is a parallelogram.
Prove: $\overline{VE} \cong \overline{TS}$ and $\overline{VT} \cong \overline{ES}$.

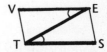

Step 1. Um, let's try drawing in $\overline{TE}$, and hey, that sure looks like we might have a pair of congruent triangles: $\triangle VET \cong \triangle STE$? Let's gather some more information.

Step 2. Our only Given (besides the diagram) is that we've got a parallelogram. And what do we know about parallelograms at this stage? Just that they have two sets of parallel lines. So, thinking of $\overline{VE}$ and $\overline{TS}$ as the (parallel) mall floors and $\overline{TE}$ as the escalator, we can use "if ||, then alt. int. ∠'s are ≅" to prove that the angles in the big "Z" are congruent: $\angle VET \cong \angle STE$. (Gimmie an "A"!)

Now, turning our heads to the right and thinking of $\overline{VT}$ and $\overline{ES}$ as the mall floors (and $\overline{TE}$ is still our escalator), we can see that they create a big backwards Z, and so the same "if ||, then alt. int. ∠'s are ≅" tells us that $\angle VTE \cong \angle SET$. (Gimmie an "A"!)

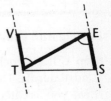

....
* See Chapter 13 if you're not sure what escalators at the mall have to do with geometry.

Okay, now some careful looking at the corresponding letters tells us that the triangles we want to prove congruent are indeed △VET ≅ △STE. (Tap those out with your pencil!) It looks like if we took the top triangle and rotated it 180° about $\overline{TE}$'s midpoint, we'd get the bottom triangle! Anyway, we still need an "S." Hey, we can use the good 'ol Reflexive Property to say $\overline{TE} \cong \overline{TE}$. (Gimmie an "S"!) Nice.

Step 3. Now, by the ASA shortcut, we know △VET ≅ △STE, and finally, CPCTC gives us both things we wanted to prove, because $\overline{VE}$ & $\overline{TS}$ are corresponding sides in our congruent triangles, and so are $\overline{VT}$ & $\overline{ES}$. Phew! Here's the two-column proof:

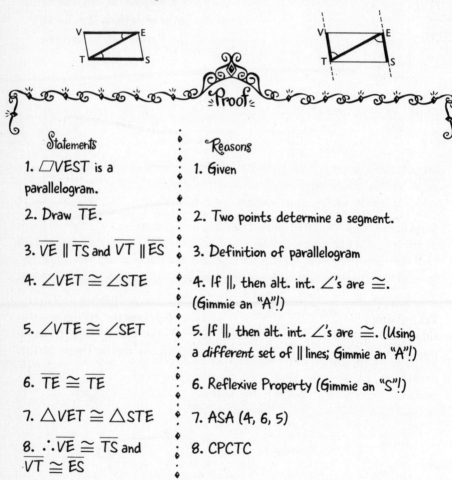

Statements	Reasons
1. ▱VEST is a parallelogram.	1. Given
2. Draw $\overline{TE}$.	2. Two points determine a segment.
3. $\overline{VE} \parallel \overline{TS}$ and $\overline{VT} \parallel \overline{ES}$	3. Definition of parallelogram
4. ∠VET ≅ ∠STE	4. If ∥, then alt. int. ∠'s are ≅. (Gimmie an "A"!)
5. ∠VTE ≅ ∠SET	5. If ∥, then alt. int. ∠'s are ≅. (Using a different set of ∥ lines; Gimmie an "A"!)
6. $\overline{TE} \cong \overline{TE}$	6. Reflexive Property (Gimmie an "S"!)
7. △VET ≅ △STE	7. ASA (4, 6, 5)
8. ∴ $\overline{VE} \cong \overline{TS}$ and $\overline{VT} \cong \overline{ES}$	8. CPCTC

And lookie there; we just proved a new Rule:

If a quad is a ||-ogram, then its two pairs of opposite sides are ≅. *

Yeah, we're good like that.

QUICK NOTE   Check out line 2 in the proof on p. 252. Yep, in a two-column proof, when drawing in a diagonal or any other useful-looking segment, we can always write something like "Draw segment $\overline{HI}$" for the Statement, and the Reason can be written as: "Two points determine a line."[†]

"I love that proud, smart feeling I get when I figure something out in math. It just feels really good to do something hard." Rebecca, 16

When *both* diagonals are drawn in a quadrilateral (as we'll see on the next page with *SKIRT*), we get long pairs of triangles like we saw in the *VEST* example above, and also *two sets* of smaller "bowtie" triangles—the side-to-side bowtie and the up-and-down bowtie. Often these triangle pairs can be proven congruent, so we'll keep an eye out for them, just in case they can be.

. . . .

* We'll see that as parallelogram property 2 on p. 259.
† See p. 389 in the Appendix for more on this.

Scene **take** *director*

Take Two: Another Example

Given: *□SKRT is a parallelogram.* **Prove:** $\overline{SR}$ *bisects* $\overline{KT}$.

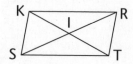

Step 1. To prove that $\overline{KT}$ is bisected, we need to first prove that $\overline{KI} \cong \overline{IT}$, right? And congruent segments can often come from CPCTC. But first we need two congruent triangles. What pair of possibly congruent triangles could have these short segments, $\overline{KI}$ & $\overline{IT}$, as corresponding congruent sides? Well, either $\triangle KIR$ & $\triangle TIS$ (the top and bottom "bowtie") or $\triangle KIS$ & $\triangle TIR$ (the side-to-side "bowtie"). And in fact, either would work for this proof! Let's pick $\triangle KIR$ & $\triangle TIS$, just because then the mall floors won't require head tilting. Notice the <u>order</u> of the letters; imagine the top triangle rotating clockwise about the center point *I*, swinging down and becoming the lower triangle. If *that's* how the triangles are congruent, then $\overline{KI}$ & $\overline{IT}$ would indeed be corresponding sides on congruent triangles, which is what we'd need in order to use CPCTC to prove that $\overline{KI} \cong \overline{IT}$. Make sense? Let's do it!

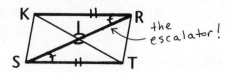

the escalator!

Steps 2 and 3. Vertical angles give us $\angle KIR \cong \angle TIS$. (Gimmie an "A"!) And the Rule "If a quad is a ||-ogram, then its two pairs of opposite sides are $\cong$" (see p. 253) gives us: $\overline{KR} \cong \overline{ST}$. (Gimmie an "S"!) *SKRT* is a parallelogram, so we of course know that $\overline{KR} \parallel \overline{ST}$. Using those as mall floors and $\overline{SR}$ as the escalator, we can use the "If ||, then alt. int. $\angle$'s are $\cong$" rule (see the big Z?) to prove that $\angle KRS \cong \angle RST$. (Gimme an "A"!) Then by AAS, we've got $\triangle KIR \cong \triangle TIS$, and by CPCTC, we've proven $\overline{KI} \cong \overline{IT}$. Finally, by the definition of *bisect*, we've proven that $\overline{SR}$ bisects $\overline{KT}$. Done!

(You'll write out the two-column proof in exercise #5 on p. 256. You can do it!)

QUICK NOTE   In the above proof, CPCTC could also tell us that $\overline{SI} \cong \overline{IR}$, right? It's just the other two corresponding sides of our two congruent triangles. And so we would conclude that the other diagonal is also bisected, which then would prove that <u>the diagonals of parallelograms always *bisect each other.*</u> We'll see that property as #4 on p. 260.

Doing the Math

Do these two-column quadrilateral proofs. I'll do the first one for you.

1. Given: ▱*COAT* is a parallelogram. Prove: ∠*O* is supplementary to ∠*C*.

<u>Working out the solution:</u> In this case, we don't need to draw any diagonals, but extending the lines can help us see the escalator and floors! We know $\overline{OA} \parallel \overline{CT}$, and if we think of $\overline{CO}$ as the transversal (escalator), then we can see that ∠*O* & ∠*C* are same-side interior angles, to which we can apply the Rule: "If ‖, then same-side int. ∠'s are supp." Done!

Statements	Reasons
1. ▱*COAT* is a ‖-ogram.	1. Given
2. $\overline{OA} \parallel \overline{CT}$	2. Definition of ‖-ogram
3. ∴ ∠*O* is supp. to ∠*C*.	3. If ‖, then same-side int. ∠'s are supp.

2. Given: ▱*COAT* is a parallelogram (see above). Prove: ∠*O* ≅ ∠*T*. *(Hint: Draw $\overline{CA}$ and find congruent triangles!)*

3. Given: *DRES* is a kite with $\overline{DR} \cong \overline{RE}$. Prove that the side angles are congruent: $\angle D \cong \angle E$. *(Hint: Figure out how to use SSS and CPCTC.)*

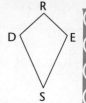

4. Given: *BOXY* is a rectangle. Do a paragraph proof to show that the diagonals are congruent: $\overline{BX} \cong \overline{YO}$ *(Hint: Rectangles are parallelograms, so opposite sides are ≅.)*

5. □*SKRT* is a parallelogram. Prove: $\overline{SR}$ bisects $\overline{KT}$. *(Hint: The strategy is laid out on p. 254. Read it again and then try doing the proof without looking!)*

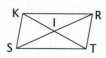

6. In *BELT*, Given: $\overline{BE} \cong \overline{EL} \cong \overline{LT} \cong \overline{BT}$. Prove: $\overline{EL} \parallel \overline{BT}$. *(This is the toothpick example I mentioned in the footnote on p. 242!)**

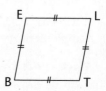

(Proof solutions at GirlsGetCurves.com)

Takeaway Tips

For *always, sometimes, never* questions about the classification of quadrilaterals, replace *is* with *satisfies the definition of*. Makes 'em much easier!

Squares satisfy all the quadrilateral definitions except for *trapezoid* and *isosceles trapezoid*.

If a proof involving a quadrilateral seems confusing, remember the golden rule: <u>Draw in diagonals</u> to create congruent △'s, an isosceles △, and/or a transversal line (a mall escalator). You can do it!

* We could repeat this proof to show that $\overline{BE} \parallel \overline{LT}$, and then we'd have proven that if a quad has four ≅ sides, then opposite sides must be parallel to each other!

Chapter 16

Reality Show Drama

Properties of Parallelograms and Other Quadrilaterals

We've all watched those reality shows with dysfunctional families. And even though we know it's mostly scripted, made-up stuff, it still seems pretty sad to watch these "real" people face public humiliation, week after week. I mean, what purpose do those shows serve? Is this really how we want to be entertained?

Well, they're about to serve a great purpose—perhaps for the first time ever. Let's face it; learning the properties of all the different quadrilaterals can be pretty intimidating. There are sides, angles, and diagonals that can be congruent, supplementary, bisected, perpendicular, you name it—and in all sorts of strange combinations! But now, a reality-show-style dysfunctional family is going to make learning them a snap.

The Quadrilateral Family Dynasty*

So there was this really wealthy girl named Paris. She was an only child at first, and totally spoiled by her mom, whom people used to call the "Queen." Paris had to have two pairs of everything—two pairs of her favorite shoes, two pairs of her favorite sunglasses, two bicycles (one pink, one purple), etc. She preferred to ride her bicycles on flat ground, though; she didn't like bicycling up or down hills' steep *angles*. When she was a few years old, her mom remarried, and Paris ended up with a stepbrother named Kit and an adopted sister named Tracy. Neither of them got the attention—or two pairs of everything—like Paris did. For example, the

. . . .

* Legal disclaimer: This is *pure fiction* and isn't meant to resemble any persons, living or dead. I totally made it up, okay?

family said Kit "only needed" one bicycle (he had fun bicycling on angled hills, though), and Tracy didn't even get a bicycle at all. Can you feel the drama already?

Once she grew up, Paris became a world traveler, and—scandal!—got herself pregnant with a little boy named Rex, to whom she gave two pairs of everything, too. To her, that was normal, after all. But as Rex became a toddler, he grew to hate all the traveling. Craving more structure in his life, Rex developed OCD.* Even as a small child, Rex liked to organize everything around him into *right angles* all the time: books, papers, you name it.

The strain of raising a child abroad finally got to Paris, so she and Rex came back home for comfort, and Paris found that comfort in the arms of . . . her stepbrother Kit. Paris and Kit ended up getting married (scandal!) and having a kid together. They couldn't agree on a name because deep inside, Kit always believed he was *right* about everything. Plus, Kit and Paris did grow up as brother and sister, so they never could agree on anything anyway. They ended up settling on the name, um, Rhonda.

Paris spoiled her little girl Rhonda with *two pairs* of everything, too, but she also tried to make sure Rhonda led a more balanced life on *all sides*, and that she could be more *flexible* than her OCD brother Rex. What parent doesn't want better for their child, after all? However, Rhonda inherited things from her dad (Kit) that her mom didn't have: a love of bicycling up steep angles (with both bicycles!) and the belief that she was *right* about everything *deep inside*. So Rhonda could be, um, stubborn sometimes.

Years later, Rhonda and her OCD half-brother Rex decided to get married and have a child of their own, Sara (major scandal!). And Sara, of course, inherited all the interesting traits from both parents: She bicycled up and down angles, was OCD (liked right angles on everything around her), *and* believed she was right deep inside.

And what ever happened to Paris and Kit's adopted sister, Tracy? There's not as much to tell! Tracy was super smart and surprisingly well-adjusted. She got married (outside the family), and had a successful career and a little girl named Isabel. In fact, Tracy didn't allow herself or Isabel to have more than one of most things, just to make sure she grew up to be appreciative and not spoiled. Isabel didn't even want a bicycle. Can you imagine that?

The rest of the family assumed Tracy was a snob because she and her daughter Isabel didn't really talk to anyone else. But Tracy just wanted to protect her daughter from all the drama. Who can blame her?

* OCD stands for obsessive-compulsive disorder.

Check out this quick sketch of their family tree,* including a "Q" to stand for the Queen matriarch. Think about this story as you do. Can you find Rhonda? Who were her parents? Where's Sara? Isabel? The love child from Europe (Rex)?

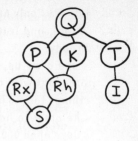

Now, can you draw this without looking? Try it! If not, read the story again and then give it another shot. Seriously, you'll thank me later.

Properties of Parallelograms

Remember how Paris had two pairs of everything, even two bicycles? Keep that in mind when reading the properties of parallelograms below. Some will seem easier than others, and that's normal. The key is to not get overwhelmed! So for each one, this is what I want you to do: Read the property slowly, look at the picture to see how it makes sense, and then say to yourself, "Yes, that makes total sense." You *have* to say or think that after each one. Deal? Okay, let's do it.

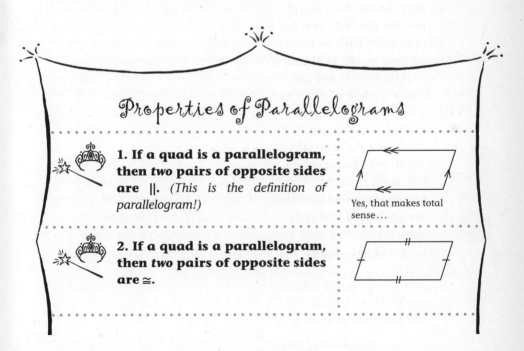

Properties of Parallelograms

1. If a quad is a parallelogram, then *two pairs of opposite sides* are ∥. *(This is the definition of parallelogram!)*

Yes, that makes total sense…

2. If a quad is a parallelogram, then *two pairs of opposite sides* are ≅.

....

* Yes, this looks exactly like the quadrilateral family tree from p. 245. I guess you've probably figured out it's not a coincidence.

3. If a quad is a parallelogram, then *two* pairs of opposite ∠'s are ≅.

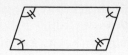

4. If a quad is a parallelogram, then the *two* diagonals bisect each other.* "Bisect" sounds like *bicycle*, doesn't it?

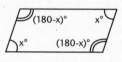

5. If a quad is a parallelogram, then *two* pairs of consecutive ∠'s are supp. *(Actually, <u>all</u> pairs of consecutive ∠'s are supp. because each side can be an escalator!)*†

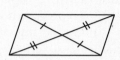

Just remember—Paris (I mean, a parallelogram) has *two* of everything: *two* pairs of ∥ sides, *two* pairs of opposite ≅ sides, *two* pairs of opposite ≅ ∠'s, *two* pairs of supp. consecutive ∠'s, and *two* diagonals bisecting each other (her two bicycles). But Paris didn't like riding up hills, so notice that there are no *angles* being bicycled (er, bisected) . . .

QUICK NOTE   **Consecutive angles** in a parallelogram are angles that are next to each other, not opposite each other. For example, in ▱*COAT* on p. 255, ∠*C* & ∠*O* are consecutive angles. Notice that any two angles in a parallelogram are either opposite or consecutive. And that means that any two angles on a parallelogram are either congruent or supplementary.

....

* We proved this property on pp. 254–5!

† And then "if ∥, then same-side int. ∠'s are supp." applies, because consecutive ∠'s on our parallelogram are same-side int. ∠'s at the mall! Review this rule on p. 218 if it helps. Now say, "Yes, that makes total sense."

Watch Out!

Although most parallelograms look like they have congru-ent diagonals, they don't! It's an optical illusion. Use the side of your finger or pencil to measure one diagonal and then the other. You'll be amazed at how different they actually are. Try it with one of the parallelograms on p. 260.

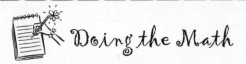

Doing the Math

Use the properties of parallelograms from pp. 259–60 to answer the questions below, referring to the parallelogram $\square JENS$. Do the following statements *have* to be true?

I'll do the first one for you.

1. $\overline{ES} \cong \overline{JN}$?

<u>Working out the solution</u>: Hmm, these are the diagonals, and there is one property that says the diagonals *bisect* each other, which means that, for example, $\overline{EA} \cong \overline{AS}$. But the properties don't say anything about the diagonals being congruent. And in fact, most of the time they aren't congruent! (See the Watch Out at the top of this page for more on that.)

Answer: No, it doesn't have to be true.

2. $\overline{JA} \cong \overline{AN}$?   6. $\overline{JA} \cong \overline{AS}$?   10. $\angle EJS \cong \angle ENS$?

3. $\overline{ES} \perp \overline{JN}$?   7. $\overline{JE} \cong \overline{SN}$?   11. $\angle EAJ \cong \angle NAS$?

4. $\angle JSE \cong \angle SEN$?   8. $\overline{SA} \cong \overline{NA}$?   12. $\angle EJN \cong \angle JNS$?

5. $\angle JSN$ supp. to $\angle ENS$?   9. $\angle JEN$ supp. to $\angle JSN$?   13. $\angle AJS \cong \angle ASJ$?

(Answers on p. 400)

Proving that a Quadrilateral Is a Parallelogram

Proving that we have a parallelogram is no big deal. If we can prove that a quadrilateral has *any* of the properties from pp. 259–60, then we've proved that it's a parallelogram. Here they are again, written as converses* and with a bonus at the end.

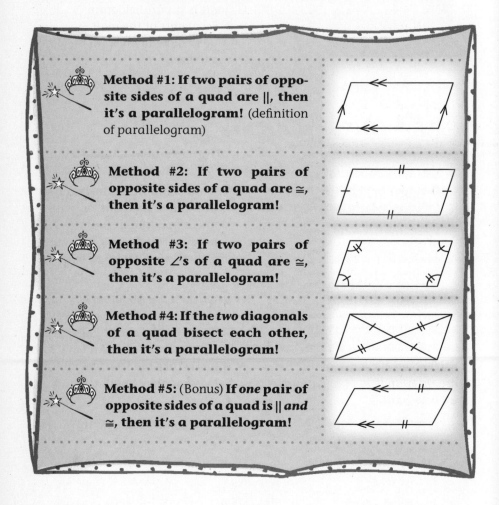

Method #1: If two pairs of opposite sides of a quad are ||, then it's a parallelogram! (definition of parallelogram)

Method #2: If two pairs of opposite sides of a quad are ≅, then it's a parallelogram!

Method #3: If two pairs of opposite ∠'s of a quad are ≅, then it's a parallelogram!

Method #4: If the *two* diagonals of a quad bisect each other, then it's a parallelogram!

Method #5: (Bonus) If *one* pair of opposite sides of a quad is || *and* ≅, then it's a parallelogram!

This last one is a bonus. It's not listed as one of the properties on pp. 259–60. So if we happen to be able to prove that one pair of opposite sides is both || *and* ≅, then we've totally proven that it's a parallelogram.

....

* As you know from Chapter 1, Rules don't necessarily have true converses (unless they're definitions). But these parallelogram properties do!

You might notice that the supplementary angle property is missing (#5 on p. 260). It could absolutely be on this list, but it would be redundant. You see, "two pairs of consecutive supplementary angles" doesn't tend to be something we can prove without *first* proving that both sets of lines are parallel, and well, if we've proved that both sets of lines are parallel, then we've already proven that we have a parallelogram, haven't we?

So when it's time to prove that a quad is a parallelogram, we just need to prove any of the items on p. 262. Escalators at the mall, isosceles triangles, and congruent triangles continue to be really helpful in doing so, and we'll keep an eye out for 'em along the way.

Step By Step

Proving that a quadrilateral is a parallelogram:

Step 1. First, we mark down everything we can on the diagram, based on what we've been given, and see which of the five methods on p. 262 might be the easiest to prove—#1: *Two* pairs of opposite || sides, #2: *Two* pairs of opposite ≅ sides, #3: *Two* pairs of opposite ≅ angles, #4: *Two* diagonals bisecting each other, <u>or</u> the bonus method, #5: One set of opposite sides that are *both* || and ≅.

Step 2. Can we find congruent triangles to help? How about an escalator at the mall (a head tilt and extending lines can help us "see" a transversal). Remember: If we have a bunch of small congruent segments, we should check to see if maybe the diagonals bisect each other. And if we've been given angles, either directly or indirectly (like via isosceles or congruent triangles, or even another parallelogram), they might tell us that we *have* parallel lines.

Step 3. If it's still not clear, just try to prove something. Anything—be creative. It could totally lead to the solution.

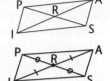

And... Action! Step By Step In Action

Given: R is the midpoint of both $\overline{PS}$ and $\overline{IA}$.
Prove: PASI is a parallelogram.

Step 1. Midpoints divide their segments into two congruent segments, so we can mark $\overline{PR} \cong \overline{RS}$ and $\overline{IR} \cong \overline{RA}$ on the diagram.

Step 2. This sure looks like bisecting diagonals—that's method #4. Nice! Now let's write it out, two-column style:

Proof

Statements	Reasons
1. R is the midpoint of $\overline{PS}$.	1. Given
2. $\overline{PR} \cong \overline{RS}$	2. A midpoint divides a seg into 2 $\cong$ seg's.
3. $\overline{IA}$ bisects $\overline{PS}$.	3. Definition of bisect
4. R is the midpoint of $\overline{IA}$.	4. Given
5. $\overline{IR} \cong \overline{RA}$	5. A midpoint divides a seg into 2 $\cong$ seg's.
6. $\overline{PS}$ bisects $\overline{IA}$.	6. Definition of bisect
7. $\therefore$ PASI is a ‖-ogram	7. If the diagonals of a quad bisect each other (Steps 3 and 6), then the quad is a ‖-ogram.

Take Two: Another Example

Given: △RPS is an isosceles triangle with base $\overline{RS}$, $\overline{PR} \cong \overline{UE}$, $\angle R \cong \angle E$ (see PURSE diagram on next page). Prove: PSEU is a parallelogram.

Step 1. Let's start marking what we know with kitty scratches. Because △*RPS* is isosceles, we know $\overline{PR} \cong \overline{PS}$ and $\angle R \cong \angle PSR$, right? Combining $\overline{PR} \cong \overline{PS}$ with the Given $\overline{PR} \cong \overline{UE}$, the Transitive Property tells us $\overline{PS} \cong \overline{UE}$. Hey, we have a pair of opposite, congruent sides on our *PSEU* quad!

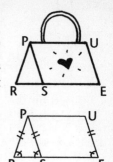

Hmm, there's no clear way to get those other opposite sides, $\overline{PU}$ & $\overline{SE}$, to be congruent. But wait—since $\overline{PS} \cong \overline{UE}$, if we can get those two sides to also be <u>parallel</u>, then we'd satisfy the "bonus" method of proving we have a ‖-ogram. Let's try that. (Plus, we haven't used all of the Givens yet!)

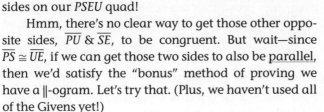

Steps 2 and 3. Combining $\angle R \cong \angle PSR$ with our Given $\angle R \cong \angle E$, the Transitive Property tells us $\angle PSR \cong \angle E$, right? Well, if we think of $\overline{RE}$ as the escalator and $\overline{PS}$ & $\overline{UE}$ as the mall floors, then we can use corresponding congruent angles to prove we have parallel lines! Here is it, two-column style:

Proof

Statements	Reasons
1. △RPS is an isosceles triangle with base $\overline{RS}$.	1. Given
2. $\overline{PR} \cong \overline{PS}$ and $\angle PSR \cong \angle R$	2. If isosceles △, then legs and base ∠'s are ≅.
3. $\overline{PR} \cong \overline{UE}$ and $\angle R \cong \angle E$	3. Given
4. $\overline{PS} \cong \overline{UE}$	4. Transitive Property (one pair opp. sides ≅!)
5. $\angle PSR \cong \angle E$	5. Transitive Property
6. $\overline{PS} \parallel \overline{UE}$	6. If corr. ∠'s are ≅, then ‖ lines. (same pair opp. sides ‖!)
7. ∴ *PSEU* is a ‖-ogram.	7. If one pair of opp. sides of a quad is both ≅ and ‖ (4 & 6), then the quad is a ‖-ogram.

Doing the Math

Based on the given info, prove that we have a parallelogram via one of the methods on p. 262. I'll do the first one for you.

1. Given $\triangle PRA \cong \triangle SRI$, prove that *PASI* is a parallelogram.

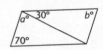

<u>Working out the solution</u>: So, we have congruent "bowtie" triangles. It might be tempting to try to get two pairs of congruent sides or angles (method #1 or #2), but these triangles are small, which means two of their sides, each, are half-diagonals of the quad. And that should make us think about bisection. In fact, since $\triangle PRA \cong \triangle SRI$, looking carefully at the order of letters, we can conclude from CPCTC that $\overline{PR} \cong \overline{RS}$ and $\overline{IR} \cong \overline{RA}$.

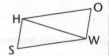

And from that point on, this proof is *exactly* like the one we did on p. 264. There's no need to rewrite it!

2. Given the parallelogram to the right, find $a°$ and $b°$.

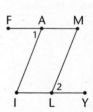

3. Given: $\triangle HSW \cong \triangle WOH$. Prove: *SHOW* is a parallelogram.

4. Given: $\overline{AM} \parallel \overline{IL}$, $\angle 1 \cong \angle 2$. Prove: *AMLI* is a parallelogram. *(Hint: Start with $\angle 2$ and $\angle M$. What do we know about them?)*

(Answer to #2 on p. 401; proofs at GirlsGetCurves.com)

What's the Deal?

Proving Two Sides Are Parallel—
On the Coordinate Plane

You might be given the coordinates of a quadrilateral's four vertices and be asked to prove that its sides are parallel (in other words, prove that it's a parallelogram). No problem! If slopes are equal, then we know the lines are parallel, so all we need to do is find the *slopes* of each of the sides, which we learned to do in algebra. Check out Chapter 11 in *Hot X: Algebra Exposed* to review how to find slope from two points, and check out GirlsGetCurves.com for an example of how it works with quadrilaterals!

The Rest of the Family

By this point, we know Paris really well, and it's time to get up close and personal with the other members of the dysfunctional quadrilateral family. The good news is, each child inherits all the traits from his or her parents, so if we remember who everybody's parents are, there's less to memorize.

For example, we know that Rex is Paris' kid, so rectangles have all the same properties as parallelograms. See what I mean? Rex has all of Paris' traits plus his OCD issue—you know, from needing more structure in his life as a young child—which ends up resulting in congruent diagonals. Who knew? (We don't know Rex's dad. He must have been some European guy.)

So that's Rex in a nutshell, and if we're given a rectangle in a problem, we can draw in diagonals and totally mark any of the above stuff that might help us. Nice.

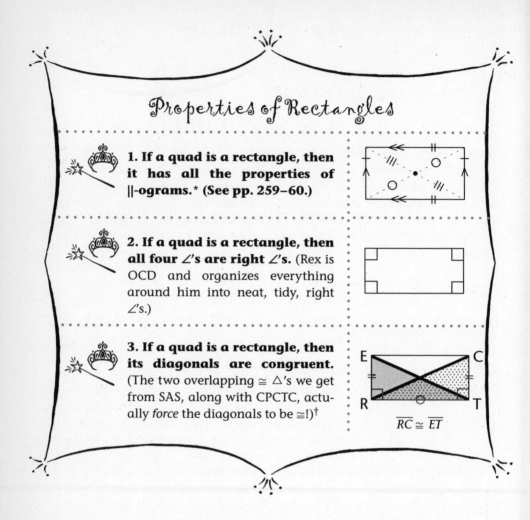

Properties of Rectangles

1. If a quad is a rectangle, then it has all the properties of ||-ograms.* (See pp. 259–60.)

2. If a quad is a rectangle, then all four ∠'s are right ∠'s. (Rex is OCD and organizes everything around him into neat, tidy, right ∠'s.)

3. If a quad is a rectangle, then its diagonals are congruent. (The two overlapping ≅ △'s we get from SAS, along with CPCTC, actually *force* the diagonals to be ≅!)[†]

$\overline{RC} \cong \overline{ET}$

How about Paris' stepbrother/husband, Kit? Remember, Kit wasn't spoiled like Paris was (for example, he only got one bicycle), and he always thinks he's *right* about everything deep *inside*.

. . . .

* I didn't mark opp. ∠'s ≅ and consecutive ∠'s supp. (2 of Paris' traits) on this row's diagram, because all that—and more—is covered in the next row: (Right ∠'s are all ≅ *and* supp. to each other!)

† Isosceles trapezoids also have ≅ diagonals for the same reason: Their legs are ≅ and their base ∠'s are ≅, so SAS & CPCTC give us ≅ diagonals. More on pp. 273–4.

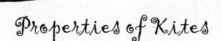

Properties of Kites

1. If a quad is a kite, then two disjoint pairs of consecutive sides are ≅. (The definition of kite! And it's what makes it look like, um, a kite that flies in the air.)

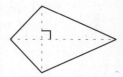

2. If a quad is a kite, then the diagonals are ⊥. (Kit feels "right" about everything *deep inside.*)

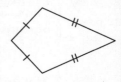

3. If a quad is a kite, then one pair of opposite ∠'s is ≅. (Paris got two; Kit only got one.)

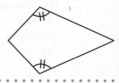

4. If a quad is a kite, then one diagonal bisects the other. (He only got one bicycle.)

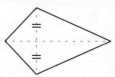

5. If a quad is a kite, then one diagonal bisects opposite ∠'s. (He *bicycled angled* hills → a diagonal *bisects angles.* Imagine folding the kite down the dotted line to "see" how the angles get bisected.)

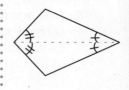

QUICK NOTE In a kite, only one diagonal is bisected. Notice that the same diagonal that *does* this bisecting, also bisects the angles it runs through. It's just one bisector (one bicycle!) doing two different types of bisection. That hard-working bisector just so happens to be the line of symmetry, which makes sense when you think about it. Segments and angles divided by a line of symmetry ought to be congruent on either side, after all.* In fact, it's really helpful to imagine folding a kite down its line of symmetry to "see" properties #1, #3, #4, and #5 from p. 269. Try it now!

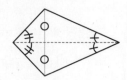

Now we're ready to get to know Paris and Kit's child, Rhonda. She gets everything they both have (two pairs of everything, Kit's inner righteousness, etc.). In fact, she gets two pairs of everything Kit has and even likes to ride *both* her bicycles up steep angles! Also, Paris wanted her to be well-balanced on all sides, remember? And she didn't end up with OCD like her half-brother Rex.

Just like with her dad Kit, it's really helpful to imagine folding rhombuses along their lines of symmetry—yep, Rhonda got two of those, too!

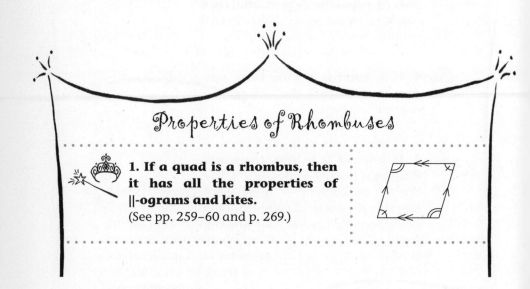

Properties of Rhombuses

1. If a quad is a rhombus, then it has all the properties of ||-ograms and kites.
(See pp. 259–60 and p. 269.)

. . . .

* See p. 112 for more on lines of symmetry.

2. If a quad is a rhombus, then all four sides are ≅. (She's well-balanced on all sides.)

3. If a quad is a rhombus, then both diagonals are ⊥ bisectors of each other. (This is really a combination of the fact that Paris' diagonals bisect each other and that Kit's diagonals create a right angle.)

4. If a quad is a rhombus, then both diagonals bisect their angles. (She *bicycled angled* hills with *both* her bicycles → *Both* her diagonals *bisect their angles*!) Notice we could use ASA to prove that the 4 right △'s that the rhombus is divided into are now congruent.

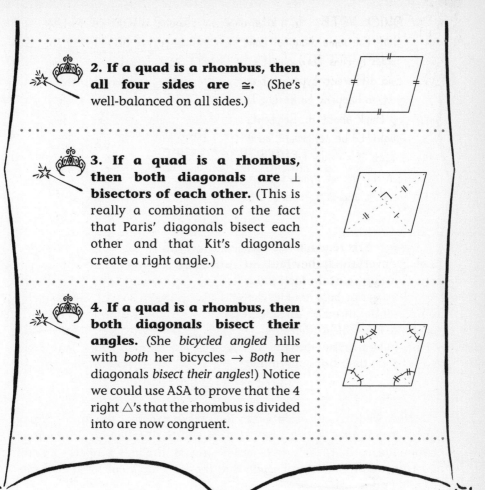

Recall that Rex and Rhonda ended up getting married and having a child of their own, Sara. She gets everything that they each had. You might think that would make it harder to memorize Sara's traits, but it's just the opposite: Squares are incredibly symmetrical, so anything you can think of involving *any* kind of symmetry for a quadrilateral will be true for squares! It's actually harder to think of what *isn't* true for a square.

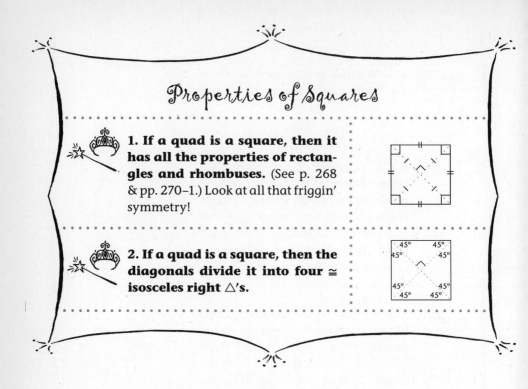

℘𝓇𝑜𝓅𝑒𝓇𝓉𝒾𝑒𝓈 𝑜𝒻 𝒮𝓆𝓊𝒶𝓇𝑒𝓈

1. If a quad is a square, then it has all the properties of rectangles and rhombuses. (See p. 268 & pp. 270–1.) Look at all that friggin' symmetry!

2. If a quad is a square, then the diagonals divide it into four ≅ isosceles right △'s.

Now we turn our attention to the other branch of the family, Tracy and her daughter Isabel. Tracy has worked hard to lead a drama-free, happy life, so there isn't much to tell. In fact, the only property we can list is the definition of trapezoid: that there is *exactly one set* of opposite, parallel sides.

Property of Trapezoids

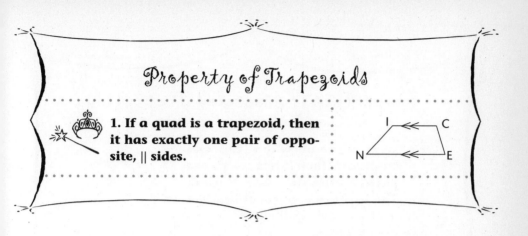

1. If a quad is a trapezoid, then it has exactly one pair of opposite, ∥ sides.

Notice that because of the parallel bases, if we think of $\overline{NI}$ and $\overline{CE}$ as mall escalators, then the Rule "If ∥, then same side int. ∠'s are supp." (see p. 218) tells us that ∠I is supp. to ∠N, and also that ∠C is supp. to ∠E.

To understand her daughter, Isabel, let's step back for a moment and think about what would happen if we sliced off the top of an isosceles triangle, parallel to its base.

The two base angles, ∠T and ∠R, would still be congruent, right? And since $\overline{AP} \parallel \overline{TR}$, if we think of the mall and $\overline{TA}$ as the escalator, the Rule "If ∥, then same-side int. ∠'s are supp." tells us that ∠A is supp. to ∠T, right? We could do the same thing using $\overline{PR}$ as the escalator, and learn that ∠P is supp. to ∠R. And if two angles (∠A and ∠P) are supp. to two congruent angles (∠T ≅ ∠R), that means they have to be congruent to each other! And that means ∠A ≅ ∠P.* Dude, we're good.

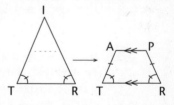

Oh hey, how about those diagonals? Let's draw them in. Hmm, it doesn't look they bisect each other or that they even bisect any angles. (Isabel never got a bike. Her life wasn't perfect.) But notice that the diagonals have created two overlapping triangles (△TAR and △RPT) that are congruent by SAS, which means by CPCTC that $\overline{AR} \cong \overline{TP}$; in other words, their diagonals must be *congruent*. In fact, this is the same exact argument we used on p. 268 for why the diagonals of rectangles are congruent!

. . . .

* This ends up meaning that in an isosceles trapezoid, *any* lower base ∠ is supp. to *any* upper base ∠. See p. 81 to review the Supplementary Angle Rule we used here!

Properties of Isosceles Trapezoids

1. If a quad is an isosceles trapezoid, then it has *exactly one* pair of opposite ∥ sides (bases). The legs are congruent and *not* ∥. (That's the definition of isosceles trapezoid!)

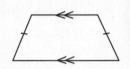

2. If a quad is an isosceles trapezoid, then the lower base angles are congruent, and the upper base angles are congruent. (It's isosceles; this makes sense!)

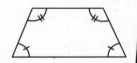

3. If a quad is an isosceles trapezoid, then any lower base angle is supplementary to any upper base angle. (Just think of the legs as escalators at the mall.)

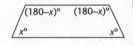

4. If a quad is an isosceles trapezoid, then the diagonals are congruent. In fact, SAS/CPCTC forces this to be true! (See the explanation on the previous page.)

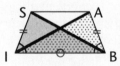

$\overline{IA} \cong \overline{SB}$

Let's see how well we know this crazy family!

Answer the questions below about quadrilaterals. For exercises 1–14, refer to the figures below: *WRLD* is a parallelogram, *KITE* is a kite (with a line of symmetry $\overline{IE}$), *HOMB* is a rhombus, *RNEG* is a rectangle, *SQUA* is a square, *TRAC* is a trapezoid, and *ISOZ* is an isosceles trapezoid. I'll do the first one for you.

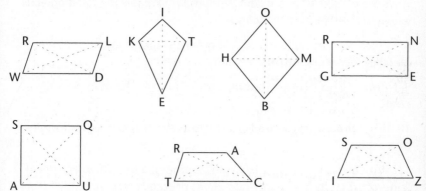

1. Name all diagonals that are bisected.

<u>Working out the solution</u>: Well, we know that Tracy and Isabel never got bicycles, so they're out. Kit only got one bicycle, and looking at the diagram, $\overline{IE}$ is the line of symmetry, and $\overline{KT}$ is the one *being* bisected—in other words, being divided into two congruent halves. Great! In terms of Paris* and all of her kids and her granddaughter (Sara), since parallelograms' diagonals are both bisected, that means all her descendants' diagonals are, too: rectangle, rhombus, square. So we'll list 'em all!

<u>Answer</u>: *KITE:* $\overline{KT}$; *WRLD:* $\overline{WL}$ & $\overline{RD}$; *RNEG:* $\overline{RE}$ & $\overline{GN}$; *HOMB:* $\overline{OB}$ & $\overline{HM}$; *SQUA:* $\overline{SU}$ & $\overline{QA}$.

. . . .

* the "WRLD" traveler . . .

2. Which shape has *only one* pair of bisected angles? *(Who has only one bicycle that rides up hills at angles?)* Name the pair of angles.

3. For which shapes are all the *angles* bisected by diagonals? *(This means both diagonals are angle bisectors: Who has two bicycles that ride up angles?)*

4. Which shapes have perpendicular diagonals? *(Who—and his or her descendants—always feels right, deep inside?)*

5. In *ISOZ*, do we know that $\triangle SZI \cong \triangle OIZ$?

6. In *WRLD*, do we know that $\triangle RWD \cong \triangle LDW$?

7. Name all sets of congruent diagonals. *(Imagine the full diagonals as sides of overlapping triangles.)*

8. Which shapes have *two sets* of opposite congruent sides? *(Who—and their kids—have two of everything?)*

9. Name the only shape that has sides that *don't* have to be congruent to any other sides.

10. What property is shared only by isosceles trapezoids, rectangles, and squares?

11. Which shapes also satisfy all the properties of kites and parallelograms? *(Which kids are descendants of both Paris and Kit?)*

12. Which shape satisfies the properties of a parallelogram, but not those of a kite (besides a parallelogram)? *(Who is Paris' kid but not Kit's kid?)*

13. What's another name for a shape that is both a rhombus and a rectangle?

14. There is one shape that doesn't have a property regarding supplementary angles. What is that shape, and why?

For the questions below, fill in the missing word(s).

(Hint: For #16–20, think about the "other parent" involved.)

15. Unlike the most general trapezoids, isosceles trapezoids have two _____ sides.

16. Unlike the most general parallelograms, a rectangle's four _____ are all congruent.

17. Unlike the most general kites, a rhombus' two pairs of opposite sides are _____ and _____.

18. Unlike the most general parallelograms, a rectangle's diagonals _____ each other. *(Careful: Remember what we know about a parallelogram's diagonals!)*

19. Unlike the most general parallelograms, a rhombus' diagonals bisect _____.

20. Unlike the most general parallelograms, a rhombus' diagonals are _____ each other.

(Answers on p. 401)

Phew, great job. So, now that we've gotten all close and snuggly with the quad properties, let's see how we can *use* them to become more powerful, problem-solving stars. (Kinda like reality TV stars, but way happier.) You can do this!

 *Doing the Math*

Answer these quadrilateral problems. I'll do the first one for you.

1. In *KITE* (see p. 275), $\overline{IE}$ is the line of symmetry, $KI = 4$, $TE = 2x + 6$, and $EK = x + 8$. Find the perimeter of *KITE*.

Working out the solution: For the perimeter, we need all four sides, right? We already know $KI = 4$. Since $\overline{IE}$ is the line of symmetry, that means $KI = IT$, which means $IT = 4$. Great! And since $TE = EK$, we can set those values equal to each other and solve for x.

Solving $2x + 6 = x + 8$, we get $x = 2$. And that means $TE = 2(2) + 6 = 10$, and $EK = 2 + 8 = 10$ (of course!). And that's a total perimeter of $4 + 4 + 10 + 10 = 28$.

Answer: The perimeter of KITE is 28.

2. If one angle of a rhombus is 120°, what are its other angles? What two shapes does the shorter diagonal of *this* rhombus divide it into? (Draw a picture!)

3. Given the kite to the right (with a vertical line of symmetry) find the angle measurements *x*° and *y*°. *(Hint: Look for isosceles triangles, and write some equations based on triangles' degrees adding up to 180°.)*

4. Given: △*RPS* is an isosceles <u>triangle</u> and *RPUE* is an isosceles <u>trapezoid</u>. Prove: *PSEU* is a parallelogram.

5. Explain why the diagonals of a trapezoid cannot bisect each other. *(Hint: This is an indirect proof; start by assuming they do bisect each other. You'll need to draw your own picture and label it. Feel free to use TRAC on p. 275 and "X" as the intersection point of the diagonals.)**

6. Given: BUCK is a parallelogram and $\overline{BC}$ bisects both ∠*UBK* and ∠*UCK*. Prove: $\overline{BU} \cong \overline{UC}$. *(Hint: Remember isosceles triangles.)* What else do we know about the sides? We can now prove *BUCK* satisfies the definition of a particular type of parallelogram. Which one is it? *(See pp. 241–3 for definitions of quadrilaterals.)*

(Answers to #2–3 on p. 401; proofs at GirlsGetCurves.com)

Memorizing the Properties

It can be challenging to memorize all of the properties about the sides, angles, and diagonals of quadrilaterals! But here's a tip: If it *looks* true, it probably *is* true. So if you're stuck on a test problem and you need to quickly find out if something is true, do a quick sketch of the quadrilateral (and its diagonals, if it's helpful) and see if it *looks* true.

Most of the time, this will lead you to the right answer. But make sure to exaggerate the angles or sides so that your picture doesn't look too much like a shape's "kids." For example, if we drew a parallelogram, we'd exaggerate the angles and sides so our picture wouldn't look too much like a rectangle, rhombus, or square. Otherwise, we might think it has more special properties than it does.

. . . .

* See Chapter 10 to brush up on *indirect proofs,* AKA *proofs by contradiction.*

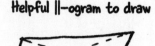 **Helpful ||-ogram to draw**

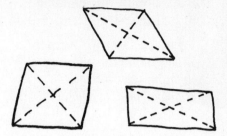 **Not helpful ||-ograms to draw!**

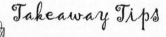

And to better "see" the properties of kites and rhombuses, imagine folding them down their lines of symmetry. If it's a rhombus you're drawing, notice that the symmetry of rhombuses is most helpful when they are drawn up-and-down, not laying on their side. (See *HOMB* on p. 275.)

Read this chapter a few times and you'll be amazed at how much you remember without even trying. Who knew a little family drama could be so helpful?

 Takeaway Tips

 Properties of specific quadrilaterals involve *sides* (they can be || and/or ≅), *angles* (they can be ≅ or supp.), and *diagonals* (they can be ≅, they can be ⊥ to each other, and they can bisect each other and/or angles).

Everyone satisfies the definitions of—and has all the properties of—their parents or grandparents.

To remember the quadrilateral properties, remember Paris and the rest of her family! Paris has *two* of everything (including bicycles, AKA bisectors) and so do her kids. Kit always feels he's *right* deep *inside* and so do his kids (perpendicular diagonals). Read the story on pp. 257–8 again and then their properties, and it'll totally sink in.

Danica's Diary
THE WHOLE POINT OF SCHOOL

Let's face it; you're learning a lot of stuff that you will never use in your life. You're learning the names of kings and queens in history class, many of which you won't remember 10 years from now, but you better know them for the test tomorrow or you won't get a good grade. Many people who take French or Spanish won't remember five words later in life if they don't practice it. Even in math and science, where the facts themselves could be very helpful depending on the career you choose, many techniques you're learning simply won't come up. If we forget so much later in life, what is the point of going to school?

When I was in the 8th grade, I was already acting on TV, so I often missed class and had to make up exams. I'll never forget approaching my history teacher, Mr. Smith, about making up a test. He said, "You know how to take tests. I'll just count your midterm for more points. You can go ahead and skip this one." I was floored. For the first time, I realized that being good at *taking tests* was its own skill, totally independent from the actual topic of the exam. And I learned something really important that day about the whole point of school: We go to school not to collect a bunch of facts. We go to school to train ourselves to be happy and productive adults by learning skills: how to focus and study (and not procrastinate!), how to become an expert at a topic in a short amount of time, how to think logically, how to express ourselves, and how to make convincing arguments.

Yep, school does serve a pretty awesome purpose. Right now, you are in training for your life as a fabulous, confident, self-sufficient young woman. Use this time in school to hone those skills and there'll be no stopping you!

Feel Great with These 8 Self-Esteem Boosters!

By being our own best friends, we can build ourselves up and fulfill our true potential—inside and out. How can we do that? Tons of ways! Here are some tips on how to boost your self-esteem and put you on the path to true power.

✳ Stop the Comparisons!   Constantly comparing ourselves to others will almost always make us feel bad—it's a losing battle! There will *always* be people in this world who have more and who have less. Sometimes we're even comparing our insides to their outsides anyway (see p. 78 for more on that). Sure, we could spend all day feeling jealous of some and feeling "superior" to others (in an unhealthy ego trip!), but why? Look, we're all different; we *all* have things we wish were different about ourselves (see p. 236 for some of mine!), and we *all* have the choice whether to waste energy feeling jealous of others or to spend that energy learning to love and celebrate our differences. The next time you realize you're comparing yourself to someone else, *stop*, and read on for some things you can focus on instead. You've got better things to do with your time!

Give Power to the Positive   When we focus on something, *we give it power in our lives.* So the next time you hear your inner voice say that you're ugly, lazy, stupid, whatever—stop right there! When you put yourself down, you're being your own worst enemy, and who needs that? Write down 5–10 things that you admire about yourself: loyal friend, hard worker, creative, etc., and look at this "happy list" at least twice a day. Take a moment to let yourself really *feel good* as you read each of them. (Really *feel* those good feelings. Ever wanted to try acting? Think of it as an acting exercise!) Then, tackle those negative thoughts and beat 'em! Ask yourself, "Where did this thought come from? Is this something I have the power to change?" (See pp. 174–5 for more on the difference between the things we can change and those we can't.) And really take the time to write out your answer in a journal; brainstorm on how you could improve the situation. You're a smart cookie, and in most cases, *you'll find that you already have solutions.* Take steps to improve whatever the negative message is—whether it's eating healthier food for better skin, exercising, adopting better study habits, whatever. The next time you hear that negative thought, you'll have earned the right to say, "I am improving every day," and feel good as you say it. Then look at your "happy list" again, and keep the good feelings flowin'!

Be Bold! The great Eleanor Roosevelt once said, "Do at least one thing every day that scares you." Did she mean we should all go home and watch scary movies? No! She was probably talking about how empowering it can be to *face our fears and overcome them*. When you're feeling scared or intimidated by something, recognize it as an opportunity to be bold and get stronger. Try out for the school play. Make a new friend at school. Raise your hand when you're feeling shy. Face down that challenging geometry proof. It's powerful to go from fear to strength. It's like we're telling the world, "I'm not gonna let you beat me!" See pp. 174–5 for more on beating insecurities.

Focus on Progress, Not Perfection Many of us feel pressure to be "perfect," whatever that means. Well, I'm going to let you in on a little secret: There's no such thing! Life is a journey, not a destination. Our journey can be a wonderful adventure, full of self-discovery and self-improvement. But if we remain focused on our lack of "perfection," it actually drags us down and makes progress much harder. Learn to feel satisfaction in the little improvements and victories along the way, and even some of your missteps: Sometimes we have to take one step back before we can take two steps forward, and that's okay, too. When the goal is progress, not perfection, we put ourselves on the path to a truly successful, fulfilling life!

Surround Yourself with Positive People Think about your friends for a moment; do they lift your spirits, or do they try to bring you down, perhaps because of their own insecurities? Letting other people walk all over us isn't exactly good for our self-respect or our self-esteem. Start by treating others the way you expect to be treated—with honest and caring considera-tion. Lift *their* spirits; be the kind of friend *you'd* like to have. And then notice if you're being treated like that in response. If not, it might be time to set some boundaries or get some new friends. See pp. 322–3 for more on setting boundaries.

Take Care of Your Body Let's face it; our bodies—including our emotions—are made up of chemicals and hormones, and we can literally *change the chem-istry of our bodies* with food and exercise! Turn up your favorite song and dance. Go play some basketball, even just by yourself, or go for a run. Even if you only do 20 jumping jacks or 5 push-ups, do it! Any amount you can stretch and move your body is great for getting your circulation going, and those endorphins make us happy. And see p. 82 for some after-school snack ideas to help you to look and feel fantastic.

Get Over Your Crush/Bad Friendship
Feeling bad about yourself because some guy doesn't like you back or because a friend hasn't been acting like a friend to you? Need to get over them? Here's how!

Remember this, like a mantra: Other people are bonuses. You don't need any particular person to make you feel good about yourself—except YOU. Can it feel great when a guy gives us attention or our friends act the way we want them to? Sure! But we don't need other people for that. In fact, the only lasting good feelings will come from ourselves. Other people will *always* disappoint us from time to time, in big and/or small ways. So the next time you're feeling bad because of someone else's behavior, think to yourself, "I have the power to make myself feel great; I don't need anyone else for that. Other people are *bonuses.*"

Do Something Nice for Others
Generosity is one of those funny things in life. Although it can seem like it requires us to "give something up," the truth is, being generous can feel *really good.* Try volunteering at a charity or even helping your parents around the house when they least expect it!* You'd be surprised how *powerful* it can feel to do nice things "for no reason." When we have pure motives—that is, we don't expect anything in return—it's great for our self-esteem. Sometimes when we're feeling down or depressed, helping *someone else* is just what the doctor ordered. It makes us feel good about ourselves for "being the kind of person who does that sort of thing." Also, helping those less fortunate (like orphans or the blind) can really put our own "problems" into perspective!

If it helps, write yourself some little reminder notes so you can have these tools at your fingertips. The great thing is, the better you feel about yourself, the more likely you are to take positive action in your life that will cause you to have even more good feelings about yourself! You just don't know, sweetie, but by being the kind of girl who takes steps toward real self-improvement, you will separate yourself from the crowd and thrive.

(Check out pp. 113–5 in Hot X: Algebra Exposed *for more great ways to gain control of our feelings.)*

....

* Your parents would probably be thrilled (and might even end up giving you more freedom when they witness this grown-up behavior), but make sure that's not your motivation. It could just be a happy bonus, that's all!

Chapter 17

Life-Sized Barbie
Similar Triangles

𝒟id you ever play with Barbie and wish you could look just like her? As it turns out, if we could magically grow Barbie to be life-sized, her body proportions would actually be kind of scary...

Similar shapes are two shapes that are absolutely identical, except that one is simply bigger than the other. Just like if you "click for a larger view" on an image or magically grew a Barbie doll: same shape, different size! In geometry, taking a shape and making it bigger (but keeping it similar) is sometimes called **augmentation** or **dilation**. Getting smaller is called **reduction**.* These are actually types of transformations, like we talked about in Chapter 7, except these obviously aren't *congruency* transformations, because our shapes end up being a different size!

We'll deal with Barbie in a few pages. For now, let's review an important concept: ratios.

. . . .

* I'm guessing mathematicians used the terms *augmentation* and *reduction* long before plastic surgeons did.

Lightning-Quick Review of Ratios, Proportions, and Cross Products*

A **ratio** is a comparison between two numbers or values, expressed in the same units and written in reduced form. For example, let's say you spend a half hour texting for every 20 minutes you spend on your homework (focus, girl!). We could convert the half hour into 30 minutes (so we have the same units), write the fraction $\frac{30}{20}$, and reduce it to $\frac{3}{2}$. Then the ratio of minutes spent texting to studying could be expressed as: $\frac{3}{2}$ or **3:2** or **3 to 2**. (Um, put your phone away?)

When we want to show that two ratios are the *same*, we can write a **proportion**, which is a statement showing the equality between two fractions, such as ratios. Since the fractions are equivalent, <u>their cross products will always be equal</u>, and we can use that to find missing values. For example, if your sister's study habits are like yours, and you know she spent 16 minutes studying, you could figure out how long she spent texting by setting up the proportion $\frac{3}{2} = \frac{?}{16}$. We could read this as, "3 *is to* 2 *as what is to* 16?" To solve this, let's use the variable t, and find the **cross product**: $\frac{3}{2} = \frac{t}{16} \rightarrow \frac{3}{2} \overset{48 \quad 2t}{\times} \frac{t}{16}$.

Now we can set the cross products equal to each other[†] and get $2t = 48 \rightarrow t = $ **24 minutes**.

And now we're ready to define similar polygons!

....

* For a full review of ratios, proportions, and cross products (and the making of the film *Attack of the 20-Foot Monster Dog*), check out Chapters 16 and 18 in *Math Doesn't Suck*.

† The cross products being equal is an *equivalent statement* to the original fractions being equal. Why? The cross-product equation is just a shortcut for multiplying both sides of the "fraction equation" by a number (the product of the two denominators). Try it!

What's It Called?

Similar polygons are polygons that have these two qualities:

1. **Their corresponding *angles* are congruent.**

2. **The <u>ratios</u> of their corresponding *sides* are equal.**

We use the wavy line ∼ to indicate similarity, and just like with congruent triangles, the order of the letters matters: Corresponding letters must match up exactly.

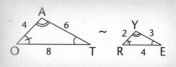

For example, if we're told that $\triangle OAT \sim \triangle RYE$, then we know:

1. $\angle O \cong \angle R$, $\angle A \cong \angle Y$, and $\angle T \cong \angle E$.

2. $\dfrac{OA}{RY} = \dfrac{AT}{YE} = \dfrac{OT}{RE}$ (In fact, they all $= \dfrac{2}{1}$.)

That set of equal ratios might look intimidating at first, but tap them with your finger or pencil on the triangle diagrams, and say "*OA* is to *RY*, as *AT* is to *YE*, as *OT* is to *RE*," and it'll be easier to see what it means. (Do this now!)

In this case, $\triangle OAT$'s sides each are *twice* the size of the corresponding sides on $\triangle RYE$. The ratios of the sides are all equal: $\dfrac{4}{2} = \dfrac{6}{3} = \dfrac{8}{4}$. They all equal $\dfrac{2}{1}$, after all!

We'll mostly deal with similar triangles, but the above definition applies to *all* polygons.

Closer to Barbie: Finding a Missing Side

If we know two triangles are similar, then we can often use cross products to find a missing side! (This is the same method we'll use to deal with Barbie on p. 292.)

For example, in LIP-LUV to the right, if all we're told is that $\triangle LIP \sim \triangle LUV$ and that $PL=12$, $UL=24$, and $LV=16$, then we can figure out LI by setting up the proportion, keeping the $\triangle LIP$ lengths on top of the fractions and the $\triangle LUV$ lengths on the bottom: $\frac{LI}{LU} = \frac{PL}{LV}$. We fill this in with our numbers to get:

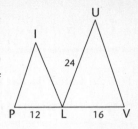

$\frac{LI}{24} = \frac{12}{16}$. Before we even do the cross products, let's make these numbers smaller with two tricks. First, we'll reduce that fraction on the right: $\frac{12 \div 4}{16 \div 4} = \frac{3}{4}$, so our problem becomes: $\frac{LI}{24} = \frac{3}{4}$. And second, we'll multiply both sides of the equation by 4, and get: $4 \cdot \frac{LI}{24} = 4 \cdot \frac{3}{4} \rightarrow \frac{LI}{6} = \frac{3}{1}$. Great! Now when we take the cross products, we won't get enormous numbers: $\frac{LI}{6} \overset{1 \cdot LI}{\underset{6 \cdot 3}{\times}} \frac{3}{1}$. And setting the cross products equal, we get: $1 \cdot LI = 6 \cdot 3 \rightarrow LI = 18$. And that's our missing side.

Answer: $LI = 18$

Watch Out!

There are a lot of places to make mistakes! When setting up the ratios, put the sides of one triangle on *top* of the fractions the whole time, and the sides of the *other* triangle on the *bottom* of the fractions, just to keep things straight. After all, $\frac{LI}{LU} = \frac{LV}{PL}$ simply isn't true.

Also, when faced with $\frac{LI}{24} = \frac{12}{16}$, it would have been tempting to "divide everything by 4" and do this: $\frac{LI}{6} = \frac{3}{4}$, but that would *change* the ratios! Just remember: We're allowed to reduce a single fraction anytime, and we're allowed to multiply both sides of an *equation* by a number, but that's it. When you're wanting to make the numbers smaller in a proportion, make sure you're doing one of those two things, and you'll be golden.

QUICK NOTE   If two triangles are congruent, they're automatically similar. It's just that the ratio of corresponding sides happens to be $\frac{1}{1}$ (or 1:1). Also, if we know two triangles are similar (so we have "AAA") and we know the ratio of one corresponding side to another is 1:1 (in other words, one set of sides are congruent), then we know the two triangles have to be congruent because we'd have ASA and also SAA, wouldn't we?

Finding a missing side in similar triangles:

Step 1. Which two triangles are similar? Pay attention to the order of the letters in the triangle names! If it helps, redraw one of the triangles so the orientations are the same.

Step 2. Keeping one triangle's info on the top and the other triangle's info on the bottom,* set up a proportion for the lengths of two sets of corresponding sides, using just the letters. To make sure the correspondences are correct, think to yourself, "This is to that, as this other thing is to that other thing" and check the letters. Now, just fill in the actual lengths (numbers) that we are given.

Step 3. If the numbers are big, try either reducing a fraction, or if the denominators have a common factor, try multiplying both sides of the equation by that factor to get smaller numbers. (We did both things in the example on p. 287.) Solve this equation for the unknown length by finding their cross products, setting them equal to each other, and simplifying. Done!

And...
Action! *Step By Step In Action*

In WINK, Given: △WNI ~ △IKN, WI = 8, IN = 5. Find NK.

Seems like there isn't enough information... Let's follow the steps and see what happens.

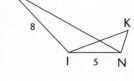

Step 1. Since △*WNI* ~ △*IKN*, just by looking at the order of the letters we've been given, we know the *W* corresponds to the *I* in the small △, right? And the *N* in the big △ corresponds to the *K* in the small △. Let's redraw these △'s side-by-side. Hmm, if we rotate the smaller △ clockwise, "lifting it up" out of the other one, then the letters match up in the right order.

. . . .

* FYI, you could also keep one △'s info in one fraction and the other △'s info in the second fraction, as long as corresponding lengths match up. For example, $\frac{LI}{PL} = \frac{LU}{LV}$ is an equivalent statement to $\frac{LI}{LU} = \frac{PL}{LV}$. The cross products will be the same.

Step 2. Ah, now things look clearer! And we see that the 5 on $\overline{IN}$ actually serves two purposes; it's the shortest side on the big △, and a longer side on the small △. Now that we see how the sides correspond, we know this must be true: $\frac{WI}{IN} = \frac{IN}{NK}$. Filling in what we know, this becomes: $\frac{8}{5} = \frac{5}{NK}$.

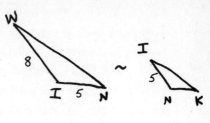

Step 3. Taking the cross products, we get $8NK = 25$. Solving, we get: $NK = \frac{25}{8} = 3\frac{1}{8}$. Great!

And that's the missing side we were looking for.

Answer: $NK = 3\frac{1}{8}$

We saw on p. 286 that all corresponding angle pairs on similar triangles are congruent; and in fact the converse is true, too! That means that if all three corresponding angle pairs on two triangles are congruent, then the triangles are similar to each other.

Shortcut Alert!

AA: If two triangles share two sets of congruent angles (AA), then the two triangles are similar.

Yep! All we need is for <u>two</u> angles to be congruent to two angles in another triangle, and the triangles are forced to be similar. Why not all three? Well, imagine if a triangle has angles measuring 30° and 80°. That forces its third angle to be 70°, since 30°+80°+**70°**=180°. And that means if some other triangle has angles measuring 30° and 80°, its third angle must also be 70°! So you can think of AA as AAA ... but that third "A" is just understood.

QUICK NOTE   Any time a segment connects two sides of a triangle so that it's parallel to the third side, it *automatically* creates two similar triangles. This is because the parallel lines become mall floors, and the two connected sides of the triangle each become their own escalator! Then we can use "If ∥, then corr. ∠'s are ≅" twice to get AA, and hence, similar triangles. We'll do an example below.

Take Two: Another Example

Given: $\overline{MG} \parallel \overline{IE}$. If MI = 4, AM = 12, and IE = 12, then find MG.

Step 1. Since $\overline{MG} \parallel \overline{IE}$, then because of "If ∥ lines, then corr. ∠'s are ≅," we know that $\angle AIE \cong \angle AMG$ and $\angle AEI \cong \angle AGM$* (see the QUICK NOTE above). So by the AA Rule on p. 289, we know that $\triangle IAE \sim \triangle MAG$.

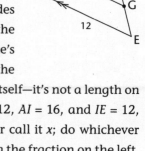

Step 2. Let's make a proportion, using the left sides and the bottom sides of the triangles, keeping the *smaller* triangle's info on top and the *big* triangle's info on the bottom: $\frac{AM}{AI} = \frac{MG}{IE}$. Filling in the values, we need to make sure *not* to use the 4 by itself—it's not a length on one of our similar triangles, after all! So AM = 12, AI = 16, and IE = 12, right? We can keep MG for the missing length or call it x; do whichever you prefer. Using x, we'd get: $\frac{12}{16} = \frac{x}{12}$. Reducing the fraction on the left, this becomes: $\frac{3}{4} = \frac{x}{12}$.

Step 3. Cross products give us: $12 \cdot 3 = 4x \;\rightarrow\; 36 = 4x \;\rightarrow\; x = 9$. Since MG = x, we're done!

Answer: MG = 9

. . . .

* And of course, $\angle A \cong \angle A$, although we didn't need it. This is the Reflexive Property for angles, by the way!

Doing the Math

Find the missing sides indicated. I'll do the first one for you.

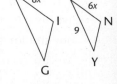

1. Given: $QUAD \sim LATS$, $QD = 5x$, $DA = 2x+1$, $LS = 2x$, $ST = x$. Find QD.

<u>Working out the solution</u>: Similar polygons work the same way as similar triangles; the corresponding sides' ratios are all equal, and that means that this is true: $\dfrac{QD}{LS} = \dfrac{DA}{ST}$, right? Now, let's fill in all this crazy x stuff and we get: $\dfrac{5x}{2x} = \dfrac{2x+1}{x}$.

Notice that we can reduce the fraction on the left by canceling an x from the top and bottom,* and we get: $\dfrac{5}{2} = \dfrac{2x+1}{x}$. Much nicer! Cross multiplying, we get $5 \cdot x = 2 \cdot (2x + 1) \rightarrow$ $5x = 4x + 2 \rightarrow x = \mathbf{2}$. But that's not the final answer; we need to find QD, which we know equals 5x. So $QD = 5(2) = \mathbf{10}$. Done!

Answer: $QD = 10$

2. Given: $\triangle BIG \sim \triangle TNY$, $BI = 8x$, $TN = 6x$, $TY = 9$. Find BG.

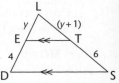

3. Your best friend is exactly 5 feet, 3 inches tall, and in a full-length picture, she appears 7 inches tall. If her head appears to be 1 inch tall, how tall is her head in real life? *(Hint: Watch the units!)*

4. In the QUAD-LATS diagram above, if we are also told that $QU = 8$, find LA.

5. Given: $\overline{ET} \parallel \overline{DS}$. If $DE = 4$, $TS = 6$, $LE = y$ and $LT = y + 1$, find LS.

(Answers on p. 401)

---

* If you're not sure why we can't cancel the x's from the other fraction, it's time to brush up on variables in fractions (and copycats). Check out Chapter 4 in *Hot X: Algebra Exposed*.

Life-Sized Barbie!

So, it's pretty widely accepted that Barbie is built at $\frac{1}{6}$ scale; in other words, the ratio of "Barbie to human" would be 1:6. By that scale, because the dolls are about $11\frac{1}{2}$ inches tall, if Barbie were life-sized, she'd be around $6 \times 11.5 = 69$ inches, in other words **5 feet, 9 inches** tall. (We also could have solved this by setting up the proportion $\frac{1}{6} = \frac{11.5}{h}$ and taking the cross products: $1 \times h = 6 \times 11.5 \to h = 69$ inches.) Measuring some of Barbie's body parts, we get: chest $= 5\frac{3}{4}$ inches, waist $= 3.5$ inches, foot $= 1$ inch, etc. And what would those numbers be if she were "life-sized"? For the chest, we'd get $\frac{1}{6} = \frac{5.75}{c} \to 1 \times c = 6 \times 5.75 \to c = 34.5$ inches, which is about a half-inch different from the circumference just below the bust (which means about a 3-inch difference on a life-sized woman). According to a bra-fitting website, that means she's a C-cup. Well, a 34-C bra size sounds fine. She's a little busty, but that's okay!

For the waist, w, we'd get: $\frac{1}{6} = \frac{3.5}{w} \to w \times 1 = 3.5 \times 6 \to w = 21$ inches. Okay, that's waaaaay too small for any real woman's waist, let alone one who's 5'9" tall. But to be fair to the makers of Barbie, they probably weren't trying to be evil and make us all wish we had impossibly small waists. They were probably much more concerned with how she would look in *clothes*. See, most doll clothes are made out of regular material, which would be <u>6 times</u> proportionally as thick as it should be. I mean, when we real gals wear clothes made of a big, thick winter coat material, our "shape" is pretty hidden, after all. So maybe they made Barbie unrealistically curvy (extra big and extra small) to compensate for that. They were more interested in how she looks in clothing than how she looks *naked* . . . well, let's hope so, anyway.

However, I'm not sure *what* to say about the shoe size. Her 1-inch foot, on a full-sized scale, becomes just 6 inches, which is a *kids'* size 8.5, often worn by, um, 2-year-olds.

Although Barbie's proportions don't make total sense on a larger scale, lucky for us, similar polygons' proportions do!

Proofs Involving Similar Triangles

Proofs involving similar triangles often require us to first prove we have similar triangles (usually with the AA shortcut from p. 289). And then we can use that similarity to prove something about proportional sides. It's unbelievable how truly helpful our little mall escalators continue to be . . .

Step By Step

Proofs involving similar triangles:

Step 1. To prove that two triangles are similar, we'll want to find two sets of congruent angles (AA—see p. 289).* And keep this in mind: Parallel lines might let us use the transversal (escalator!) rules from p. 218, which can give us congruent angles.

Step 2. Once we have two congruent angles (AA), we've proven the two triangles similar! If that's what we needed to prove, we're done. Otherwise, we're probably being asked to prove a proportion or a product. At any point, try redrawing the triangles into the same orientation. It'll really help!

Step 3. We set up the proportions we know are true (being careful about the order of letters) and maybe find their cross product, if that's what we're being asked for. Done!

Remember altitudes from p. 156? Well, for *any* right triangle, if we draw in an altitude from the right angle's vertex, we end up creating three similar right triangles! Below is part of the proof (which you'll totally understand), and you'll do the rest of it on p. 298.

. . . .

* There are actually two more methods for proving similar triangles, called "SSS~" and "SAS~", but they don't come up very often. Check out GirlsGetCurves.com if you want to learn more about 'em.

Given: △PAW is a right triangle, with right angle ∠PAW and altitude $\overline{AS}$. Prove: △PAW ~ △ASW.

Step 1. We already have right angles in both triangles; all we need to do is come up with one more set of congruent angles. First, let's redraw the triangles separately, in

matching orientations: Well guess what; they both have ∠W.* Nice.

Steps 2–4. Now we have AA and have proven △PAW ~ △ASW, which is what the problem was asking for! Hmm, that wasn't so bad, was it?

Proof

Statements	Reasons
1. △PAW is a right triangle, with right angle ∠PAW.	1. Given
2. $\overline{AS}$ is an altitude.	2. Given
3. ∠ASW is a right angle.	3. Altitudes create right ∠'s.
4. ∠PAW ≅ ∠ASW	4. All right ∠'s are ≅. (Gimmie an "A"!)
5. ∠W ≅ ∠W	5. Reflexive Property (Gimmie another "A"!)
6. ∴ △PAW ~ △ASW	6. AA

You totally understood that. You're such a smart cookie.

. . . .

* So, why can't we say "they both have ∠A"? Because in this diagram, there is more than one angle at the vertex A. In fact, "∠A" doesn't mean anything here—it's too vague!

QUICK NOTE   If we're being asked to prove something like
AB · CD = EF · GH, that usually means it's a similar–triangle
proof, so don't panic! We'll do an example below.

Take Two: Another Example

See the VIOLET diagram to the right. Given: $\overline{VI} \parallel \overline{LE}$ and $\overline{IO} \parallel \overline{TE}$.
Prove: $VI \cdot TL = VO \cdot LE$.

Um, how the heck are we supposed to prove that two
products are equal, based on a couple of pairs of parallel lines? Well, this is where YOU are about to become
more powerful. Because in a few minutes, you'll have
yet another trick up your sleeve . . .

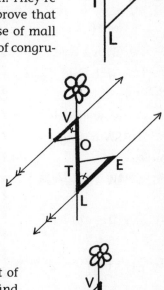

Step 1. Remaining calm, we look at our diagram and
see two triangles, which is always a good sign. They're
clearly not congruent, but perhaps we can prove that
they are similar with AA? And, hey! Because of mall
escalators, parallel lines can lead to all sorts of congruent angles. To make this easier to see, let's
extend just one set of mall floors, $\overline{VI} \parallel \overline{LE}$
(doing both sets can look confusing!). Now
we can see that the stem is the escalator.
Now, <u>only looking at the mall floors and
escalator</u>, are there any pairs of angles
<u>inside our two triangles</u> (those are the only
ones we care about, after all) that must be
congruent? Well, remember the "oh-no-he-
didn't-Z-snap" from p. 215? Yep! The big Z
is a set of alternate interior angles:
$\angle IVO \cong \angle ELT$. (Gimmie an "A"!)

We can do the same thing for the other set of
parallel lines, $\overline{IO} \parallel \overline{TE}$, and this time, we'll find
alternate exterior angles in our triangles:
$\angle VOI \cong \angle LTE$. (Gimmie another "A"!) Pant,
pant! This is incredible brain exercise! Read this
stuff a few times until you can "see" it, even if
just for a moment. I'm seriously so proud of you.

Step 2. And how about that? We have <u>two sets</u> of congruent angles, so we can prove that our triangles are similar, being careful to correspond the letters in the correct way: $\triangle VIO \sim \triangle LET$. It helps to use those skills we developed in Chapter 7 to "un-rotate" the big triangle so we can see the correspondence more clearly.

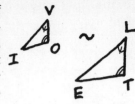

Step 3. Now that we have two similar triangles, we know that corresponding sides are proportional, right? Somehow that's going to get us to our last step, $VI \cdot TL = VO \cdot LE$. Working backwards and using these same four sides, how would we put together a proportion that we know is true for our two triangles? Looking at the side-by-side triangles we drew above, $\overline{VI}$ corresponds with $\overline{LE}$ and $\overline{VO}$ corresponds with $\overline{TL}$. That means we know this is true: $\dfrac{VI}{LE} = \dfrac{VO}{TL}$. The cross products of equivalent fractions are always equal, and that gives us: $VI \cdot TL = VO \cdot LE$. Ta-da! We did it! And here it is, in a two-column proof:

Proof

Statements	Reasons
1. $\overline{VI} \parallel \overline{LE}$	1. Given
2. $\angle IVO \cong \angle ELT$	2. If $\parallel$ lines, then alt. int. $\angle$'s $\cong$. (Gimmie an "A"!)
3. $\overline{IO} \parallel \overline{TE}$	3. Given
4. $\angle VOI \cong \angle LTE$	4. If $\parallel$ lines, then alt. ext. $\angle$'s $\cong$. (Gimmie an "A"!)
5. $\triangle VIO \sim \triangle LET$	5. AA (Great, now we have similar triangles!)
6. $\dfrac{VI}{LE} = \dfrac{VO}{TL}$	6. If two $\triangle$'s are $\sim$, then corr. sides are proportional.
7. $\therefore VI \cdot TL = VO \cdot LE$	7. Cross products of equivalent fractions are equal.

For that last Reason, instead of saying "Cross products of equivalent fractions are equal," some people will just say "algebra," which I think is pretty funny. It's like, "My Reason? Algebra. Yes, just all of algebra. No, I don't think I need to be more specific than that." I mean, by that same logic, shouldn't we be able to write "geometry" for every Reason and be done with it? I'm kidding, of course. Sort of.

QUICK NOTE   Just like we did in the VIOLET problem above, it can be really, really helpful to draw the triangles side by side to see the correspondence. Another trick is to fill in pretend angle measurements. We'll do this in #1 below.

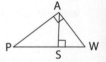

Do these problems on similar triangles. I'll do the first one for you.

1. In the PAWS diagram on p. 294, we proved that △*PAW* ~ △*ASW*. There is a third triangle within the diagram that is also similar to these two. Write the name of that triangle, with the *letters in the correct order* to complete the statement: △*PAW* ~ △*ASW* ~?

<u>Working out the solution</u>: Well, △PAS is the only other △ in this diagram; it's the one on the left. But what is the correct order for the letters? Let's try filling in some pretend measurements. What if ∠W = 70°? Then looking at the big △, this would force ∠P to be 20°. See why? Now fill those in! Looking at the smallest △, it would then *have* to be true that ∠SAW = 20°, and then because those two top angles add up to 90°, it would force ∠PAS to be 70°. And now it's much easier to see the correspondence! In order to match up with △PAW, for example, we want to write the letters of △PAS in 20°–90°–70° order,* so that would be △PSA.

Answer: △PAW ~ △ASW ~ △PSA

. . . .

* Instead of starting with ∠W = 70°, we could have used ∠W = x, and then ∠P would equal (90° – x), etc. Everything works out the same way; but using pretend numbers is usually easier for "seeing" what's going on at this stage in your math career!

2. In #1, we noticed that $\triangle PSA \sim \triangle PAW$. Time for the two-column proof. Given: $\triangle PAW$ is a right triangle, with right angle $\angle PAW$ and altitude $\overline{AS}$. Prove: $\triangle PSA \sim \triangle PAW$. (Hint: This is very similar to the proof on p. 294!)

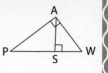
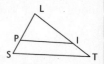

3. See the SPLIT diagram to the right. Given: $\overline{PI} \parallel \overline{ST}$. Prove: $\dfrac{PL}{SL} = \dfrac{IL}{TL}$ in a two-column proof.*

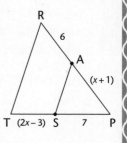

4. **The Midpoint Theorem** says that if a segment connects two sides of a triangle at their midpoints, then that segment is parallel to the third side and is half the length of the third side. Given: $RA = 6$, $AP = (x+1)$, $TS = (2x-3)$, $SP = 7$, and S is the midpoint of $\overline{TP}$. Use a paragraph proof to show that $\overline{TR} \parallel \overline{SA}$ and that $\overline{SA}$ is half the length of $\overline{TR}$. (Hint: If S is the midpoint of $\overline{TP}$, then what two segments must be congruent? Use that to find x, and then find AP. What does this tell us about A? Now use the Midpoint Theorem!)

(Proof solutions at GirlsGetCurves.com)

By the way, in #2 above, you finished the first part of a proof of the Pythagorean Theorem. Check out GirlsGetCurves.com to see how it works!

* This is actually the *beginning* of a proof of something called the "Side-Splitter Theorem," which in this case would tell us $\dfrac{PL}{PS} = \dfrac{IL}{IT}$: A parallel line splits the sides of a triangle proportionally. Check out GirlsGetCurves.com for the rest of the proof, and more!

Reality Math

There's this guy in your class who likes to show off. A lot. His family owns a local store that just put an enormous Christmas tree out front, and he's been telling everyone how he single-handedly carried the 100-foot-tall tree. You take out a tape measure and ask him, "Are you sure it's 100 feet tall?" He smirks and says, "You can take my word for it. What, you really think you're going to drag a 100-foot ladder over here and measure it?" You just smile politely, noticing that the overhead sun is peeking through the winter sky just enough to cast shadows. You mark the ground where the top of the tree's shadow hits the ground, then mark the ground where your feet are, and also where the top of your head's shadow hits the ground. You know that because the sun's rays are nearly parallel, they hit the ground at a nearly consistent angle (which of course varies depending on the time of day).

Then you easily use similar triangles to fill in the missing length:
$\frac{5.5}{h} = \frac{4}{36} \rightarrow \frac{5.5}{h} = \frac{1}{9} \rightarrow 9(5.5) = h$. A little multiplication on the spot reveals that **$h = 49.5$ feet.** Hmm. Who's smirking now?

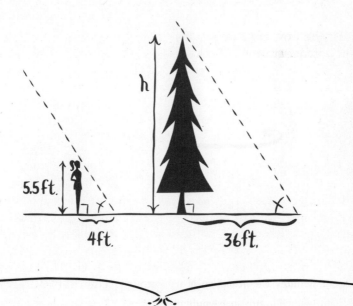

Takeaway Tips

Similar triangles (and other similar polygons) have these two qualities: Their corresponding angles are congruent, and the ratios of their corresponding sides are equal.

The most common way to prove that two triangles are similar is with the AA shortcut.

Try redrawing similar triangles to make the correspondence easier to see.

If a proof involves parallel lines and a proportion (or a product) and you have no idea where to start, try looking for two triangles that might be similar—and remember the transversal (escalator) theorems from Chapters 13 and 14. They could be the key!

TESTIMIONIAL

Jessica Keramas
(Los Angeles, CA)
<u>Before</u>: Overweight and shy
<u>Now</u>: Fit and Fabulous Owner of an Organic Pizzeria!

In high school, I was the "scholarly" type. I didn't play sports, and I didn't really seek out friends. I spent my time studying and going to math team meets, which I loved. Not only did they sharpen my math skills, but one year I even placed 18th in all of Maine, which meant I got to go to the nationals on an overnight trip! It was awesome.

But there was something missing from my life. I wasn't taking care of my body. Alright, I'll say it: I was fat.

Throughout high school, it bothered me, but I just kept ignoring it. After all, it wouldn't last forever, right? Well, there I was, graduating from high school. I suddenly realized that if I didn't make some changes, well, things

"Math skills helped me get in shape."

weren't going to change! Being overweight just wasn't how I saw myself as a young woman, and I needed to take action.

Luckily, with all the time I'd spent studying math, my brain was super sharp—and those same problem-solving math skills helped me lose weight and get in shape. See, instead of feeling overwhelmed and frustrated with my body, I was able to break down the problem into its parts and use logic to solve it. I never starved myself; I always ate when I was hungry. The huge difference would be *what* I ate: I researched and learned that I should avoid greasy, sugary, artificial, and processed foods, and I read about the importance of organic food. Every day that summer, I worked the problem. I made charts to keep track of the food I ate, which helped me not to cheat! I ate no more than 20 grams of "bad" fat per day, and I only drank water—no more sugary or so-called diet sodas, which can also lead to weight gain. I started biking every day, a little farther and a little faster each week (usually by 10 percent), and did light exercise at home, like push-ups, squats, and leg lifts. Nothing extreme. I made graphs to track my progress in both weight loss and bike-riding stamina. At the same time, I researched and measured the best angles for leg lifts (sometimes 45°, sometimes 90°, depending on the positioning of the leg), which helped me get the most out of my exercise time.

By the end of the summer, I had lost nearly all my extra weight, and in a healthy way. The progress was so slow and steady that I didn't realize how much even my face had changed, but a good friend walked right by me that September. He honestly couldn't believe it was me!

I continued my new healthy lifestyle in college, where I became a math major. And today I'm in better shape than ever. Now I use math to help *others* stay healthy and fit. I own a restaurant called PizzaSalad, where we only use fresh, organic ingredients. It costs more to use truly fresh ingredients, so math is essential for the competitive analysis required for us to be a profitable business—from price setting to comparing the areas of my pizzas (which are rectangular) to others (which are usually circular), setting proportions of ingredients in our salads, and more.

Running a restaurant can be challenging, but studying math taught me how to tackle any challenge that might lie ahead. It's so great to see my happy customers discovering a healthier lifestyle, and looking and feeling great!

Team Spirit

The results are in: Sports are great for girls!

*"The more physically active girls are, the greater their self-esteem and the more satisfied they are with their weight, regardless of how much they weigh. In fact, 83% of very active girls say that physical activity helps them feel good about themselves."**

Team activities are good for our bodies and self-esteem, too. Don't forget they're also a great way to meet new friends—girls we might not ordinarily hang out with. On sports teams, girls learn and grow in an environment where there's no pressure to impress guys, and sports teach us cooperation and give us an outlet for healthy competition. Plus, playing a sport increases our confidence in our bodies for more than superficial reasons. So lace up your running shoes, hit the court, and be proud!

* *The Girl Scout Research Institute,* The New Normal? What Girls Say About Healthy Living (2006)

Chapter 18

The Birds and the Bees . . . and Puppy Dogs, Too

Circles

People used to talk about the "birds and the bees" to explain to kids about the universality of, um, that thing married couples do to have babies. Well, birds and bees have a way of teaching us about circles, too! Also, if you like drawing animal faces, then you know that circles are a good place to start—especially for birds and puppies, which we'll get to in a few moments. But first, let's get some basics under our belts!

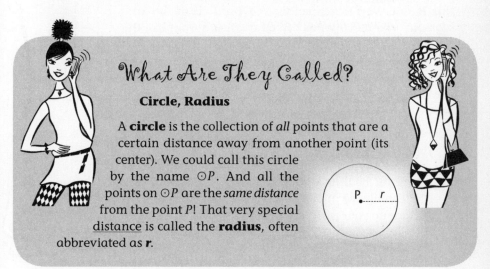

What Are They Called?

Circle, Radius

A **circle** is the collection of *all* points that are a certain distance away from another point (its center). We could call this circle by the name $\odot P$. And all the points on $\odot P$ are the *same distance* from the point P! That very special distance is called the **radius**, often abbreviated as **r**.

The plural of *radius* is "radii," and it's pronounced like if you saw a giant eyeball headed toward your two friends Ray and Dee, and you shouted, "Ray, Dee! . . . eye!!" (We'll actually see a giant eyeball in a few pages.)

Here's a simple theorem, and it will be your <u>best friend</u> during circle proofs.

 All radii of a circle are congruent.

As soon as we see two or more radii on a diagram, we can mark 'em congruent with kitty scratches! Just as a quick example, check out $\odot I$ to the right. Do you see all four radii? Without being told anything about this diagram, we could mark the radii congruent with kitty scratches, and then because of vertical angles and SAS, we'd actually be able to prove those two skinny triangles congruent: $\triangle QIU \cong \triangle KIC$.

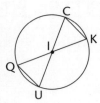

Chords

I love hearing a talented piano player work her magic, don't you? The melodies, the chords—I have so much respect for all the practice it takes to get good! Most people would agree that the most beautiful music comes from the piano, violin, banjo . . . wait. Maybe not banjo. But this is one way to remember the word *chord*: It's any segment whose endpoints are on a circle, sort of like the "circle" part of a banjo.

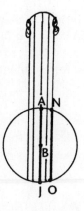

In our BANJO picture, $\overline{ON}$ is a **chord** in $\odot B$, and $\overline{AJ}$ is a very special chord called a **diameter** (often called **d**) because it passes through the center, B. The diameter is twice the length of the radius. In other words, **d = 2r**.

 Here are some more examples of chords. This banjo looks, um, a little harder to play.

So how do chords help us with circle proofs? It all comes down to . . . puppies!

Puppy Faces: Doodle Time!

Simple puppy faces are adorable and pretty easy to draw. We'll start with a circle and a nose in the center, then add two eyes and floppy ears. Then we can draw a line down from the nose and connect it to a mouth.

Now let's do some bulldog puppies with wide, flat mouths and a little tooth jutting out. They're so cute, aren't they? And I'll let you in on a little secret: Each "tooth" is actually a little right angle marker and shows that the nose-to-mouth segments are *perpendicular* to their "mouth" chords.

Notice that the <u>shorter</u> the nose-to-mouth segment is, the <u>wider</u> the mouth will be, because it gets to stretch out to a bigger part of the circle. Now it's time for some theorems about puppy mouths! Or, you know, chords.

What's It Called?

Distance from center to chord; Symmetrical Bulldog Puppy Mouth (radius-bisecting-chord) Theorems

The **distance from the center of a circle to a chord** is defined to be the length of the *perpendicular* segment connecting the nose to the mouth (connecting the center to the chord, like $\overline{ON}$ below). This also happens to be the shortest distance from the center to the chord, which makes sense, because any other segment we'd draw from the nose to the mouth would be unnecessarily long, and it just wouldn't be a puppy face at all.

In this diagram, $\overline{ON} \perp \overline{BE}$, so we know that $\overline{ON}$ is the *shortest segment* we could draw from the O to anywhere on the chord . . . and that's why we call the length of $\overline{ON}$ the "distance"!*

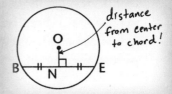

. . . .

* The distance between any point and line is defined to be the length of the ⊥ segments connecting them.

This little perpendicular nose-to-mouth segment will always cut the "mouth" chord in half (symmetry is nice in puppy faces), so if we're told that $\overline{ON} \perp \overline{BE}$, we automatically know that $\overline{BN} \cong \overline{NE}$, and we could draw in the little kitty scratches.* The converse is also true: If we're told that $\overline{BN} \cong \overline{NE}$, we automatically know that $\overline{ON} \perp \overline{BE}$. Yep, there's that puppy tooth.

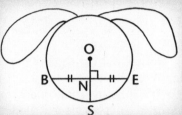

Now imagine if $\overline{ON}$ continued past the chord to the rim of the circle—we'd get a radius! (Hey, who drew a line down the middle of the puppy's chin?!)

In fact, here are two theorems about this radius:

 If a radius $(\overline{OS})$ is $\perp$ to a chord $(\overline{BE})$, then it bisects the chord.

 Converse: If a radius bisects a chord, then it's $\perp$ to the chord.†

We could call these the "Symmetrical Bulldog Puppy Mouth" Rules, because a symmetrical mouth goes hand-in-hand with having one bottom tooth poking out!

Once we have a radius and a chord: If we are given a tooth (right angle), then we know the mouth is symmetrical. If we are given a symmetrical mouth, then we know we can draw in a tooth. Not so bad, right?

QUICK NOTE   Notice that if a radius is $\perp$ to a chord, it must be the $\perp$ bisector. And if a radius *bisects* a chord (other than a diameter), it must be the $\perp$ bisector. It can't be just one and not the other; it has to be both.

. . . .

* Sometimes puppies end up with kitty scratches on their faces.

† . . . unless the "chord" happens to be the diameter. The center bisects the diameter already, so the radius could be drawn in any direction, and wouldn't have to be $\perp$ to anything! ☺

It's amazing how powerful these little chord theorems are. They often give us enough info to create congruent triangles, and then we can prove lots of stuff! For example, in ⊙*T* below, let's say *all we're told* is that $\overline{TO}$ bisects $\overline{PN}$.

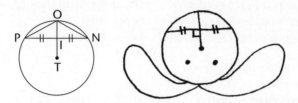

I see an upside-down puppy mouth! Since $\overline{TO}$ bisects $\overline{PN}$, the second theorem on p. 306 automatically tells us that $\overline{TO} \perp \overline{PN}$, so we can draw in a little right-angle box. All right angles are congruent, so this means ∠OIP ≅ ∠OIN. And by definition of *bisect*, we know $\overline{PI} \cong \overline{IN}$, right? Hmm, lookie there, because of the Reflexive Property ($\overline{OI} \cong \overline{OI}$) and SAS, we now have two congruent triangles: △*POI* ≅ △*NOI*. And using CPCTC, we could then prove tons of things: $\overline{PO} \cong \overline{NO}$, ∠P ≅ ∠N, $\overrightarrow{OT}$ bisects ∠PON, and more!

Notice that instead, *if all we were told* was that $\overline{TO} \perp \overline{PN}$, we could still prove all the same stuff. Our first step would be to say, "Ah, since $\overline{PN}$ is a chord that is perpendicular to a radius, we can draw in a tooth (right-angle box), and then by the Symmetrical Bulldog Puppy Mouth Theorem, that automatically means $\overline{TO}$ bisects $\overline{PN}$." We'd make our congruent-segment kitty scratches and go from there.

We'll see how to use these theorems with the Pythagorean Theorem on p. 310 and actually find distances and lengths of chords and radii. Where there are right angles, there are right triangles, after all . . .

Two Puppies Are Better Than One: Congruent-Chord Theorems!

Here's what can happen with *two* chords . . .

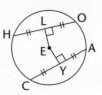

As it turns out, if two chords are the same distance from the center, then the two entire chords must be congruent. So if we're told that $\overline{LE} \cong \overline{YE}$, it must be true that $\overline{HO} \cong \overline{CA}$.

 Congruent-Chord (Twin Puppy Dog) Theorem: If two chords are equidistant* to the center, then the two chords are congruent.

I call this the "Twin Puppy Dog" Theorem, because if we were going to try to draw two puppy mouths on the same circle (or two circles with the same radius), if the distance from the nose to one mouth is the same as the distance to the other mouth, then the mouths will have to be the same size, too! So you'd get two of the same puppy face.

 Converse of Congruent-Chord (Twin Puppy Dog) Theorem: If two chords are congruent, then the two chords are equidistant to the center.

(So if we're told that $\overline{HO} \cong \overline{CA}$, it must also be true that $\overline{LE} \cong \overline{YE}$.) If we draw the same size mouth (chord) on two different parts of the same circle or on circles with the same radius, the distance from the nose to each mouth will be identical.

Let's do another puppy proof! Before we even start, can you spot the chords with distances to the center marked on the diagram?

Given: △RHA is an isosceles triangle with base $\overline{RA}$, $\overline{RH} \perp \overline{BE}$, $\overline{HA} \perp \overline{ET}$, and ⊙H. Prove: △BET is isosceles with base $\overline{BT}$.

Whoa, complicated-looking diagram. But that's okay; we're tough cookies and we can handle it. Let's skip to the end for a moment and work

· · · · · · · · · · · · · · · · · · ·

* *Equidistant* means "the same distance." See p. 158 for more!

backwards. In order to prove that $\triangle BET$ is isosceles with base $\overline{BT}$, we'd either need to prove that $\angle B \cong \angle T$ or that $\overline{BE} \cong \overline{ET}$, right? Well, because $\overline{BE}$ and $\overline{ET}$ are chords with distances to the center drawn in, hmm . . . if we knew that the distances ($\overline{RH}$ and $\overline{HA}$) were congruent, then our Twin Puppy Dog Theorem would tell us that $\overline{BE} \cong \overline{ET}$! Can we prove that $\overline{RH} \cong \overline{HA}$? Sure! They're the legs of the tiny isosceles triangle we're given to begin with. Now that we've worked backwards, let's do this sucker.

Proof

Statements	Reasons
1. ⊙H, $\overline{RH} \perp \overline{BE}$ and $\overline{HA} \perp \overline{ET}$	1. Given (The $\perp$'s tell us that $\overline{RH}$ and $\overline{HA}$ are <u>distances</u> from H to the chords, and now we know they're puppy faces.)
2. $\triangle RHA$ is an isosceles triangle with base $\overline{RA}$.	2. Given
3. $\overline{RH} \cong \overline{HA}$	3. If a $\triangle$ is isosceles, then its legs are $\cong$. (Great, now we know that the distances from the center to the chords are the same!)
4. $\overline{BE} \cong \overline{ET}$	4. If two chords are equidistant from the center, then the chords are $\cong$. (Twin Puppy Dog Theorem!)
5. $\therefore \triangle BET$ is isosceles.	5. If a $\triangle$ has two $\cong$ sides, then it's isosceles.

In this proof, the key was to first realize we had two *distances from the center to chords* marked on the diagram. The fact that $\overline{RH}$ and $\overline{HA}$ are drawn from the center and are <u>perpendicular</u> to the chords tells us these are *distances*, and not just some random lines drawn from the center to the chords. Then, because of the isosceles triangle $\triangle RHA$, we knew these two distances were in fact congruent!

Watch Out!

It's easy to glaze over while reading proofs and not really absorb everything. But don't let that happen, or you'll end up thinking you don't understand them! These strategies and proofs are meant to be read slowly. For each angle or segment, look on the diagram and actually *find* it, even marking the diagram as you go. I mean, imagine trying to understand a short story by only reading a word at the top of each page and then saying, "I didn't understand it. I just don't *get* this stuff." Of course you didn't understand it—nobody could without really reading it! And in geometry, that means reading slowly and letting the logic sink in at each stage. Keep your brain on while you read these and give yourself a chance, okay?

Now we'll see how these chord theorems can help us find actual distances . . . by mixing it up with some Pythagorean Theorem. Sometimes we'll be given the drawing, and sometimes we'll have to draw the entire thing ourselves. The trick is to create a <u>right triangle</u> using a chord, a radius, and the distance from the chord to the center. I'll show you how it's done in #1 below, and then it's your turn!

Doing the Math

Answer these questions about chords. I'll do the first one for you.

1. The diameter of a circle is 8 ft across (see above). A chord measures 6 ft. How far is the chord from the center?

<u>Working out the solution:</u> The distance from the chord to the center is the ⊥ segment connecting them, right? Let's draw that in (including the right-angle box), and label that distance f. Okay, now a diameter of 8 ft means the radius is 4 ft. Well, in order to build a right triangle, the most helpful radius to <u>draw</u> is one that connects to an endpoint of

our chord. Notice that because we have a ⊥ segment from the center to the chord, the Symmetrical Bulldog Puppy Mouth Theorem on p. 306 tells us that it must *bisect* the chord! So half the chord would be 3 ft, right? Great! So we've created a right triangle where the hypotenuse is 4 ft, and one leg is 3 ft. That means we can figure out the other leg, f, by using the Pythagorean Theorem: $f^2 + 3^2 = 4^2 \rightarrow f^2 + 9 = 16 \rightarrow f^2 = 7 \rightarrow f = \sqrt{7}.$* Done!

Answer: The distance from the center to the chord is $\sqrt{7}$ ft (approximately 2.65 ft)

2. The diameter of a circle is 6 ft across. A chord measures 4 ft. How far is the chord from the center? *(Hint: Where would it be most helpful to draw the radius?)*

4 ft.

3. A circle has a chord whose distance from the center is 6 inches, and the radius is 10 inches. What is the length of the chord? *(Hint: Draw the circle, chord, and distance first. Then fill in the radius to make a right triangle.)*

4. In ROMANCE, $\overline{RM} \cong \overline{EN}$. $OA = 2x + 5$, $AC = 4x + 1$, and $RM = 8x - 3$. Find EC. *(Hint: Use a Congruent-Chord (Twin Puppy) Theorem and the first two equations to find x, then evaluate RM, and then find EC.)*

5. For the POINT diagram, Given: $\odot T$ and $\overline{TO} \perp \overline{PN}$. Prove $\angle P \cong \angle N$ in a two-column proof. *(Hint: Reread p. 307 for strategy tips!)*

6. For the JOKING diagram, Given: $\odot G$, $\overline{GO} \perp \overline{KJ}$, $\overline{GI} \perp \overline{KN}$, $\overrightarrow{KG}$ bisects $\angle JKN$. Prove: $\overline{KJ} \cong \overline{KN}$. *(Hint if you want one: Draw in $\overline{KG}$. How can we prove those two little triangles are $\cong$? (Clue: We don't need HL.) Then, which two lengths can CPCTC prove $\cong$ so that we can use the Congruent-Chord "Twin Puppy" Theorem?)*

7. For the SMILE diagram, Given: $\odot I$ and $\overline{SE} \parallel \overline{ML}$. Prove $\angle MSE \cong \angle LES$. *(Hint: Use congruent radii, isosceles triangles, and transversals. You can do it!)*

(Answers to #2–4 on p. 401; proofs at GirlsGetCurves.com)

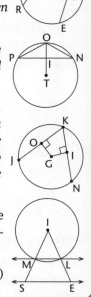

* We already knew f was positive, so we didn't need the negative solution, $f = -\sqrt{7}$. Check out p. 178 for more.

In Your Circle of Friends: How Do You Handle It: Text, Call, or Face-to-Face?

Even though there are so many ways to communicate with each other, it can be tempting to text everything because, well, it's usually faster and easier . . . and much less confrontational! But we all know it's not always best, and that often, a more personal touch is the way to go. Here are a few suggestions on how to send that next message:

Text: *Tell* a friend you'll be at her party a few minutes late.

Call: *Tell* a friend about your latest crush.

Face-to-Face: *Tell* a friend why something she did or said upset you.

Text: *Let* Mom know what time to pick you up after practice.

Call: *Let* your neighbor know you'll have to cancel your baby-sitting job.

Face-to-Face: *Let* your crush know you've liked him for a long time.

Text: *Ask* a friend what's up for the weekend.

Call: *Ask* your boss for a day off.

Face-to-Face: *Ask* someone out on a date.

Text: *Give* directions to your house.

Call: *Give* your best friend advice about family problems.

Face-to-Face: *Give* a hug!

She's Like, So Distracted: Tangents

Do you know what it means to "go off on a tangent"? Imagine you're tutoring someone, and you say, "Okay, we're going to convert miles into feet. Oh, that reminds me, have you smelled my feet? They really smell nice." You've just gone off on a tangent, getting distracted and talking about something sort of related, but totally off-point.

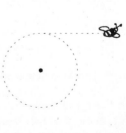

And speaking of points, imagine a bee is flying around in a circle, very focused, and then gets distracted by an amazing flower (or some smelly feet). At this "point," our bee leaves the circular path to fly in a straight line toward her new interest. She, too, has gone off on a tangent!

What's It Called?

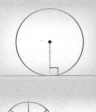

A line is **tangent** to a circle if the line touches the circle at exactly one point. As it turns out, if we draw in a *radius* to that point, it will always be perpendicular to the tangent line! It's just like our bicycle wheel from Chapter 7, only now it's rolling along the ground, and whichever spoke (radius) touches the ground (the tangent line) is guaranteed to make a right angle with the ground. Can you picture it?

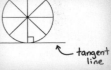

tangent line

Bicycle Wheel Rule: A tangent line to a circle will always be perpendicular to the radius that touches the tangent point. (I like to call this the Bicycle Wheel Rule; it just makes it easier to remember.)

A **tangent segment** is just a segment that touches a circle and is part of a tangent line to that circle. For example, $\overline{YA}$ is part of the line $\overleftrightarrow{YA}$, which is tangent to the circle because it only touches it at one point. So $\overline{YA}$ is a *tangent segment* to the circle.

However, $\overline{NO}$ is part of the line $\overleftrightarrow{NO}$, which isn't a tangent line, so $\overline{NO}$ isn't a tangent segment.

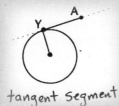

tangent segment

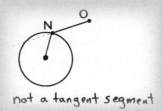

not a tangent segment

Just because a segment has an endpoint on a circle doesn't mean it's a *tangent* segment after all!

Bird Beaks: The Two-Tangent Theorem

Here's an easy way to draw a little bird head: Just draw a circle with its center dot and then pick any point outside the circle. Draw two tangent lines, and there's the beak!

Oh, there's a theorem about this, too. See, two tangent segments starting from a single point and whose endpoints are tangent to the circle will always be congruent! (Symmetry is nice for bird beaks, I think you'll agree.) For example, if all we're told is that $\overline{BI}$ and $\overline{BR}$ are tangent segments to $\odot D$, then we'd know it *must* be true that $\overline{BI} \cong \overline{BR}$.

 Two-Tangent "Bird Beak" Theorem: If a circle has two tangent segments drawn from a single external point, they will always be congruent.

We'll prove this theorem on pp. 318–9, and you'll totally understand the logic; it'll be awesome.

Watch Out!

Just because a circle has two tangent segments touching it doesn't mean the segments are necessarily congruent. The two tangent segments both have to come from a single point. Check out these two circles that each have two tangent segments.

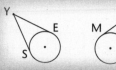

Yes! $\overline{YS} \cong \overline{YE}$. But are $\overline{MA}$ and $\overline{BY}$ congruent? Who knows? Maybe. Notice, however, that if we extended $\overline{MA}$ and $\overline{BY}$ so that they reached their intersection point (up to the right, past the A and Y), then those *longer* tangent segments *would* be congruent. See what I mean? Just look for the sharp bird beaks, and you'll do great!

Two circles can actually be tangent to each other, too. If they're externally tangent, they're "kissing"* at a single point, and if they're

. . . .

* I'm not kidding; some textbooks call these *osculating circles*, and "to osculate" means "to kiss"!

internally tangent, well, they look like a giant eyeball . . . which is staring at the kissing circles. I'm sorry, but that's just creepy.

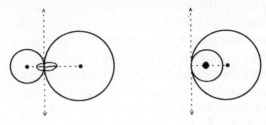

Externally tangent circles   Internally tangent circles

Doing the Math

Do these circle problems. I'll do the first one for you.

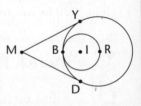

1. In *MYBIRD*, ⊙*I* and ⊙*R* are internally tangent at *B*. $\overline{MY}$ and $\overline{MD}$ are tangent to ⊙*R*, as shown. If *MY* = 5*y* – 4, *MD* = 3*y* + 6, and *BI* = *y* – 2, what is the radius of ⊙*I*?

Working out the solution: No need to panic; what do we know? Because of the Two-Tangent Theorem, *MY* = *MD*, right? And that means we can set up the equation: 5*y* – 4 = 3*y* + 6, and solve for y right away. Adding 4 and subtracting 3y on both sides gives us 2*y* = 10 → **y = 5**. Great! Now that we've figured that out, let's turn our attention to what the problem asked. We want the radius of ⊙*I*; in other words, we want the length of *IR*, or the length of *BI*. They're both radii, after all! And since *BI* = **y – 2**, we can use substitution to find that *BI* = **5 – 2 = 3**. Done!

Answer: The radius of ⊙*I* is 3.

2. In TUCAN,* $\overline{TU}$ and $\overline{TN}$ are tangent to $\odot A$. The radius of $\odot A$ is 6 cm. $TU = 2x + 2$, and $TN = 3x - 1$. Find TN, TA, and TC. (Hint: After finding x, use the Pythagorean Theorem!)

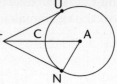

3. In the double scoop (DBLSCOP) figure to the right, given that $\overline{DO}$ & $\overline{LO}$ are common tangents to $\odot B$ & $\odot C$ at points D, S, L & P, prove $\overline{DS} \cong \overline{LP}$. (Try it on your own, and then use this hint if needed).[†]

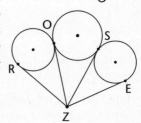

4. In the ROSEZ diagram, all segments are tangent to the circles they touch. Is it necessarily true that $\overline{RZ} \cong \overline{EZ}$? Explain one way or the other in a paragraph proof. (Hint: Think about the Transitive Property.)

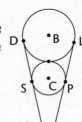

(Answer to #2 on p. 401; proofs at GirlsGetCurves.com)

Time to summarize what we've learned so far, so you can handle any 'ol circle problem they throw your way.

Step By Step

Tackling "How-the-Heck-Am-I-Supposed-to-Do-That?" circle problems:

Step 1. No matter what we're supposed to do, whether it be proving something or just finding a length—always, always <u>draw in radii from the center to any points that are marked on the rim of the circle</u>, and then mark all radii congruent with kitty scratches. Don't think about it; just do it!

Step 2. If the circle has a tangent line, draw in the radius that touches the tangent (if you didn't already do this in Step 1), and since we know *that*

. . . .

* A *toucan* is a kind of a bird, by the way.

† Use the "Bird Beak" Two-Tangent Theorem twice and also the Subtraction Property!

radius is automatically ⊥ to the tangent line (Bicycle Wheel Rule), we mark the right angle!

Step 3. If the circle has two tangent segments that come from a single point, then mark 'em congruent. (See p. 314 for a diagram of the "Bird Beak" Two-Tangent Theorem.)

Step 4. If a chord is involved, remember the bulldog puppies! Try using the Symmetrical Puppy Dog Mouth Theorem: Draw in the ⊥ bisector that goes through the center of the circle, if it's not already drawn in. Then mark the new midpoint on the chord, the two new congruent segments with kitty scratches, and of course, the right angles with a little bulldog tooth. Twin Puppy Face Theorem: Two chords in a circle will be congruent if they are equidistant to the center, and vice versa. (See p. 306 and p. 308 to review the puppy theorems.)

Step 5. It's amazing how much more info we have about the diagram now, huh? If we're looking for distances, try using the Pythagorean Theorem (see p. 176). If this is a proof, try looking for congruent triangle pairs we might have created (look for SSS, SAS, ASA, AAS, and HL),* and go from there. Remember, we're always allowed to connect any two points with a segment, so if that helps to create congruent triangles, draw that segment in, too. Also keep an eye out for possible quadrilaterals. You can do this!

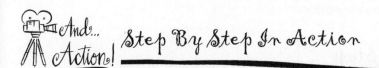

And... Action! Step By Step In Action

In the BEAK diagram below, Given: ⊙E and $\overline{BA} \cong \overline{KA}$. Prove: $\overrightarrow{AE}$ bisects ∠BAK.

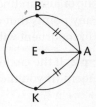

Um . . . how the heck are we supposed to do that? Let's follow the steps and see!

Step 1. Well, because we have a circle with some points on its rim, the first thing we want to do is draw in the missing radii . . . and voilà!

....

* To review the HL Theorem for congruent right triangles, check out p. 143.

All radii are congruent, so we know that $\overline{BE} \cong \overline{KE}$, and of course the Reflexive Property gives us the shared side, $\overline{EA}$, and we were given $\overline{BA} \cong \overline{KA}$. As if by magic (or maybe by SSS), we suddenly have a couple of congruent triangles. Along with CPCTC, we've got this covered. Hey, we didn't even need the rest of the steps! Nice.

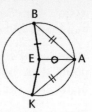

Proof

Statements	Reasons
1. ⊙E, $\overline{BA} \cong \overline{KA}$	1. Given (Gimmie an "S"!)
2. Draw $\overline{BE}$.	2. Two points determine a line.
3. Draw $\overline{KE}$.	3. Same as above.
4. $\overline{BE} \cong \overline{KE}$	4. All radii are congruent. (Gimmie an "S"!)
5. $\overline{EA} \cong \overline{EA}$	5. Reflexive Property (Gimmie an "S"!)
6. $\triangle BAE \cong \triangle KAE$	6. SSS (1, 4, 5)
7. $\angle BAE \cong \angle KAE$	7. CPCTC
8. $\overrightarrow{AE}$ bisects $\angle BAK$.	8. If a ray divides an $\angle$ into two $\cong$ $\angle$'s, then it bisects the $\angle$.

Take Two: Another Example

Let's prove the "Bird Beak" Two-Tangent Theorem from p. 314. In other words:

Given: ⊙O and tangent segments $\overline{NS}$ and $\overline{ES}$.
Prove: $\overline{NS} \cong \overline{ES}$.

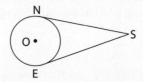

Wow, this seems strange! Let's just follow the steps and see what happens.

Step 1. Are there any points on the rim that we can draw radii to? Yep! We'll draw in radii $\overline{ON}$ and $\overline{OE}$, making sure to mark them as $\cong$ on the diagram with kitty scratches.

Step 2. Because these radii touch tangent segments, we know that they are also $\perp$ to those segments, so we can draw in the right angle markings for the newly created angles $\angle N$ and $\angle E$. Nice! (See p. 313 for the Bicycle Wheel Rule.)

Step 3. (We have to skip this step, because that's the theorem we're proving!)

Step 4. Hmm, there are no chords on this diagram, so we can skip this step, too.

Step 5. We're not finding a distance, but we are looking to prove $\overline{NS} \cong \overline{ES}$. So the question becomes: Can we find two congruent triangles in this diagram? Yes! Let's draw the segment connecting O to S, and then we can see that the two $\cong$ triangles could be $\triangle NOS$ and $\triangle EOS$. Using the Reflexive Property and HL, we can totally prove those triangles congruent, and then CPCTC tells us that $\overline{NS} \cong \overline{ES}$. Ta-da!

And you, little missy, will write out this proof (two-column style) in #4 on p. 321!

QUICK NOTE Remember, any time perpendicular things are involved (or could be drawn in, like with distances to chords or tangent lines to circles), we should always check to see if the Pythagorean Theorem might be helpful.

Doing the Math

Do these circle problems, using the Step By Step on pp. 316-7. Remember to draw in radii! I'll do the first one for you.

1. In the "Hide Me" diagram, $\overline{HE}$ is tangent to $\odot I$, $HM = 6$, $HI = 3\sqrt{5}$, $DE = 2$. Find ME.

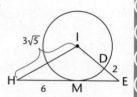

<u>Working out the solution:</u> First step, always look for radii! Well gosh, there's one on the figure already: $\overline{ID}$. Also, if we connect I to M, we'll get another radius: $\overline{IM}$. So let's fill that in and give kitty scratches to show that $\overline{IM} \cong \overline{ID}$. Since the radius $\overline{IM}$ touches a tangent line to its circle, we've got a right angle, so we'll mark that, too. Lookie there, we've got ourselves a couple of right triangles: $\triangle IMH$ and $\triangle IME$. We can find the radius by using the Pythagorean Theorem on the left triangle: $(IM)^2 + 6^2 = (3\sqrt{5})^2$. Simplifying,* we get: $(IM)^2 + 36 = (3^2 \cdot \sqrt{5}^2) \rightarrow (IM)^2 + 36 = 9 \cdot 5 \rightarrow (IM)^2 + 36 = 45 \rightarrow (IM)^2 = 9 \rightarrow IM = \mathbf{3}$. Great—now we have the radius! And this tells us that $IE = 3 + 2 = \mathbf{5}$. Now that we have two sides of the right triangle (the smaller one on the right), we could easily apply the Pythagorean theorem again, or we could notice that $\triangle IME$ is a 3-4-5 right triangle.† And that means $ME = \mathbf{4}$.

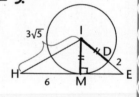

Answer: $ME = 4$

2. In the EAR‡ diagram, $\overline{EA}$ is tangent to $\odot R$, $EA = 2\sqrt{6}$, and the radius of $\odot R$ is 2. Find RA.

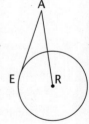

....

* To brush up on how exponents distribute over multiplication and simplifying radicals, check out Chapters 17 and 19 in *Hot X: Algebra Exposed*.

† For Pythagorean triples, check out p. 178.

‡ Looks kinda like one rabbit ear, right?

3. In PAWS, concentric* circles both have center W, with radii of 2 and 4. Find PS. (Hint: Neither radius has been drawn in yet.)

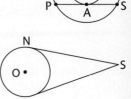

4. Write out the two-column proof of the Two-Tangent Theorem we proved on pp. 318–9. In the NOSE diagram, given $\odot O$ and tangent segments $\overline{NS}$ & $\overline{ES}$, prove that $\overline{NS} \cong \overline{ES}$. Try it on your own before looking at pp. 318–9 for help!

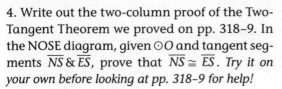

(Answers to #2–3 on p. 401; proof at GirlsGetCurves.com)

Takeaway Tips

All radii of a circle are $\cong$, and as soon as we see two or more radii on a diagram, we can mark 'em with kitty scratches. Chords should make us think about bulldog puppies!

Bisected Radii "Symmetrical Bulldog Puppy Face" Theorem: If a radius is $\perp$ to a chord, then the radius bisects that (non-diameter) chord, and vice versa.

Congruent-Chord "Twin Puppies" Theorem: If two chords in a circle are $\cong$, then the chords are equidistant to the center, and vice versa.

Tangent "Bicycle Wheel" Theorem: A tangent line to a circle will always be $\perp$ to the radius that touches the tangent point.

Two-Tangent "Bird Beak" Theorem: If a circle has two tangent segments drawn from a single point, they will always be $\cong$.

* Concentric circles are two circles that have the same center point.

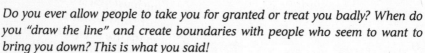

Confessional

Drawing the Line—Respect!

In geometry, drawing lines is often the key to solving challenging proofs, like #4 on p. 321. In life, "drawing lines" is even *more* important: Setting boundaries can be the key to being treated with respect from others. However, it doesn't always come easy. In a perfect world, we would all have total self-respect and always stand up for ourselves no matter what. But often, we're afraid that if we stand up for ourselves, we won't be liked or accepted.

Do you ever allow people to take you for granted or treat you badly? When do you "draw the line" and create boundaries with people who seem to want to bring you down? This is what you said!

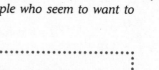

"I used to let people take advantage of me because I didn't want to hurt anyone's feelings. My moral 'lines' were being crossed and blurred until I finally learned to say a very important word: 'no.' Telling people 'no' isn't being rude or disrespectful, it's simply knowing where your boundaries are, and it's helped me to be a happier person." —Tori, 15

"If you don't have respect for yourself, no one else will—and that includes any guys you might end up dating." —Lola, 18

"If someone is trying to bring me down, they do not belong in my life. The people that I would like to surround myself with are those who are confident, happy, enjoy my company, and like me for who I am." —Jade, 17

"A lot of girls my age face the problem of being objectified, or looked at as pretty faces. One of the things I hate the most is when boys patronize me, especially in math and science. I take pride in my intelligence, and I won't let anyone treat me as if I'm not smart." —Victoria, 16

"I once became friends with a great guy, but as I got to know him, he did everything in his power to try to bring me down. He would tell me that 'nobody can make it as a writer, especially you,' and he would tell me that I was wasting my time caring about the environment. It almost seemed as though everything that I cared about or was striving toward, he would bash down in some way. Needless to say, we are no longer friends." —Lia, 17

"Women with self-respect are strong and confident: They live life for themselves and tend not to care about what other people think. The strongest and most beautiful of girls, inside and out, are those who have enough respect for themselves to say 'no' if they're not comfortable, and they'd rather be alone than be with a guy who wants to disrespect them or push them into something they're not ready for."

—Bailey, 17

"To get self-respect, first you must learn to accept yourself for who you are and not try to be a different person to fit in." —Brock, 15

"One must first see the big picture and decide that all humans are of equal worth. It's simple from there—if no one is better than you, then you won't take abuse from anyone." —Minerva, 16

Having the courage to stand up for ourselves often requires the confidence that comes from true self-respect. If you just don't feel like you have self-respect, don't worry—you can build it. How? *By doing things you believe to be respect-worthy.* Treat others like you want to be treated, even when you're in a bad mood. Be kind to yourself: Eat foods and do activities that make you feel healthy and good, and resist bad habits. Keep trying when you feel frustrated by a geometry proof. Volunteer at a local charity. Study in all your classes so you feel confident on test days,* knowing you're building a better future for yourself.

Changing behaviors and thought processes isn't easy, but it's much easier when you make notes to yourself—on your phone, in your locker, at home—wherever you might see them. By taking one tiny positive action at a time, you can build true self-respect and pride in yourself. Self-respect is essential for choosing the right career and guy later in life. And believe it or not, you have the power to steer yourself in a healthy direction *right now.* You can do it!

. . . .

* Check out pp. 305–11 in *Kiss My Math* for the "Math Test Survival Guide." It seriously makes test taking way easier.

Chapter 19

The Warrior Princess
Arcs and Arc Length

$\mathscr{J}$oan of Arc was a brave warrior. At the age of 14, she had visions of divine lights that inspired her to lead the French to many victories in the Hundred Years War. And now, in her honor, we shall conquer *arcs*.

What's It Called?

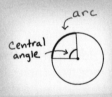

An **arc** is just part of a circle. More specifically, an arc is two points on a circle and all the points on the perimeter between them, too. If an arc takes up half the circle, then it's called a **semicircle**. If an arc is smaller than a semicircle, it's called a **minor arc**. If the arc is bigger than a semicircle, it's called a **major arc**. In the circles below, the angles are all called **central angles**, because each vertex (armpit) is at the *center* of its circle.

$\overset{\frown}{HT}$ is a *minor arc*, $\overset{\frown}{FR}$ is a *semicircle*, and $\overset{\frown}{SBY}$ is a *major arc*.

The distance around the *entire* circle is called its **circumference**.

Notice that for the major arc, $\overset{\frown}{SBY}$, we needed to stick "B" into its name or someone could get confused, thinking we really meant the minor arc, $\overset{\frown}{SY}$. No biggie. (We'll usually just deal with minor arcs and semicircles.)

Imagine for a moment that you're playing touch football, and a cute guy throws a ball (making an *arc* in the air) that is intended for someone on the opposite team, but you cut off that person and catch the "arc" instead! There's a name for that; it's called an *interception*. Angles intercept arcs, too: When an angle "cuts off" part of a circle with its sides, the arc that it creates is called an **intercepted arc**. On p. 324, $\overset{\frown}{HT}$ is the arc *intercepted* by the angle ∠HOT. Nice move, hottie!

So, how do we measure an arc? Yes, we could get out a tape measure and find the actual length of the curved segment in inches or feet or whatever. But even if we don't know anything about the size of the circle—it could be the size of your ring or a planet—we can still measure the arc in terms of *degrees*, just like an angle.

What's It Called?

The **measure of an arc** is *defined* to be the measure of the *central angle* created by the radii that connect to the endpoints of the arc!

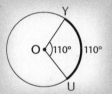

In this case, because ∠YOU = 110°, we define the intercepted arc's measurement to also be 110°, and we could say: $m\overset{\frown}{YU}$ = 110° or just $\overset{\frown}{YU}$ = 110°. And in the circles on p. 324, we could say that $\overset{\frown}{HT}$ = 50°, $\overset{\frown}{FR}$ = 180°, and $\overset{\frown}{SBY}$ = 290°, with the same measurements as the angles that intercept them! There are other ways to figure out an arc's measure, even if we don't have a central angle, and we'll see those on p. 333.

QUICK NOTE   Just like with angles, I prefer to leave off the *m* for arc measurement notation, but we'll do it both ways in this book, just to remain flexible. Be sure to ask your teacher what she/he prefers!

By the way, in the "BOYS" circle on p. 324, the *minor* arc $\overset{\frown}{SY}$ has a measure of 70°. Do you see why? There are 360° total in a circle, and because the major arc is 290°, the minor arc using the same endpoints must be 360° – 290° = **70°**. Makes sense, right?

And since a circle has a total of 360°, we can figure out the fractional amount of the circle that the arc covers, too. How? By making a fraction with 360 as the denominator and reducing! So if an arc measures 36°, then it's $\frac{36}{360} = \frac{1}{10}$ **of the circle**.

In the case of $\overset{\frown}{HT}$ from p. 324, since we know it equals 50°, we also know that it takes up $\frac{50}{360} = \frac{5}{36}$ of the circle. Then, let's say someone told us that the total circumference of the circle is 9 feet. How long is the arc? Well, it would be $\frac{5}{36}$ of 9 feet, right? And that's just: $\frac{5}{36} \times 9 = \frac{5}{36} \times \frac{9}{1} = \frac{45}{36} = \frac{5}{4} = 1\frac{1}{4}$ feet; in other words, **1 foot, 3 inches.***

This is the "formula" we just used:

Arc length = (fractional part of circle) × total circumference

. . . where the "fractional part of the circle" is: $\dfrac{central\ angle\ of\ arc}{360}$. Nice trick, huh?

Is it me, or does this SIR diagram look like a guy with a big moustache?

Given: ⊙I, ∠S = 20°. Find the measure of $\overset{\frown}{SR}$.

Whoa, how are we supposed to do that? Well, what should we always, always do with every circle diagram before we even start a problem? We should look for radii and mark 'em congruent! Here, that means $\overline{IS} \cong \overline{IR}$. And that means △SIR is isosceles! Because of the "if sides, then angles" rule from p. 148, we now know ∠S ≅ ∠R, which means that ∠R = 20°, right? And to find the third angle in our triangle, we just do: 20° + 20° + ∠I = 180° → ∠I = **140°**. Since ∠I is a central angle, the arc it intercepts has the *same measure*.

Answer: $\overset{\frown}{SR}$ = 140°

And if we were now told that the circumference of ⊙I were 90 feet, we could totally find the arc length of $\overset{\frown}{SR}$ like this:

. . . .

* For a refresher on reducing fractions, check out Chapter 6 in *Math Doesn't Suck*.

$$\text{Arc length} = (\text{fractional part of circle}) \times \text{total circumference}$$

$$\text{Arc length of } \overset{\frown}{SR} = \frac{140}{360} \times 90 \text{ ft}$$

With a little simplifying, we'd get **35 feet**. Nice!

Congruent Arcs . . . and Angles and Chords

For two arcs to be considered congruent, they must have the same measure *and* also be on the same circle (or on two congruent circles). After all, a 20° arc on a penny will be quite a different size from a 20° arc on a dinner plate. Unless it's a Barbie dinner plate, I guess. But you don't have those lying around anymore . . . right?

If we're given any two points on a circle, we could draw a chord between them, a central angle, and the arc that's intercepted by the angle. I mean, why not, right? (Don't answer that.)

For example, in ⊙*I* to the right, if someone gave us the points *M* and *S* on the circle, we could fill in the central angle ∠*MIS*, the intercepted arc $\overset{\frown}{MS}$, and the chord $\overline{MS}$.

Now let's say we have *two* central angles on a circle that are congruent. We'd also have *two* intercepted arcs and *two* chords, right? If the two central angles are congruent (like if both were 50°), obviously their intercepted arcs would be congruent, too (they'd be, um, 50°), and vice versa! With me so far?

☺  ☺

Now, we might not have any idea what the lengths of those *chords* are, but as it turns out, they'd have to be congruent to each other. In fact, *they're all tied together*: central angles, intercepted arcs, and chords. If we know one pair of things is congruent, the other pairs of things must be congruent, too!

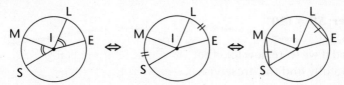

Angle-Arc-Chord Congruent Rules*

 If two central ∠'s are ≅, then their intercepted arcs are ≅, and vice versa.

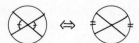

 If two intercepted arcs are ≅, then their corresponding chords are ≅, and vice versa.

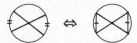

 If two central ∠'s are ≅, then their corresponding chords are ≅, and vice versa.

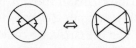

For example, in the SMILE diagram on p. 327, if we were told that $\overarc{MS} \cong \overarc{LE}$ and also that the length of $\overline{MS}$ is 13 inches, then we'd automatically know that $LE = 13$ inches, too! Since *all* the combos work, the above chart can seem more complicated than it is. But it's actually pretty simple for a sharp cookie like you, isn't it?

"*The girls I admire the most are the ones who make choices for themselves and don't care what anyone else thinks. They're independent, strong, and confident.*" Macy, 16

* Note: These Rules work for a single circle or for two congruent circles.

Doing the Math

Answer these questions about arcs; I'll do the first one for you.

1. If a circle has an arc measuring 135° with length 6 inches, what fraction of the circle is it, and what is the circumference of the circle?

__Working out the solution:__ Hmm, if the arc is 135°, then it's $\frac{135}{360}$ = $\frac{135 \div 9}{360 \div 9}$ = $\frac{15}{40}$ = $\frac{3}{8}$ of the circle. Notice that we didn't need to use the 6 inches to know that! Now the question becomes: If 6 inches is $\frac{3}{8}$ of the circumference, how many inches is the entire circumference? Let's call the circumference c and translate this English sentence into math*: 6 inches is $\frac{3}{8}$ of the circumference → $6 = \frac{3}{8} \times c$. Multiplying both sides by $\frac{8}{3}$, we get: $\frac{8}{3} \cdot 6 = \frac{8}{3} \cdot \frac{3}{8} c$ → $\frac{48}{3} = \frac{24}{24} c$ → $16 = c$. Does it make sense for the circle to be a total of 16 inches around, if 6 inches is $\frac{3}{8}$ of it? Sure does!

__Answer: The arc is $\frac{3}{8}$ of the circle; the circumference is 16 inches.__

For questions #2–11: In the MUSICAL diagram to the right, there are two concentric circles with center S, $\overline{UI}$ & $\overline{MC}$ are their diameters, and $\angle LSC = 100°$.

2. What is the measure of $\overset{\frown}{AI}$? How about $\overset{\frown}{LC}$?

3. What is $\angle USA$? How about $m\overset{\frown}{ML}$?

4. Name a major arc in the smaller circle starting with A and give its measurement.

5. True or false: $\angle USA \cong \angle MSL$?

6. True or false: $\overset{\frown}{UA} \cong \overset{\frown}{ML}$?

....

* Remember: When "of" is immediately surrounded by two values, it means multiplication. For tons of easy tips on translating English into math, check out Chapter 11 in *Kiss My Math*.

7. What are the three radii of the bigger circle?

8. What is the measure of ∠M? *(Hint: See "SIR" on p. 326.)*

9. What fraction of the circumference is $\overset{\frown}{LC}$?

10. If the circumference of the big circle is 24 feet, what is the length of $\overset{\frown}{LC}$?

11. If the length of $\overset{\frown}{UA}$ is 2 feet, what is the circumference of the smaller circle?

12. In the PETAL diagram, a regular pentagon is inscribed in a circle, meaning that P, E, T, A, & L are evenly distributed around the circle. What fraction of the circle is $\overset{\frown}{PE}$? What is the measure of each arc?

13. If a regular hexagon were inscribed in a circle, what would each of the six arcs measure? *(Draw a picture if it helps!)*

14. As we saw in Chapter 12, a regular *n*-gon is a polygon with *n* sides. Imagine if a regular *n*-gon were inscribed inside a circle. What would each of its *n* arcs measure?

15. In the SARONG* diagram, Given: ⊙A, $\overline{SG} \cong \overline{RO}$. Prove: $\overset{\frown}{SO} \cong \overset{\frown}{RG}$.

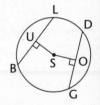

16. In the BULDOGS diagram, Given: ⊙S, $\overline{US} \perp \overline{BL}$, $\overline{SO} \perp \overline{DG}$, $\overline{BL} \cong \overline{GD}$. Prove: $\overline{US} \cong \overline{SO}$. *(Hint: Remember the bulldog puppies from Chapter 18?)*

(Answers to #2–14 on pp. 401–2; proofs at GirlsGetCurves.com)

----

* Sarongs can be tied low around the hips like a long skirt, often showing the belly button. Can you see it in the diagram?

Art Class Gone Wrong: Secant Lines

Hey, remember tangent lines from the last chapter? Say you were asked to draw one by holding a pencil between your toes. Despite your protests, you'd give it a try and probably manage to get your segment to pass through the circle . . . but, um, not in a tangent way! You'd then say, "See? Can't be done!" Chances are, you've drawn a **secant** line, which is pronounced "See? Can't!"

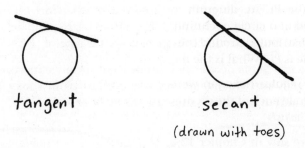

tangent   secant

(drawn with toes)

We'll deal with secant lines in a few pages . . .

Where's My Armpit? More Angle-Arc Relationships

As it turns out, even when the vertex of an angle isn't at the center of a circle, if we know the angle measurement, then we can still totally figure out its intercepted arc measurement, and vice versa. The key is to realize that it's all about the location of the armpit (er, vertex). Think about it; if you have a circle, there are four different kinds of places we could put a point: on the center, on the rim, floating around aimlessly inside, or totally outside the circle! And once we know the location of the vertex of an angle, if we know the measurement of either the intercepted arc *or* the angle, we can figure out the other one.

But before we get around to learning all the Angle-Arc Rules, let's reflect for a moment on four different types of people . . .

And now, here they are: The four types of angle-arc relationships. Betcha they're not as hard to remember as you might've thought!

Types of Angle-Arc Relationships: Peaceful, Reckless, Lazy, Outcast

Peaceful: The vertex is the *center* of circle

Angle = intercepted arc†

What a nice, *balanced* equation!

$$\angle E = \overset{\frown}{ZN}$$

....

* Check out pp. 85–7 for a self-proclaimed outcast who became a sports star and entrepreneur.

† Technically, an angle can't *equal* an arc; it's their *measurements* that are equal. So we "should" say things like "angle measure = intercepted arc measure" and "$m\angle E = m\overset{\frown}{ZN}$," but I wanted to keep this chart easy to read. Check your teacher's preferences!

Reckless: Vertex on circle rim (*edge*)

Angle = $\frac{1}{2}$ intercepted arc

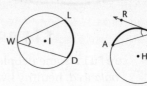

Remember, this *edgy* behavior could cut someone's life span in $\frac{1}{2}$ . . . !

(Note: ∠*W* is called an **inscribed angle.*** And ∠*S* is called a **tangent-chord angle**, because its sides are made from, yep, a tangent and a chord!)

WILD: $\angle W = \frac{1}{2}\widehat{LD}$

RASH: $\angle S = \frac{1}{2}\widehat{AS}$

~~~~~~~~~~~~~~~~~~~~

Lazy: Vertex *floating around* inside circle

Angle = $\frac{1}{2}$ the sum of the two intercepted arcs

This is the *average* . . . Like lazy people!

(Note: ∠*EOD*—and also the congruent ∠*BOR*—are called **chord-chord angles**, because they're made up of two chords intersecting.)

$\angle EOD = \dfrac{\widehat{ED} + \widehat{BR}}{2}$

~~~~~~~~~~~~~~~~~~~~

Outcast: Vertex *outside* circle

Angle = $\frac{1}{2}$ the **difference** between the two intercepted arcs

Because outcasts are usually *different*, and maybe *half* are losers but *half* are cool . . .

(Note: ∠*T* is called a **tangent-tangent angle**, ∠*Y* is called a **secant-tangent angle**, and ∠*G* is called a **secant-secant angle**, because of where the sides of the angle touch the circle. No biggie.)

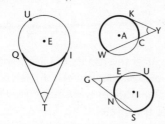

QUIET: $\angle T = \frac{1}{2}(\widehat{QUI} - \widehat{QI})$

WACKY: $\angle Y = \frac{1}{2}(\widehat{KW} - \widehat{KC})$

GENIUS: $\angle G = \frac{1}{2}(\widehat{US} - \widehat{EN})$

. . . .

*An angle is *inscribed* in a circle when the vertex <u>and</u> a point from each side of the angle are all on the rim, too.

The strangest one is the outcast (big surprise!).*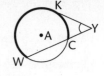
In geometry, we only deal with positive angles,†
so we always subtract the smaller arc from the
bigger arc. For example, in the WACKY dia-
gram, if we're told that $\overset{\frown}{KW} = 170°$ and $\overset{\frown}{KC} = 70°$, then
we could figure out $\angle Y = \frac{1}{2}(\overset{\frown}{KW} - \overset{\frown}{KC}) \rightarrow \angle Y = \frac{1}{2}(170° - 70°) \rightarrow$
$\angle Y = \frac{1}{2} \times 100° \rightarrow$ **$\angle Y = 50°$.**

By the way, in the BORED example on p. 333, those vertical angles
mean that $\angle BOR \cong \angle EOD$, which is good, because it ends up being the
same formula to find both! And what if we wanted to find one of the "side"
angles in BORED, like $\angle BOE$? We could do it; our formula would give us
$\angle BOR$, and after all, $\angle BOE + \angle BOR = 180°$; they're supplementary angles!

Watch Out!

When filling in the "lazy" and "outcast" angle-arc equations,
make sure you use only the *intercepted* arcs' measurements, not
some other arc. In the WACKY example we just did, if the prob-
lem had told us that $\overset{\frown}{KW} = 170°$ and $\overset{\frown}{CW} = 120°$ and asked us to
find $\angle Y$, it could have been tempting to use 170° and 120° for the
subtraction—oops! Instead, we'd need to think, "Hmm, $\overset{\frown}{CW}$ isn't
actually intercepted by $\angle Y$, but $\overset{\frown}{KC}$ is. So what is $\overset{\frown}{KC}$? Since there
are 360° around every circle, $\overset{\frown}{KC} + 170° + 120° = 360°$, which we
can solve to get $\overset{\frown}{KC} = 70°$." And then we'd continue as we did!

Step By Step

Angle-arc problems:

Step 1. Don't worry about how complicated the diagram looks. For now,
answer this one question: Where's the armpit? Figure out if the vertex is
peaceful (center of circle), reckless (on the edge of the circle), lazy (floating inside), or an outcast (totally outside the circle).

. . . .

* Hey, some of the greatest minds in history were considered outcasts . . . Joan of
Arc, hel-lo!
† In trigonometry and beyond, we see negative angles, too.

Step 2. Write down the relationship that applies:

Peaceful: $\angle$ = **arc**   Lazy: $\angle = \frac{1}{2}$ **(sum of two arcs)**

Reckless: $\angle = \frac{1}{2}$ **arc**   Outcast: $\angle = \frac{1}{2}$ **(difference of two arcs)**

And then write it with the actual letters from the diagram. For example, if it's a "lazy" vertex, it might look something like: $\angle A = \frac{1}{2}(\overset{\frown}{BC} + \overset{\frown}{DE})$.

Step 3. Plug any numbers we know into the equation, making sure to use the correct *intercepted* arcs only, and solve for the unknown.

Step 4. Make sure you've answered what the problem was actually asking for. Done!

And...
Action! Step By Step In Action

In the GENIUS diagram, Given: $\angle G = 20°$, $\overset{\frown}{EN} = 35°$. Find: $\overset{\frown}{US}$.

Step 1. This is definitely an outcast; look at G all the way out there!

Step 2. So we'll use this formula: $\angle = \frac{1}{2}$ (difference of two arcs), which becomes: $\angle G = \frac{1}{2}(\overset{\frown}{US} - \overset{\frown}{EN})$.

Step 3. Filling in what we know, we get: $20° = \frac{1}{2}(\overset{\frown}{US} - 35°)$. Time to simplify! Multiplying both sides by 2, this becomes $40° = \overset{\frown}{US} - 35° \rightarrow \overset{\frown}{US} = 75°$.

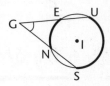

Step 4. Yep, that's what we were asked to find. Nice!

Answer: $\overset{\frown}{US} = 75°$

QUICK NOTE   Let's use the "m" notation for this next problem. Every teacher will have different preferences on this, and it's good to be flexible!

Take Two: Another Example

Given the circle to the right, prove that △BAR ~ △IAD.

This one is tricky! There are so many angles and arcs, it's hard to know which we'll need. It doesn't seem like we've been given enough information, either. Hmm. We'll let go of the steps for a moment and think about what the problem is asking. If we want to prove that the two triangles are *similar*, we only need to find two pairs of congruent angles (see p. 289 for a refresher), so let's see what we can do.

Since we have vertical angles, we know ∠**BAR** ≅ ∠**IAD**, right? And what about another pair of corresponding angles? Looking at ∠R and ∠D, we might notice that they each have their vertex on the edge of the circle, so they're the "reckless" type (with, you know, $\frac{1}{2}$ the life span), and—hold the phone—they both intercept the *same* arc, $\widehat{BI}$! That means that $m\angle R = \frac{1}{2}m\widehat{BI}$* (find that on the diagram) <u>and</u> $m\angle D = \frac{1}{2}m\widehat{BI}$, right?

The *right* sides of both of those equations are identical, so that means the *left* sides must be equal! In other words, <u>$m\angle R = m\angle D$</u>. And, of course, that means ∠**R** ≅ ∠**D**. Now we have two pairs of congruent angles, so by AA ~, we've proven that △*BAR* ~ △*IAD*.

∴ △**BAR** ~ △**IAD**

Ta-da!

By the way, we also could have shown ∠B ≅ ∠I because both angles also intercept a common arc, $\widehat{RD}$.

QUICK NOTE   We used a great trick in that last example: noticing that two of *the same type* of angles intercepted the same arc (in this case they were both "reckless"). It's something to look out for, just like vertical angles!

. . . .

* Remember, we're using the *m* notation for this problem, just to remain flexible. We also could have written this formula as ∠$R = \frac{1}{2}\widehat{BI}$, for example.

Joan of Arc was a heroine of France . . . and she was burned at the stake for the victories she led. Guess *some* people weren't ready for a kick-ass girl to flex her muscles. Talk about an outcast who changed the world! Personally, I think she'd be proud to see <u>you</u> conquering arcs. Ready or not, here we come . . .

Doing the Math

Do these angle-arc problems based on the given figures. I'll do the first one for you.

1. In the CRAZY ONES diagram, Given: $\overline{RO}$ is tangent to $\odot E$, $\overset{\frown}{CR} = 22°$, $\overset{\frown}{ZY} = 26°$, $\overset{\frown}{ON} = 45°$. Find: $\overset{\frown}{OS}$.

<u>Working out the solution:</u> Okay, we're not going to panic. Notice that the point A is the vertex of the angle $\angle 1$. With respect to the small circle, A is floating around the inside—a "lazy" one—but with respect to the big circle, A is totally outside—an "outcast"! We already know that $\overset{\frown}{ON} = 45°$, so just looking at the *big* circle, if we knew $\angle 1$, then we'd have a typical outcast problem and could use the "half the difference" formula to find $\overset{\frown}{OS}$, right? And how can we find $\angle 1$? By using the "lazy" formula on the small circle! That's the average, so we get: $\angle = \dfrac{1}{2}$ (sum of two arcs)

$\rightarrow \angle 1 = \dfrac{1}{2}(\overset{\frown}{CR} + \overset{\frown}{ZY}) \rightarrow \angle 1 = \dfrac{1}{2}(22° + 26°) \rightarrow$

$\angle 1 = 24°$. Great progress! Now we can turn our attention to the big circle/outcast part of this problem, and do:

$\angle = \dfrac{1}{2}$ (difference of two arcs) $\rightarrow \angle 1 = \dfrac{1}{2}(\overset{\frown}{OS} - \overset{\frown}{ON}) \rightarrow$

$24° = \dfrac{1}{2}(\overset{\frown}{OS} - 45°)$. Solving for $\overset{\frown}{OS}$, we multiply both sides by

2 to get: $48° = \overset{\frown}{OS} - 45° \rightarrow \overset{\frown}{OS} = 93°$. And voilà!

Answer: $\overset{\frown}{OS} = 93°$

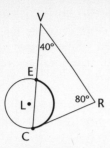
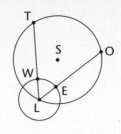
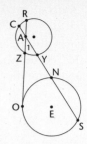

2. In CLEVR, Given: $\odot L$ is tangent to $\overline{CR}$ at the point C, $\angle V = 40°$, $\angle R = 80°$. Find $\overarc{CE}$. *(Hint: First, find $\angle C$.)*

3. In the TOWELS diagram, Given $\odot S$ & $\odot L$, $\overarc{TO} = 132°$, and L lies on the circumference of $\odot S$. Find $\overarc{WE}$.

4. Looking at the CRAZY ONES diagram, use these Givens *instead*: $\overline{RO}$ tangent to $\odot E$, $\overarc{ON} = 47°$, $\overarc{OS} = 99°$, and $\overarc{CR} = 25\frac{1}{2}°$. Find $\angle 1$ and $\overarc{ZY}$.

5. In RITE, Given: $\overline{RE}$ is a diameter of $\odot T$. Find $\angle I$. *(Hint: How many degrees of the circle is $\angle I$ intercepting? And what "personality type" is the point I?)*

6. In NOPE, Given: $\overline{PO}$ and $\overline{PE}$ are tangent segments to the circle. Explain why $\overline{OE}$ can't be a diameter, in a paragraph-style indirect proof. *(Hint: Assume $\overline{OE}$ is a diameter. What would that mean for $\overarc{OE}$? Now set up the "outcast" formula for $\angle P$.)*

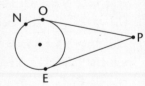

7. In the BOWTIE diagram, Given: $\odot E$, $\overarc{BO} = \overarc{IT}$, and $\overarc{OT} = 32°$. Do a paragraph proof to demonstrate that $\overline{OW} \cong \overline{TW}$, using congruent triangles and CPCTC. *(Hint: Use our new angle-arc relationships with $\angle B$ & $\angle I$ and also the Angle-Arc-Chord Rules from p. 328. You can do it!)*

(Answers to #2–5 on p. 402; proofs at GirlsGetCurves.com)

Takeaway Tips

 In the same circle (or two ≅ circles), congruent *central angles* mean congruent *intercepted arcs*, which also mean congruent *corresponding chords*. If we have *any* of those ≅ pairs, we know the other two pairs are also ≅.

 Look for the location of the vertex of the angle intercepting the arc(s) on a circle, and remember the four types of people when determining the angle-arc formula we need: peaceful, reckless, lazy, or outcast!

Keep an eye out for when two different angles both intercept the *same* arc. If both angles' vertices are the same "personality type," the angles must be congruent, after all. And if they're different "personality types," we can often still figure out one from the other!

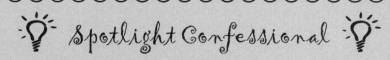

Spotlight Confessional

Dealing with Eating Disorders and Academic Pressures: One 16-Year-Old's Story

Nobody's life is perfect. Many of us come from families that are broken in some way, have been bullied in school, or just feel "less than." So many of us believe no one could possibly understand what we're going through. We feel alone, and the darkness begins to lead us into self-hatred.

I know. I know because I've seen it happen to friends. I know because it nearly destroyed me.

For as long as I can remember, my home life was stressful. My parents never particularly got along, and I wanted more than anything to fix their marriage . . . the strategy I settled on? Academic excellence, and I threw myself into it wholeheartedly. Maybe if I brought home perfect grades, my parents would be so overjoyed that they would stop arguing and we could be one big, happy family. But of course it didn't work; I didn't have the power to change them, no matter how much I wanted to. If perfect grades didn't impress my parents enough to make everything better somehow, then maybe something else would . . .

In sixth grade, my mother commented that I was beginning to look a bit pudgy in my uniform. At the time, I thought nothing of it, but the seed had been planted. It grew in my head, and by the 7th–8th grade, I was convinced that she was silently judging me. Then I began to compare my weight to that of other, more athletic girls in my class, and I became more and more unhappy with myself. But rather than take positive action like join a sports team or adjust my eating habits in healthy ways, I became self-destructive. At meals, I started eating less and less. Then later, starving, I would fill up on the wrong foods, which is actually a recipe for gaining weight . . . and which is exactly what happened.

My next move was extreme: I decided I would only eat raw, vegan food. That meant no bread, no meat—just raw fruits and vegetables. And I'm not saying that raw vegans can't be healthy, but I used it as an excuse to starve myself: I was only eating about 100 calories a day. I lost 35 lbs in two and a half months and got down to 95 lbs—about 20 lbs under my ideal weight. On so few calories a day, I was constantly getting lightheaded, I couldn't concentrate on schoolwork, and of course I thought about food obsessively. I was miserable. My mother tried to enroll me in an eating disorder program at a local clinic, but I never followed their advice. Who did they think they were? Why should I listen to them anyway?

But after months of my mom begging me to eat more, I finally did . . . and then I didn't stop. I was so angry that my plan hadn't fixed my life, my parents, my happiness, that I went to the other extreme. Maybe I'd find happiness in food. And sure, white bread, French fries, and ice cream sometimes felt good in the moment, but in the next moment my self-hatred would get the best of me . . . which just made me eat more. It's almost as if I was proving to myself just how loathsome I was. And I felt worse every day.

The problem with eating disorders, or even looking in the mirror and criticizing yourself, is that thought patterns develop that are very difficult to break. Myself, I was spinning down a drain. I ate for comfort, all day long. Incredibly depressed, I wouldn't move if I didn't have to. I would spend weekends and school breaks in bed or on the couch, just watching movies or reading. And I gained 80 lbs in less than 9 months.

I hated my body so much that I convinced myself I honestly didn't care about it. It was a glorified jar of fluids for transporting my brain. But that created its own, constant pressure. Suddenly I was terrified of getting a wrong answer on a test. I had decided that my brain was my entire identity, so I was absolutely nothing without perfect grades. I felt trapped. I lived in these extremes, and I felt like there was no way out. And then, thankfully, the clouds began to part.

Summer break came, and I went to camp. There, away from school, my parents, and the rest of my stressed-out life, I was able to begin to relax. Most importantly, I found friends I could talk to, and especially one camp counselor who really listened and made me realize I wasn't alone. It meant so much to me that someone cared enough to listen and tell me I had the strength to get through this. To my amazement, the comfort I craved could be found in talking, not food! The weird thing is, it didn't matter how much my mom tried to get me to talk about things, it never helped in the same way. I guess I just needed a more objective opinion and someone who wasn't so closely tied to my home life. Learning that there are other people out there who care and have been through the same struggles made me realize I wasn't alone. I finally felt motivated to help myself, for real this time. I started changing my thinking and even started following the meal plan from the eating disorder clinic, and you know what? It's been working. By eating five small, healthy meals a day, I'm losing weight at a reasonable rate with the goal of finding an equilibrium set point for my weight—and keeping it there.

Sometimes I still look in the mirror and think, "If only I weighed twenty fewer pounds," but instead of "I will sacrifice every single other aspect of my life to achieve this goal," my thinking is, "I'll look a bit better if/when I lose a few pounds. So what. I've got other priorities in my life. If I really want this, I'll make an effort to exercise a bit more." I've lost the extremes in my thinking, and frankly, it's helped me so much.

My parents are divorced now, and I'm at peace with the fact that I never did have any control over their happiness. I guess we all wish for power we don't have. I've been learning to focus on the power I actually do have . . . the power to make positive changes in my life. Not for my parents and not for the counselors, but because I want to be healthy for me.

My biggest mistake was thinking I had to suffer alone, and the best advice I can give to others is to find a trusted adult to talk to. Life can be painful, but there is always hope; there is always tomorrow, and remember that a good listener is the best kind of friend you could hope for.

Hope can be misplaced, but it is never lost. And when it is dark enough, you can see the stars. I'm still not sure if I'm out of the darkness completely, but I do know that I'm not alone in it. Nobody is.

Tori C., 16

Danica Says: *Please talk to a nurse or a counselor if you have unhealthy thought patterns developing. Take action early!*

Chapter 20

Suddenly in the Mood for . . . Pie

Finding Perimeter and Area

$\mathcal{L}$ook, I love to eat healthy—organic, lots of vegetables and whole grains and all that. But what fun is being good if you can't splurge from time to time? Plus, my mom makes an amazing pie . . . and as it turns out, a little splurge here and there can really help us remember those geometry formulas for circumference and area.

We'll find the perimeters and areas for all sort of shapes, but I feel like cooking, so let's start with circles. Have you ever made a pie 100% from scratch? First, we make the crust, which starts off as pie dough.* Yep, **Crust = pie dough**. Say it out loud, like a chant: "Crust is pie dough! Crust is pie dough!" What is crust? Pie dough! Got it?

We could abbreviate that like this: **C = π*d***, so I guess you just memorized the formula for the circumference of a circle, didn't you?

Anyway—back to our pie. Time to figure out what to put *inside*. You know what I love? Fresh red raspberries. And some research says that red raspberries can help prevent cancer, so that sounds pretty healthy. You have to agree, no matter what kind of crust we use, it's the *area inside the pie* that determines what kind of pie it is, and red raspberries is the winner. "The **Area** inside is **red raspberry pie**." Say that one a few times out loud . . . it has kind of a nice rhythm to it, and it even sort of rhymes.

---

* I like to make mine with whole-wheat pastry flour.

Abbreviated, we could write: **A = rr**π, in other words: **A = πr^2**. The two *r*'s stand for <u>r</u>ed <u>r</u>aspberry, get it? And yep, you've just memorized another formula. Nice.

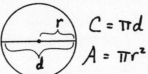

QUICK NOTE   Remember that the radius is just half of the diameter, so *d = 2r*. Sometimes you'll also see *C = 2πr*, but it's the exact same thing as *C = πd*, isn't it?

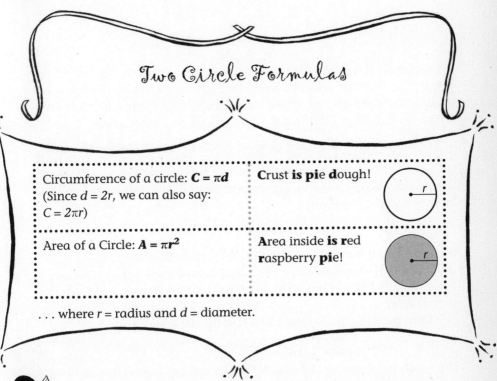

Two Circle Formulas

Circumference of a circle: **C = πd** (Since *d = 2r*, we can also say: *C = 2πr*)	**C**rust **is** **p**ie **d**ough!
Area of a Circle: **A = πr^2**	**A**rea inside **is** **r**ed **r**aspberry **pi**e!

. . . where *r* = radius and *d* = diameter.

QUICK (REMINDER) NOTE   The letter "π" (pronounced "pie") is an irrational number that equals approximately 3.14159 and shows up in lots of formulas, including those for circles. And why not? Pies look kinda like circles, don't they?

Watch Out!

Because π is an irrational number, its digits go on forever with no repeating pattern.* If you've got a calculator handy while you're doing circumference and area for circles, you might be tempted to punch in "3.14159 . . ." for π in order to get your final answer. But most of the time, the answer in the back will be written with the π left on, and your answer will look nothing like it! Unless you're asked to approximate your answer, it's really best to keep the π, because it's *exact* . . . and as a bonus, it still sounds like *pie*.

Just the Crust, Please

Some people like to eat just the *crust* of pie. I don't get it; I mean, all the good stuff's inside. But whatever. In Chapter 19, we found some *arc lengths*, based on the size of the central angle that intercepted that arc—which is kind of like finding the *length of crust* on a piece of pie.

Arc length = (fractional part of circle) × total circumference

We found the fractional part of a circle by putting the measure of the arc (or central angle) over 360, remember? Finding the *area* of a slice† of a circle works the same way.

<u>Area</u> of "slice" = (fractional part of circle) × total <u>area</u>

For example, if our pie has a radius of 4 inches (so the diameter is 8) and if we cut a piece that measures 40°, we can easily find the length of crust and also the area of the piece.

First, we'll find the fractional part of the circle we're dealing with: $\frac{40}{360} = \frac{4}{36} = \frac{1}{9}$. Great! The piece is $\frac{1}{9}$ of the entire pie. "Crust is pie dough," so the total circumference of this pie is: $C = \pi d = \pi(8) = \mathbf{8\pi}$. And that means the length of crust of *just our piece* is: $\frac{1}{9} \times 8\pi = \frac{\mathbf{8\pi}}{\mathbf{9}}$ **inches**, which is approximately 2.8 inches of crust. What a nice, small piece of pie. Perfect.

. . . .

* For more on irrational numbers, check out p. 282 in *Hot X: Algebra Exposed*.

† The real name for "slice" is **sector**, and they both start with s, so that's helpful.

For the area, since we already know the fractional part of the circle is $\frac{1}{9}$, we just find the total area, which is A = $\pi r^2 = \pi(4)^2 = \mathbf{16\pi}$, multiply it by $\frac{1}{9}$, and we get $\frac{1}{9} \times 16\pi = \dfrac{\mathbf{16\pi}}{\mathbf{9}}$ **square inches**. And that's approximately 5.58 square inches of pie. Mmm.

QUICK NOTE **Square units** When we multiply a length X length to get an area, we're also multiplying the units together. So 3 ft x 4 ft = 12 square feet. We can also write this as 12 sq ft, or 12 ft².

Reality Math: Tricky Pizza Man

*L*et's say you're ordering pizza for a group of friends. The pizza restaurant menu has three sizes: 8-inch for $8, 12-inch for $12, and 16-inch for $16. At pizza restaurants, the measurements refer to the diameters.

You figure the large should be enough for the five of you, and after you order your large veggie pizza, the guy behind the counter says, "Hey miss, we're having a special on our 8-inch pizzas—two for $14. It's a better deal than getting one large!"

Is he, like, full of it?

As you know, a much better measure of how much pizza we'll get is the *area*, not the diameter. So you say, "Excuse me one moment," and grab a napkin to jot down this quick calculation. Remembering to use radii instead of diameters, you'd do this:

$$\text{Area of large} = \pi r^2 = \pi(\mathbf{8})^2 = \mathbf{64\pi}\ \mathbf{in.^2}$$

If you didn't have a calculator, a rough approximation of the large pizza area is: 64(3) = <u>192 in²</u>. Next, you'd do this:

$$\text{Area of two smalls} = 2 \times \pi r^2 = 2 \times \pi(\mathbf{4})^2 = 2 \times 16\pi = \mathbf{32\pi}\ \mathbf{in.^2}$$

And a rough approximation of the two smaller pizzas' areas is: $32(3) = \underline{96 \text{ in.}}^2$. Whoa—the area of the big pizza is twice as much as *two* of the 8-inch ones?? And the price difference isn't very much. But with a calculator app, we could figure out exactly how much better a deal the bigger pizza is by figuring out the *unit price per square inch of pizza*:*

Big pizza: $\dfrac{\$16}{192in^2}$ = \$0.083—so just a bit more than **8 cents** per square inch of pizza.

Two small pizzas: $\dfrac{\$14}{96in^2}$ = \$0.14583—so more than **14 cents** per square inch of pizza. That's almost twice as expensive!

Yeah, you can tell the pizza guy to shove his "deal." But . . . you wouldn't do that, because he might not have understood the math, and it could have been an honest mistake on his part. So instead, you'll probably smile politely and simply say, "No thanks. I'll stick with the large."

. . . It sure is nice to be savvy, isn't it?

Now that we're total experts on the areas of circles (and, um, pies), let's talk about the areas of some familiar polygons.

Areas of Rectangles, Squares, and Triangles

A rectangle's area is the easiest area to find; it's just the base times the height.

$$\text{Area} = b \times h \quad h$$
$$b$$

Rectangle Area = $b \cdot h$

This same formula works for squares, but because *base* <u>equals</u> *height* for squares, we'll often see it written like this, where **s** is a side of the square:

Square Area = s^2

. . . .

* To brush up on unit rates, check out Chapter 17 in *Math Doesn't Suck*.

Rectangles also lead us to triangles. Why? Because every single triangle is just *half* a rectangle! Let's see how this works. We'll take any 'ol triangle, lay it on its side, and draw in the triangle's altitude and a big rectangle surrounding it.* Do you see how the area of this big rectangle is $b \cdot h$?

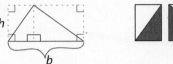

When we split our diagram up, it's easy to see that in each of these two rectangles, the shaded triangle part represents exactly half of the area, right? And that means when we put 'em together, the triangle part is still *half of the total area*. We've just shown that the area of any triangle is one half of the product of its base and its altitude (height). Nice!

$$\text{Triangle Area} = \frac{1}{2}(b \cdot h)$$

Notice that the height is the same as the *altitude*; it must make a right angle with the base. For example, to find the area of $\triangle RED^\dagger$ below, there's no altitude drawn to $\overline{RD}$, so $\overline{RD}$ isn't a good base to use. But guess what? This is a right triangle, so the two legs *are* altitudes. (They each go from a vertex to make a right angle with the opposite side, after all!)

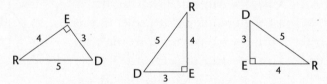

So we can either think of $\overline{ED}$ as the base and $\overline{RE}$ as the height, or vice versa. Either way, we get the same answer:

$$\text{Area} = \frac{1}{2}(b \cdot h) = \frac{1}{2}(4 \cdot 3) = \frac{12}{2} = 6$$

Notice that since we have the area, we can now *find the length of the altitude* going from E to $\overline{RD}$. How? Well, let's draw it in (we'll call the new point "A"), and, using $\overline{RD}$ as the base and $\overline{EA}$ as the height, we can write this expression for the area of $\triangle RED$:

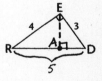

$$\text{Area} = \frac{1}{2}(b \cdot h) = \frac{1}{2}(5 \cdot EA) = \frac{5}{2}EA$$

. . . .

* FYI, the Parallel Postulate (see p. 389 in the Appendix) is what "allows" us to draw these sides. Notice that each side passes through a vertex of the triangle and is ∥ to either the base or the altitude of our triangle. So the rectangle sides are indeed ⊥ to each other.

† I guess I'm still thinking about red raspberries. So fresh, so tangy . . .

But we already know the area equals 6! So that means we can write: $6 = \frac{5}{2}EA$. Multiplying both sides by 2, and then dividing by 5, we get: $12 = 5EA \rightarrow EA = \frac{12}{5} = 2\frac{2}{5}$ or **2.4**, and that's the length of the altitude we drew! Pretty sneaky, huh?

Areas of Regular Polygons

Polygons, as we saw in Chapter 12, can be divided into triangles by drawing in diagonals from a *single vertex* to all the other vertex points. If we want to find *area*, however, it's usually more helpful to divide a polygon into triangles by drawing lines from the *center point* to its many vertex points. Let's do that for this regular pentagon, and because there aren't any right angles yet, we'll draw in an altitude.

Notice that the area of the triangle $\triangle REC$ is $\frac{1}{2}bh = \frac{1}{2}(RC) \cdot a$, which we could rewrite as: $\frac{1}{2}a \cdot$ *(one side of pentagon)*. Why not? It's a free country.

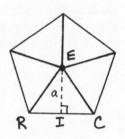

But that's just the area of *one* triangle. In this pentagon, there are 5 triangles total, so the total area would be $\frac{1}{2}a \cdot$ **(one side of pentagon)** • **5**. But hey, what's 5 times one side of a pentagon? It's the entire *perimeter*, isn't it? So we can rewrite this formula as $\frac{1}{2}a \cdot$ **(perimeter)**. And guess what? Since we could have used *any* number of sides, this formula works for *any* regular polygon:

Area of Regular Polygon: $\frac{1}{2}a \cdot p$

...where a is our little altitude, and p is the perimeter of the polygon.

QUICK NOTE   Truth be told, that little altitude we drew in the polygon above has a special name: *apothem*. That's technically what the little "a" stands for in the area formula above, but I like to think of it as "altitude." So sue me, okay?

Hexagons Are Special

Regular hexagons are actually pretty special, and here's why: They're made up of six equilateral triangles—meaning that every angle equals 60°! This makes sense, because we saw on p. 206 that each interior angle of a regular hexagon equals 120°, and those are getting split in half.

And if we look at the center point, the 360° gets split into six parts, right? (And after all, 360° ÷ 6 = 60°!)

Now if we draw in little altitudes for each of them, we get 12 little 30°-60°-90° triangles, which as we know, are *really* special.

QUICK (REMINDER) NOTE   If we know <u>one</u> side of a 30°-60°-90° or 45°-45°-90° triangle, we can find the other sides! (See pp. 187–9 to review this.)

Let's see some of this hexagon business in action...

***In the YAMS diagram, if YS = 4 inches, what is the perimeter <u>and</u>
the area of the regular hexagon?***

If we knew the length of $\overline{AS}$, then we'd just multiply it by 6 to get the full perimeter, right? Well, this
is a hexagon, so that means it's made up of equilateral triangles, one of them being △*YAS*! So because
we know *YS* = 4, that means *AS* = 4, too. Nice. A regular hexagon has six of these equal sides, so the
perimeter must be 6 × 4 = **24 inches.**

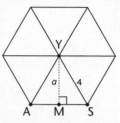

Now let's figure out the area. Notice the equilateral triangle △*YAS*
above has been split into two 30°-60°-90° triangles. (That's what happens
when we draw an altitude in an equilateral triangle.) Let's focus on the
half that makes up △*YMS*. We know the hypotenuse of △*YMS* equals 4, so
the shortest leg must equal 2, right? And so the longer leg must be $2\sqrt{3}$.
(See p. 189 to review this.)

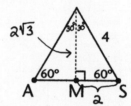

This means we've found the altitude of the equilateral triangle
△*YAS*: **$a = 2\sqrt{3}$** . Yay! We also know the base (*AS* = 4), so we can now find
the area of △*YAS*: A = $\frac{1}{2}(b \cdot h)$ = $\frac{1}{2}(4 \cdot 2\sqrt{3})$ = $\frac{1}{2}(8\sqrt{3})$ = **$4\sqrt{3}$**. And since
there are six of these triangles that make up the hexagon, the **total area**
= 6 · **$4\sqrt{3}$** = **$24\sqrt{3}$ square inches.**

By the way, after we found the altitude *a*, we could have used our new
polygon formula: Area = $\frac{1}{2}ap$, but it's really the same amount of work.

Watch Out!

Regular hexagons are the *only* regular polygons that can be
divided into equilateral triangles like we saw above. So just
because we know the sides of a regular polygon doesn't mean
we can assume anything about the distance from the center to
a vertex, or the distance from the center to the middle of a side.
And we certainly can't count on getting 30°-60°-90° triangles.
Remember why regular hexagons can get split up like that, and
you'll do great!

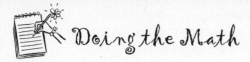

Doing the Math

Do these area problems for circles, triangles, and polygons. Leave π in your answers, wherever applicable. I'll do the first one for you.

1. Pizza A has twice the radius of Pizza B. What is the ratio of their areas?

<u>Working out the solution</u>: Hmm, no numbers, eh? Well, let's label Pizza B with radius b, and since Pizza A's radius is twice as long, we'll call its radius $2b$. Now, what's an expression for the area of Pizza B? Area $= \pi r^2 = \pi b^2$, right? And we could write Pizza A's area as: $\pi r^2 = \pi(2b)^2 =$ **$4\pi b^2$**. Then the ratio of their areas is: $\frac{\text{Area of Pizza A}}{\text{Area of Pizza B}} = \frac{4\pi b^2}{\pi b^2}$. We can reduce this fraction to $\frac{4}{1}$, in other words, 4 : 1. Done!

Answer: The ratio of their areas is 4 : 1.

2. In the diagram SLICE, $\overline{LC}$ and $\overline{IE}$ are altitudes of the triangle $\triangle SCI$. If $SI = 12$ mm, $SC = 9$ mm, and $LC = 3$ mm, find the area of $\triangle SCI$ and the length of $\overline{IE}$. (Hint: See pp. 347–8.)

3. A rectangular pizza's dimensions are 4 inches by 16 inches. A circular pizza has the same area; what must its radius be?

4. The radius of Pizza A is three times the radius of Pizza B. What is the ratio of their areas?

5. If a 14-inch-diameter pizza is cut into 10 equal slices, how much area is *each slice*?

6. Find the area of a regular hexagon with a perimeter of 60 feet. (Hint: Draw a picture. What is the length of each side? Drop an altitude and fill in the rest of the 30°-60°-90° triangle you create.)

7. The diagonals of a square measure $5\sqrt{2}$ cm each. What is the area of the square? (Hint: Use 45°-45°-90° triangles to find the square's sides.)

8. If the distance from the center of a regular hexagon to one of its sides is $7\sqrt{3}$ miles, then what are the perimeter *and* the area of the hexagon? (Hint: Draw a picture!)

9. The arc length of the crust on a slice of pizza is π
inches.* If a protractor tells us the angle of this slice is
30°, what is the length of each *side* of the slice? *(Hint: This
is the same as the radius. Use the arc length formula to find
the diameter and then the radius.)*

10. The arc length on a (different) slice of pizza is π inches. If a
protractor tells us the angle of this slice is 60°, what was the area of
the entire pizza? What is the area of just this slice? *(Draw your own
picture this time!)*

(Answers on p. 402)

More Quadrilateral Area Formulas: Squishing a Box

The rest of the quadrilateral family's areas can be figured out by, well,
drawing rectangles and triangles!
   Let's take a rectangle with base 10 and height 4. Now let's imagine
that the corners are hinges that bend so it becomes a parallelogram. The
more we squish it down, the smaller the area gets. (See the first two diagrams
below.)

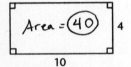 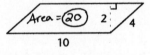 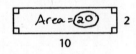

Believe it or not, the <u>formula</u> for the area of a parallelogram is actually the same as for a rectangle: $b \cdot h$. The big difference is that we have
to make sure we really use the <u>height</u> for "h" (and
not the length of a side).
   The middle parallelogram above whose height
is 2 has the *same exact area* as the rectangle with
height 2. Just imagine that we cut off the triangle
on the right and stuck it on the left. Voilà!

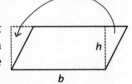

Parallelogram Area = $b \cdot h$

Because Rhonda is Paris' kid, we can use this formula for the area of
rhombuses, too.[†]

. . . .

* Remember, π is just a number! So π inches is just a little more than 3 inches.

† And, of course, this formula also works for Rex and Sara. See Chapter 16 if you
don't know who any of these people are.

How about a kite? Let's draw a kite with diagonals called d_1 and d_2, and then fill in a rectangle around it, using the same lengths: d_1 and d_2.
The area of the big rectangle is just $d_1 \cdot d_2$, right? Do you see how the kite has exactly *half* the big rectangle's total area? (Just look at the top half, and you'll see a similar diagram to our triangle-in-a-rectangle from p. 347. Same thing with the bottom half!)

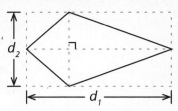

$$\text{Kite Area} = \frac{1}{2}(d_1 \cdot d_2)$$

Since Rhonda is Kit's kid, we can also use this formula for rhombuses, or even squares!*

And how about trapezoids? Let's draw a trapezoid with bases called b_1 and b_2. Now we'll draw a median,† cut off the bottom triangles, rotate 'em, stick 'em up top, and poof! We get a rectangle that has the <u>same area</u> as the trapezoid. So now we just need to figure out the area of

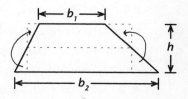

the rectangle. Well, the height of the rectangle is the same as the height of the trapezoid, right? And the base of the rectangle is the same as the median we drew, and *that* length is just the average of the two bases: $\frac{b_1 + b_2}{2}$. So that means the area of the rectangle, and also of our trapezoid, is:

$$\text{Trapezoid Area} = \frac{b_1 + b_2}{2} \cdot h$$

And this makes sense, because using the smaller base would result in an answer that would be too small, right? And using the bigger base would result in an area that would be too big. The average of the two bases? Perfect! And of course, this works just fine if the trapezoid happens to be isosceles. So we've covered all of our special quadrilaterals. Nice.

Recap: All of Paris' descendants (kids and grandkid) can use the regular 'ol **b • h** as long as h is really the *height*, meaning that it's perpendicular to the base. A triangle is just half of a rectangle, so that's $\frac{1}{2}b \cdot h$. Kit and his descendants have those wonderful perpendicular diagonals, which means we can use 'em as the "base and height" for the

. . . .

* For squares, the two diagonals are equal, so $d_1 = d_2$, but the formula works just fine.

† Trapezoids have medians, too, and they divide their sides into two congruent segments, just like the medians of triangles. (See p. 155 for medians of triangles.)

triangles inside that make them up: $\frac{1}{2}d_1 \cdot d_2$. Tracy's (and Isabel's) area is almost like Paris' clan, except that she uses an average of the bases, just to make sure her life doesn't have any extremes: $\frac{b_1 + b_2}{2} \cdot h$. Not so bad, right?

⟶❧⟵

Keep in mind *why* these formulas are true, and they won't be hard to learn. You're a total rock star, by the way. Have I mentioned that lately?

QUICK NOTE   Notice that the key to areas of triangles, quadrilaterals—and all polygons, really—is that we need *right angles*. Most of the time that means multiplying a *base* times an altitude (*height*), or two diagonals that cross at right angles. So if we want to find the area of a shape that doesn't have a helpful right angle, then we can draw in an altitude, or diagonals (but only if we know they're perpendicular).*

Complicated Shapes . . . and Frankenstein

Maybe it's because I'm an actress (so I'm partial to playing different characters), but I really love dressing up for Halloween. What's your favorite costume? A superhero of some sort? The good 'ol black cat? Or maybe even Frankenstein?

I'm going to guess your answer isn't "Frankenstein." But just pretend that it is, okay? You see, Frankenstein is made up of a bunch of smaller parts, all stitched together, and that's the best way to look at the area of complicated shapes: We unstitch the complicated shape into smaller, easier shapes. And who doesn't like smaller and easier?

We've already done a bunch of area problems so far, so here's a nice little step-by-step strategy that summarizes what we've been doing and incorporates our new Frankenstein† perspective.

· · · · · · · · · · · · · · · · · · ·

* Kit and his kids have ⊥ diagonals (kite, rhombus, square).

† I had these two girlfriends in high school whose last names were Frank and Stein, and we used to say, "Look! It's Frank 'n Stein!" They didn't mind. In fact, I think they're the ones who came up with it.

Finding the area of complicated shapes:

Step 1. Within the complicated-looking shape, look for easy shapes like rectangles, triangles, slices of circles, etc. There are two basic types of area problems: ones where we subtract a shape from another shape, and the Frankenstein ones where we divide up the figure into smaller shapes and then add 'em all up at the end. Try drawing in altitudes or other straight lines (like diagonals) to separate out the shapes and help find right triangles.

Step 2. If there are still key pieces of information missing, try using subtraction for rectangle info, look for right triangles (especially 30°-60°-90° and 45°-45°-90° triangles—see p. 187–8 to review), and/or use the Pythagorean Theorem to fill in any missing lengths.

Step 3. One at a time, find the areas of the Frankenstein parts. (Make sure to use radii for the areas of circles, and not diameters!)

Step 4. Add 'em up! Or if there's subtraction to be done, subtract off the parts we don't want. Done!

And... Action! Step By Step In Action

Let's say we found this picture of a lipstick tube in a magazine. It's made up of a 25-mm by 18-mm rectangle and 40 degrees of a circle. If we wanted to cut it out and fill it in with glitter, what total area, in square millimeters, would we be filling?

Steps 1 and 2. It's Frankenstein time! We've got a rectangle and a "slice" of a circle. No problem.

Step 3. The area of the rectangle is just $18 \times 25 =$ **450 mm^2.*** The area of the partial circle is just the *total* circle area, times the *fraction* of the circle that it is. Notice that the width of the rectangle is the radius of the circle sector. So, because the radius is 18, the total circle area would

. . . .

* I did that in my head. You know how? I pretended the 25 was 100, and I got 18 × 100 = 1800. Easy, right? But I knew my answer was 4 times what it should be. First I divided by 2 (and got 900), and then I divided by 2 again and got 450.

be $A=\pi r^2 = \pi(18)^2 = 324\pi$. Since we've only got 40° of the circle, the "slice" is $\frac{40}{360} = \frac{1}{9}$ of the entire circle's area, right? So the area of the slice is: $\frac{1}{9}\times 324\pi = \frac{324\pi}{9} = \mathbf{36\pi}$.

Step 4. Add 'em together, and we get: **450 + 36π**.

Answer: 450 + 36π mm²

By the way, if we wanted to approximate this answer, we could substitute 3.14 for π and get approximately 563.04 mm², or 5.63 cm².*

Take Two: Another Example

In the WALNUT diagram, find the total area of the trapezoid WANT.

Steps 1 and 2. Hmm, we can't use the trapezoid formula right off the bat; we'd need the height. Either *AL* or *NU* would work, but we don't have them. Let's attack this, Frankenstein-style! We'll divide this

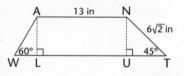

shape into two triangles and a rectangle.† And we should notice that both triangles are special; △*WAL* has to be a 30°-60°-90° triangle,‡ and △*NUT* is a 45°-45°-90° triangle. The hypotenuse of △*NUT* is $6\sqrt{2}$, so we know that each leg must be 6, and that means *NU* and *UT* = 6. Great! We have our height, and we can fill that in for *NU*, *UT*, <u>and</u> *AL*. Since *AL* = 6, that means $WL = \frac{6}{\sqrt{3}}$, so we can fill that in, too. Also notice that because ANUL is a rectangle, *LU* = *AN* = 13.

Step 3. Time to find some areas! The rectangle is easy: $bh = 13 \times 6 = \mathbf{78}$. And △*WAL*: $\frac{1}{2}bh = \frac{1}{2}\cdot\frac{6}{\sqrt{3}}\cdot 6 = \frac{36}{2\sqrt{3}} = \frac{18}{\sqrt{3}}$. But answers shouldn't have square roots in the denominators, so we'll multiply this times the copycat

. . . .

* Wondering why the conversion isn't 56.3 cm²? See pp. 359–60.

† We know that $\overline{AN} \parallel \overline{LU}$ (the definition of trapezoid), so because "if ∥, then alt. int. ∠'s are ≅" (or even because "if ∥, then same side int. ∠'s are supp."), we could prove the other two angles must be 90°, too, which means ANUL is a rectangle!

‡ Remember, triangles' angles must add up to 180°. So, for example, we already see 60° and 90°, and that means ∠*WAL* has to be 30°.

fraction* $\dfrac{\sqrt{3}}{\sqrt{3}}$ and get: $\dfrac{18\cdot\sqrt{3}}{\sqrt{3}\cdot\sqrt{3}} = \dfrac{18\sqrt{3}}{3} = \mathbf{6\sqrt{3}}$. Nice! The area of the other triangle, $\triangle NUT$, is $\dfrac{1}{2}bh = \dfrac{1}{2}\cdot 6 \cdot 6 = \dfrac{36}{2} = \mathbf{18}$.

Step 4. Adding up our Frankenstein parts: $78 + 6\sqrt{3} + 18 = \mathbf{96 + 6\sqrt{3}}$. Done!

Answer: Area of the trapezoid is $96 + 6\sqrt{3}$ square inches.

Once we got the height, **6,** and the full length of $\overline{WT}$, $\dfrac{6}{\sqrt{3}} + 19$, we could have plugged those into the trapezoid area formula on p. 353. Try it. You'll see that it involves a complex fraction. Personally, I think this was the easier way to do it! Plus, we got to see Frankenstein in action.

"*I* will need this type of thinking in my future career as a businessperson, whether in building houses or anything else." Branden, 15

Take Three: Yet Another Example

Find the missing side and the area of this trapezoid.

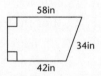

58in
34in
42in

On p. 184, we drew in an altitude and used subtraction and good 'ol Pythagoras to find that the missing side is **30 inches**. Since the missing side is also the height, now we can find the area! We can either do it Frankenstein-style as a rectangle and triangle, or we can use the trapezoid formula from p. 353, using 30 for h. Both give the same answer.

Trapezoid area $= \dfrac{b_1 + b_2}{2} \cdot h = \dfrac{(58 + 42)}{2} \cdot 30 = 50 \cdot 30 = \mathbf{1500}$.

Answer: Missing side = 30 in.; Area = 1500 in.^{2}

. . . .

* See the footnote on p. 186!

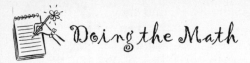

Doing the Math

Do these area problems. Leave π in your answers, where applicable.
I'll do the first one for you.

1. A regular hexagon whose perimeter is 24 inches is
inscribed* in a circle. Find the shaded area.

<u>Working out the solution</u>: We'll need the area of the cir-
cle and the area of the hexagon, and then we'll subtract
them. If the perimeter of the hexagon is 24, then each
side must be 4, right? We found the area of this hexagon on p. 350; it's **24√3 in².** To find the area of the
circle, let's draw in a radius to a vertex. Since hexagons
are made up of equilateral triangles, the radius is 4! So
the circle's area is A = πr^2 = $\pi(4)^2$ = **16π,** which means the
shaded area is: circle area − hexagon area = **16π − 24√3.** There's
no way to simplify that, so we're done!

Answer: 16π − 24√3 square inches

2. Find the total area of the lopsided house
figure, ALMOND. *ALND* is a rectangle, and
LMON is a trapezoid.

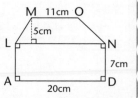

3. For the square/circle diagram to the right, if the <u>diag-
onal of the square</u> measures $2\sqrt{2}$ inches, what is the
shaded area?

4. A rhombus's area is 25 cm², and one of its diagonals is twice the
length of the other. What are the lengths of the diagonals? *(Hint:
Draw a picture, label the shorter diagonal "x" and the longer one "2x,"
and use the kite formula on p. 353.)*

. . . .

* When a regular polygon is inscribed in a circle, all of its vertices lie on the circle's
rim.

5. For the FISH diagram to the right, if the short diagonal of the kite, $\overline{IS}$, measures 3 in. and the diameter of the circle, $\overline{FH}$, measures 10 in., then what is the shaded area? *(Hint: This is multistep; keep your brain on!)*

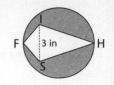

6. Find the area of the parallelogram PEAS. *(Hint: Draw in the altitude from E down to $\overline{PS}$.)*

7. Find the area of this trapezoid. *(Hint: We saw this diagram on p. 185.)*

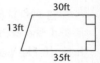

8. A circle is drawn inside a regular hexagon so that each side is tangent to the circle. The circle's radius is 3 cm. What is the distance from the center to a *vertex of the hexagon*? What is the area of the shaded region? *(Hint: This is different from #1! Draw one of the equilateral triangles like we did on p. 349 to see what's going on.)*

(Answers on p. 402)

(Answers on p. 402)

What's the Deal?

Converting Square Units

To go from millimeters to centimeters, we just divide by 10, right? For example, 50 mm = 5 cm. No big deal. However, *square* units are a different story. To convert from square millimeters to square centimeters, we'd have to divide by 10^2; in other words, we'd divide by <u>100</u>. Imagine we had a rectangle 30 mm tall and 20 mm wide. The area would be 30 mm × 20 mm = **600 mm²**, right? Then, if we wanted to convert the answer to square centimeters, it would be **6 cm²**, not 60 cm². To see why this is true, just convert from mm to cm *before* we multiply: 3 cm × 2 cm = 6 cm², after all!

When we leave the metric system, it gets more complicated. Converting from inches to feet, we'd divide by 12, right? For example, 24 in. = 2 ft. But to convert from square inches to square feet, we'd divide by 12^2; in other words, we'd divide by <u>144</u>. Yikes! So, if you know you'll need to convert your answer at the end, I highly recommend converting your units *before* doing the multiplication and you can avoid the whole "converting square units" thing.

Takeaway Tips

 Remember the circle formulas like this: Crust is pie dough: $C = \pi d$, and the Area inside is red raspberry pie: $A = \pi r^2$.

 The area of parallelograms (including rectangles) is: $A = bh$, where the height (h) must always be $\perp$ to the base (b). The area of a trapezoid is like a parallelogram, but using the *average* of the bases: $A = \dfrac{b_1 + b_2}{2} \cdot h$.

 A triangle is *half* a rectangle, so its area is: $A = \dfrac{1}{2}bh$.

 The area of any quad whose diagonals are $\perp$ can be found using: $A = \dfrac{1}{2}(d_1 \cdot d_2)$.

 The area of any regular polygon is $A = \dfrac{1}{2}ap$, where p is the perimeter, and a is the altitude from the center to a side (in other words, the *apothem*).

 Regular hexagons are special: They can be divided into six equilateral triangles, or twelve 30°-60°-90° triangles!

 For complicated areas, draw in altitudes and look for 30°-60°-90° and 45°-45°-90° triangles.

TESTIMONIAL

Fernmarie Brady

(Seattle, WA)
<u>Before</u>: Constantly moving schools
<u>Now</u>: Designs airplanes—and floats in zero gravity, too!

I had a pretty rough childhood. But I do recall a loving mother who always supported me and my brothers as she worked hard to raise four children by herself.

"I struggled to keep up."

However, in an attempt to seek a better lifestyle for our family, my mom had to move every few months. For much of my childhood, I changed schools at least twice every school year.

Changing schools so frequently meant that I often missed important topics; I struggled to keep up! I remember being in a trigonometry class and converting fractions into decimals in order to solve my test problems, because I didn't know how to reduce fractions (I eventually did learn). Having to work twice as hard as my classmates in order to keep up may have been difficult at times, but no matter how hard things got, I held onto a secret wish I'd had since I was 6 years old: I dreamed of someday floating in space, like astronauts do. All throughout school, I believed that if I worked hard enough in math and science, I could make that dream a reality.

I ended up graduating from college as an engineering major, and today I am an Avionics Systems Engineer Lead! I design and develop improvements for the next generation of airplanes, and I use math to simulate aircraft trajectories, create devices that take aerial imagery, and more. When we consider how an airplane will respond to wind, for example, we need geometry and trigonometry to calculate the amount of perpendicular wind speed (called crosswind) created by wind hitting the airplane at an angle.

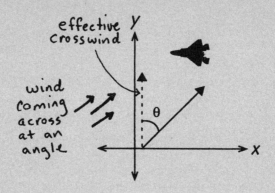

Effective crosswind = Wind Speed × cosΘ*

    I love helping society by creating new and exciting technologies; it makes me feel smart and powerful! And did the 6-year-old girl get her wish? Yep! I work with a recreational company that allows me to float in zero gravity several times a year!

    We get onboard a 727 aircraft, fly up to 35,000 feet, and use math and physics to fly a path of vertical parabolas, which generate centrifugal forces that cancel gravity, allowing the passengers (and me) to float inside the plane. It's so much fun!

    Having a dream helped me get where I am today—along with the ability to keep motivated when things got tough. Those two ingredients are the keys to success, and we <u>all</u> have the power to give those things to ourselves. In fact, nobody else *can* give them to us. What are your dreams? I can attest that having a dream worked for me, and I am certain that it will work for you, too!

* Don't worry if that equation doesn't make sense to you now. After you take trigonometry, it will!

Chapter 21

It's a Wrap!

Surface Area and Volume

I love wrapping presents, don't you? The shiny paper and sparkly bows, knowing we're about to give someone that exciting anticipation of wondering what's inside . . . Finding the surface area and volume of 3-dimensional objects involves a whole bunch of formulas—and wrapping paper is going to help us remember them!

What Are They Called?

Surface area, volume

The **surface area** of a 3-dimensional solid is the <u>sum of the areas</u> that enclose the solid. In other words, it's the sum of all of the areas that would be covered by wrapping paper. Remember, surface area is *area*: It's a 2-D measurement on a 3-D object. So we use square units like in.^{2}, cm^2, ft^2, etc.

On the other hand, the **volume** of a 3-dimensional solid is the space *inside* the solid—all the stuff inside that's being wrapped up! Volume is a 3-D measurement, so we use cubic units like in.^{3}, cm^3, ft^3, etc.

You know those foil-wrapped chocolate Easter bunnies? The best kind are solid chocolate (not hollow), and the *volume* of the shape tells us how much actual chocolate we're getting.* The *surface area* of the shape would tell us the outside area that is covered by the shiny foil. For now, we'll focus on surface area. To find the surface area of *any* 3-D object, just remember this:

Surface area of a solid = Area of wrapping paper needed to cover it

For example, to find the surface area of a **cube** whose edges each measure 3 inches, we'd just find the area of one of the squares and multiply it by 6, because there are 6 square faces total (just think about dice!). As we saw in Chapter 20, the area of one square = s^2. Here, that becomes $3^2 = $ **9**. So, the total surface area = $9 \times 6 = $ **54 in.^{2}**

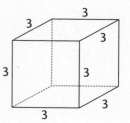

Reality Math

So you've gotten an enormous ice-cream cone for your sister—well, it's her birthday and you're just a sweetie like that. You're innocently waiting for her to join you when this snotty girl from school walks up to you and says, "Um, you better hurry up and eat that. My friend got that same exact ice-cream cone yesterday. She didn't eat it fast enough, and it totally overflowed, filling the whole cone and spilling everywhere." She walks away with her nose in the air. Luckily, you brought your ruler. A quick measurement shows that the radius of the ice cream is 3 inches, the radius of the cone is 2 inches, and the height of the cone is 5 inches. Is the snotty girl telling the truth? Let's get some more math under our belts, and we'll solve this on p. 378.

Time to learn some 3-D solids!

. . . .

* Beauty tip: Instead of eating the entire thing, share your chocolate bunny with friends. Your skin will thank you.

What Are They Called?

Prism, Cylinder, Pyramid, Cone, Lateral face, Slant height

A **prism*** is a solid defined by two congruent, parallel Bases† and a height. The Bases can be *any* polygon—a square, rectangle, pentagon, hexagon, etc.

EXAMPLES OF PRISMS

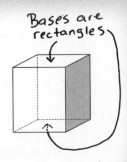

Bases are rectangles

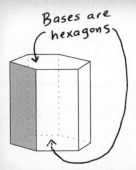

Bases are hexagons

All of the non-Base faces of a prism are called the **lateral faces**. (Two examples of lateral faces have been shaded.) By the way, a cube is an example of a prism whose Bases *and* lateral faces are all squares!

A **cylinder** is just like a prism, except that its Bases are *circles*. This is the general shape of most soda cans. A cylinder has no lateral faces because it's sort of got just one big, curved lateral "face," doesn't it? (See p. 370 for more.)

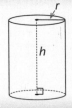

* Worst joke ever: Q: Where do bad math students go? A: To *prism*. Get it? Sounds like *prison*. Yeah, it's pretty bad . . .

† We'll use a capital B for the 2-dimensional Bases that have area, so we don't confuse them with the 1-dimensional *bases* (line segments) we've seen up till this point. By the way, for a formal definition of *parallel planes*, see p. 388 in the Appendix.

A **pyramid** is a solid defined by a single Base and a height that is ⊥ to the Base. The Base can be *any* polygon. It also has *triangular lateral faces* and a pointy top! (Two of the lateral faces have been shaded on the pyramids below.)

EXAMPLES OF PYRAMIDS

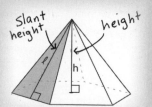

The **slant height** on pyramids is the, um, slanted height. And it also happens to be the *altitude* of the triangle faces on pyramids, which means it's ⊥ to the edge of the Base. That's what the little slanty right-angle box means! We use the letter *l* to denote slant height. Probably because it's l̲eaning, y'know?

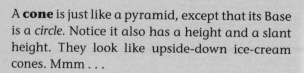

A **cone** is just like a pyramid, except that its Base is a *circle*. Notice it also has a height and a slant height. They look like upside-down ice-cream cones. Mmm . . .

(We'll look at **spheres** on pp. 377–9.)

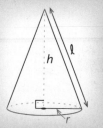

QUICK NOTE   Unless I say otherwise, all the prisms, pyramids, cylinders, and cones will be *right*, which means if we draw its height from the center of the bottom Base, it will pass through the center of its top Base (or through the top vertex, in the case of pyramids and cones), and it'll make a <u>right angle</u> with the <u>center</u> of the bottom Base, as you can see on the previous page and also at the bottom of this page. All of the solids we've seen so far in this chapter have been *right*.

Just for comparison's sake, the two solids below are *not* right; in fact, they're called *oblique* (pronounced, "Oh, bleak").*

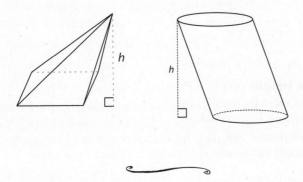

QUICK NOTE   For right prisms and cylinders, the slant height (*l*) is equal to the height (*h*), which makes sense, right?

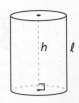

To find the *surface area* of prisms and pyramids, we just add up the areas of the Bases and lateral faces (the rectangles or triangles)—all the area that would be covered by wrapping paper. That's it! Here's the Step By Step.

* In oblique prisms and cylinders, the Bases are still parallel to each other, though!

CHAPTER 21: IT'S A WRAP!   367

Step By Step

Finding the surface area of prisms and pyramids:

Step 1. If it's not given, sketch the object, and also draw the shape "spread out" into its different flat parts. Be sure to label the info we've been given.

Step 2. Find the area of the Base(s), using the techniques we learned in Chapter 20.

Step 3. Find the area of one of its lateral faces. If it's a prism with a polygon base, then these lateral faces will be rectangles, whose area, **bh**, is found using the height of the prism for h and an <u>edge</u> of the prism's Base for b. If it's a pyramid, then the lateral faces will be triangles, whose area is found using the *slant height* (l) of the pyramid and an <u>edge</u> of the pyramid's Base for b, so we'd use $\frac{1}{2}bl$ for each triangular face. It can help to draw the open pyramid like a flower! (See p. 374 for an example.)

Step 4. If we're dealing with a right, regular prism or pyramid, then once we have the area of one of its lateral faces,* we multiply that times the *number* of lateral faces on the solid to make sure we don't leave any surface unwrapped on this package.

Step 5. Add 'em all up. Done!

QUICK NOTE   The lateral surface area of a solid is all of the surface area that *doesn't* come from a Base.

. . . .

* If it's not a right, regular pyramid or prism, that means the lateral faces aren't necessarily identical to each other, so we'll just find each lateral face individually. We do an example of this next!

And... Action! Step By Step In Action

Find the surface area of a prism with a height of 7 ft and whose Base is a scalene triangle with lengths 3 ft, 4 ft, and 5 ft.

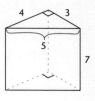

Steps 1 and 2. Let's draw this! Noticing that we have a 3-4-5 triangle, we know there's a right angle opposite the "5" side. That makes the area of the Bases easy to find; they're each $= \frac{1}{2}bh = \frac{1}{2} \cdot 3 \cdot 4 = \textbf{6}$. There are two Bases on a prism, so that's $2 \cdot 6 = \textbf{12}$ for the area of our two Bases. So far, so good?

Steps 3 and 4. To find the lateral surface area, we can either find the three rectangle areas separately and add them up (in this case, $21+28+35 = \textbf{84}$), or we can even imagine the single, big rectangle of "wrapping paper" that would do the same job (we'd add $3+4+5 = 12$, and then multiply $bh = 12 \cdot 7 = \textbf{84}$).

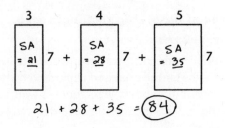

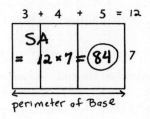

Either way, that's **84** for the lateral faces.

Step 5. Now we add the total Base area, 12, to the total lateral area, 84, and we get a grand total of $12+84=\textbf{96}$. Done!

Answer: The total surface area of the prism is 96 ft^2. That's a lot of wrapping paper!

Toilet Paper and Ice Cream: Cylinders and Cones

That combination sounds . . . messy. Hmm. Anyway, let's think about the surface area of a cylinder. Let's say you wrapped up the cardboard center from a roll of toilet paper for your brother's birthday. How special. To cover it in wrapping paper, you'd need two circles for the two Bases, right? And then the rest of it would be just one big rectangle whose height is the exact same as the original cylinder. Notice that the *width* of the rectangle would be the same as the <u>entire circumference</u> of a circle Base:

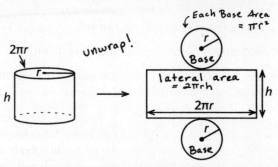

Total S.A. of cylinder = 2(area of Base) + big rectangle area
$$= 2(\pi r^2) + 2\pi rh$$

There's no need to memorize this "formula." Seriously, as long as we remember our red raspberry pie formulas for the area and circumference of a circle, we'll be fine. The key is understanding how the rectangle's width is the same as the circumference of the Base, $2\pi r$,* and then it's easy to re-create!

On the other hand, if we unwrap a cone (like those drumstick ice-cream cones), then we get something that looks like a big fan. Interestingly, this "fan" has the *same area* as a triangle with Base $2\pi r$ and height l, which is $A = \frac{1}{2}bh = \frac{1}{2}(2\pi r)l = \pi rl$. (To see the proof of this "fan" formula, see GirlsGetCurves.com.)

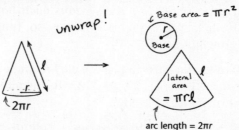

Total S.A. of a Cone = area of Base + area of fan
$$= \pi r^2 + \pi rl$$

....

* Which as you know, is the same as πd.

The Base part of the formula, πr^2, is just the red raspberry pie—easy, *right*? And since π is spelled "pi," the "fan" part of the formula, $\pi r l$, sort of spells "pirl," which sounds like "pearl"! Just another way to remember it. I'd sure love to have a fan decorated with pearls; wouldn't you?

Step By Step

Finding the surface area of cylinders and cones:

Step 1. Draw a picture if there isn't already one given.

Step 2. Find the area of the circular Base, using the area formula $A=\pi r^2$. A cylinder has two of these; a cone has only one.

Step 3. If it's a cylinder, then the lateral surface area is one big rectangle, whose long base is the same length as the circumference of the circle: $2\pi r$! So the area of the big rectangle is: $A=bh=2\pi rh$. If it's a cone, then the lateral surface area is a fan, whose area is $\pi r l$; remember the pearl fan . . .

Step 4. Add together the lateral surface area and the area of the Base(s). Done!

And... Action! Step By Step In Action

Find the surface area of this cylinder with diameter = 12cm and height = 8cm.

Steps 1 and 2. Okay, if the diameter of the Base is 12, then the radius is 6, right? And that means each Base area$=\pi r^2=\pi(6)^2=36\pi$. There are two bases, so that's $2(36\pi)=$ **72π** for the Bases.

Step 3. The lateral surface area is just the big rectangle whose *width* is the circumference of the Base, $2\pi r=2\pi(6)=$ **12π**. Its height is 8, so the lateral surface area is just: $bh=(12\pi)8=$ **96π**.

Step 4. Total S.A. = Area from Bases + Lateral S.A., and that's: $72\pi+96\pi=$ **168π**. Done!

Answer: 168π cm^2

Take Two: Another Example

Find the surface area of a cone with height 9 cm, whose Base has a diameter of 12 cm.

Steps 1 and 2. We've got a picture, and this problem starts like the previous one: The area of the Base is **36π cm^2**. There's only one Base, so this step is done.

Step 3. In order to get the lateral surface area of this cone, we need the <u>slant</u> height. Since the height of the cone makes a right angle with the Base, let's draw the right triangle formed by the height, slant height, and radius of the Base, and use good 'ol Pythagoras to *find* the slant height. We know that $9^2+6^2=l^2$, right? So $l^2= 81+36=117 \rightarrow l=\sqrt{117}$, which can be simplified: $\sqrt{117} = \sqrt{9 \cdot 13} = 3\sqrt{13}$. Now that we have l, we can find the area of the "fan" (remember the pearls): $A = \pi r l = \pi \cdot 6 \cdot 3\sqrt{13} = 18\pi\sqrt{13}$ cm^2.

Step 4. Putting it all together, Total S.A.= Base area+fan area = **$36\pi+18\pi\sqrt{13}$ cm^2**.

Answer: The total surface area of the cone is $36\pi+18\pi\sqrt{13}$ cm^2.*

Watch Out!

One of the most challenging parts of finding surface area is just keeping everything straight! In other words, this is *not* the time to skip steps. Write out every formula you're using before you even stick in the values from the problem, and label everything you find, like "Area of Bases," "Lateral Area," and "Total S.A." Otherwise, believe me, it's easy to get confused and write the "answer" too soon—before all the parts have been put together. Also, because π is sometimes used for straight lengths now, it's easy to think that something you've written as the area for a rectangle is really the area of your circle Base. Just pay attention, label everything clearly, and you'll do great.

---

* For a "neater" answer, we could factor out 18π from both terms to get $18\pi(2+\sqrt{13})$, but it's fine to leave the answer the way it was, too.

Christmas Ornaments: Surface Area of Spheres

You know how the area of a circle is πr^2? Well, a **sphere** with the same radius has a surface area that's exactly four times the circle's area! In other words:

Surface Area of Sphere $= \mathbf{4\pi r^2}$

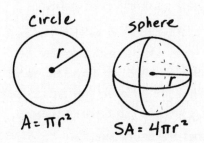

circle
sphere

r

$A = \pi r^2$
$SA = 4\pi r^2$

So if the radius of a sphere is 2, then its surface area is $4\pi r^2 = 4\pi(2)^2 = \mathbf{16\pi}$. Notice that if we put glitter on a circle to make a cute but simple Christmas ornament, we'd need exactly four times that much glitter to cover a *spherical* ornament with the same radius. Good to know!

☺

Doing the Math

Find the **total surface area** of the shapes described below. Leave π in your answers when applicable. I'll do the first one for you.

1. A pyramid with a slant height of 8 inches, and with a regular hexagon Base whose perimeter is 24 inches.

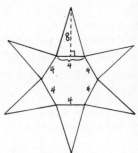

Working out the solution: Let's draw this: The Base is a hexagon, which has 6 sides, so each side must be $24 \div 6 = 4$ inches long, and it's a pyramid, so we can imagine opening it like a flower! Now we can easily see that it's just a hexagon and 6 skinny triangles.

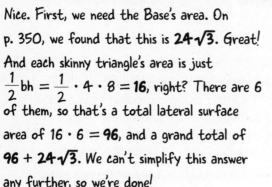

Nice. First, we need the Base's area. On p. 350, we found that this is $24\sqrt{3}$. Great! And each skinny triangle's area is just $\frac{1}{2}bh = \frac{1}{2} \cdot 4 \cdot 8 = 16$, right? There are 6 of them, so that's a total lateral surface area of $16 \cdot 6 = 96$, and a grand total of $96 + 24\sqrt{3}$. We can't simplify this answer any further, so we're done!

Answer: $96 + 24\sqrt{3}$ square inches

2. A pyramid with slant height 5 in., whose Base is a square with sides of length 5 in.

3. A pyramid with *height* 4 meters and whose Base is a square with sides of length 6 meters. See the picture. *(Hint: First use the shaded triangle to find the slant height. See p. 372 for an example with a cone.)*

4. A sphere with diameter 10 mm.

5. A prism with height 7 inches, whose Base has a total perimeter of 8 inches, and each Base area is 4 square inches.

6. The cylinder to the right, whose Base has a diameter of 5 meters and diagonal (as depicted) of 13 meters. *(Hint: First find the height of the cylinder.)*

7. The oversized crayon to the right, whose measurements are in feet. Remember, surface area is just the total area we need to cover with wrapping paper! *(Hint: Don't blindly use formulas.)*

(Answers on p. 402)

Turn Up the Volume

How do we measure the 3-dimensional space inside a solid object? With **volume**! In a lot of ways, finding volume is easier than finding surface area (and not just because we already know how to find the volume button on a phone). In fact, finding the volume of any prism or cylinder is pretty simple; it's just the area of the Base times the height:

Volume of prism or cylinder = Area of Base × height

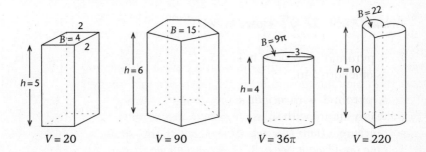

In other words:

Volume of prism or cylinder = *Bh*

As evidenced by the heart-shaped solid, the Bases don't even have to be polygons or circles; as long as we *somehow* know the area of the Base, *B*, the formula ***V=Bh*** still works.

Even more amazing is that these solids don't need to be *right*. Oblique prisms and cylinders use the same formula, **$V = Bh.$** Remember from p. 352 how we saw that the area of a parallelogram is just *bh*, the same formula as for a rectangle?

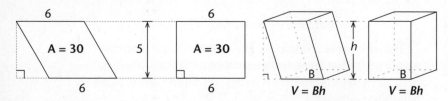

Well, the same kind of thing is true for oblique prisms and cylinders: If the Base area is *B* and the height is *h*, then no matter how slanted they are, the volume will always be *Bh*! Imagine that the above oblique solid were a really tall, slanted stack of paper (yes, it would probably fall over, but let's pretend it wouldn't). Then imagine if we were to push it back so that it's perfectly straight up and down. Think about it: The height wouldn't have changed, and the volume of actual paper wouldn't have changed, either!

That means if we find the *volume* of the straight stack of paper, we've also found the volume of the slanted stack. And voilà! The slanted stack's volume is *Bh*. Crazy, huh?

Our Pointy-Headed Friends: Pyramids and Cones

Imagine if a cone were inscribed in a cylinder so that it had the same Base and height as the cylinder. And imagine if a pyramid were inscribed in a prism in the same way. Obviously, our pointy-headed friends would have a smaller volume than their flat-top counterparts. But how much smaller?

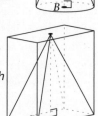

Looking at the cone/cylinder picture, it almost sort of looks like we could squeeze two upside-down, cone-like shapes on either side of the cone, right? And that sort-of-kind-of makes it look like three cones would fit inside the cylinder. Sort of. See what I mean? And even though this was a weird way to look at it, as it turns out, the cone's volume is indeed *exactly* $\frac{1}{3}$ of the cylinder's volume!*

....

* Once you take calculus, you'll have no problem proving that this formula works, but for now you'll just have to take my word for it.

This works for pyramids and prisms, too; a pyramid with the same Base and height as a prism will have exactly $\frac{1}{3}$ of that prism's volume.

Volume of pyramid or cone $= \frac{1}{3} \times$ **(Area of Base)** $\times$ **height**

In other words:

Volume of pyramid or cone $= \frac{1}{3} Bh$

Remember this cone from p. 372? It has a height of 9 cm and a base with a diameter of 12 cm. In order to find its surface area, we had to first find the slant height. But for volume, we're all set! Recall that the Base area was 36π cm^2. So the volume of the cone is just $\frac{1}{3} Bh = \frac{1}{3}(36\pi)(9) = 12\pi(9) = \mathbf{108\pi}$ **cm^3**. Nice!

Oh, by the way, just like with the stacked paper example we saw on p. 376, this formula works for *oblique* pyramids and cones, too. Does the madness ever end? (Not really, no.)

QUICK NOTE   When working with pyramids and cones, the slant height might be really helpful for surface area, but we don't use slant height in volume formulas—just regular 'ol height because we need the height to be *perpendicular to the Base.*

Birdseed, and the Volume of Spheres

Last but not least, here's the formula for the volume inside a sphere.

Volume of a Sphere $= \frac{4}{3} \pi r^3$

Hmm, how can we remember this? Well, all good formulas for circles or spheres somehow involve π, so that's no surprise. Also, notice that volume must use 3-dimensional units, and r has an exponent of 3.* So far, so good? Now we just need a way to remember the $\frac{4}{3}$ part. Well, have you ever looked really closely at birdseed?† Some pieces look like tiny spheres.

. . . .

* Semi-advanced footnote: Our other volume formulas also have a **degree** of 3, which also gives 3-D units. For example, for a prism with height h and a square base with sides s, the volume would be: $V = Bh = s^2 h^1$, and $2 + 1 = 3$. See Chapter 20 in *Hot X: Algebra Exposed* to review adding exponents and degrees.

† I sort of hope you answered "no."

And who is birdseed for? It's for birds. Yep, *for birds*, which rhymes with *four-thirds*.

We could also imagine cutting up some pie into tiny cubes for the birds (I bet they'd like that), and then we could think, "For birds, pie are cubed!" Look at the formula $V = \frac{4}{3}\pi r^3$ while saying that out loud. (Yes, now.) I mean, the grammar's not great, but that's not really the point, is it?

Let's say an orange has a radius of 2 inches. Then its volume is just $V = \frac{4}{3}\pi r^3 = \frac{4}{3}\pi(2)^3 = \frac{4}{3}\pi \cdot 8 = \frac{32}{3}\pi \text{ in.}^3$ (which is approximately 33.5 in.^{3}). If a grapefruit has a radius of 3 inches, its volume would be $V = \frac{4}{3}\pi r^3 = \frac{4}{3}\pi(3)^3 = \frac{4}{3}\pi \cdot 27 = 36\pi \text{ in.}^3$ (which is approximately 113 in.^{3}).

We only changed the radius by 1 inch, and yet the volumes are different by about 80 in.^{3} Ah, a reminder to never underestimate the power of birdseed, I mean, cubing . . .

Remember the snotty girl from p. 364? It's time to see if she was telling the truth. If the cone's height is 5 inches and the radius is 2 inches, then the volume of the cone is $V = \frac{1}{3}Bh = \frac{1}{3}[\pi r^2]h = \frac{1}{3}[\pi \cdot 2^2] \cdot 5 = \frac{20}{3}\pi \text{ in.}^3 \approx 21 \text{ in.}^3$

So far, so good? The ice cream is the same size as the grapefruit above, so that volume is $36\pi \text{ in.}^3 \approx 113 \text{ in.}^3$

Um, yeah, 113 in.^{3} is just a little bigger than 21 in.^{3}; so if it melted, it would totally spill everywhere. Maybe we shouldn't have judged that girl so quickly. Looks like she was trying to be helpful after all . . .

Circle/Sphere Formulas at a Glance!

Circumference of a Circle	$C=2\pi r$	Since $2r=d$, this is the same as "Crust is pie dough" from p. 342: $C=\pi d$. Notice that length is 1-dimensional and this formula uses r, which has an (invisible) exponent: 1.
Area of a Circle	$A=\pi r^2$	Remember the red raspberry pie from p. 343? Area is 2-D, and uses r^2, which has an exponent of 2.
Surface Area of a Sphere	$S.A.=4\pi r^2$	Surface area of a sphere is exactly 4 times more than a circle's area of the same radius. Surface area is 2-D,* and this uses r^2, which has an exponent of 2.
Volume of a Sphere	$V=\dfrac{4}{3}\pi r^3$	Volume is 3-D, and this uses r^3, which has an exponent of 3. And remember the birdseed—it's *for birds* (which sounds like *four-thirds*)!

Time to practice all this volume stuff . . .

* Even though the surface area of a sphere couldn't exist without 3 dimensions, the surface area itself is considered to be "locally" 2-dimensional—it can be flattened into wrapping paper, after all. You can study more about this in a college math class called Topology. (See pp. 141–2 to read about a fabulous topologist.)

Doing the Math

Answer these volume problems. Leave π in your answers, where applicable. *Note: Drawing pictures will be very helpful.* I'll do the first one for you.

1. What is the volume of a pyramid whose lateral edge is 8 inches, with a regular hexagon Base whose perimeter is 24 inches?

<u>Working out the solution</u>: The Base has the same shape as in #1 on p. 374, so we know the area of the Base is **$24\sqrt{3}$**. Now we can use the formula $V = \frac{1}{3}Bh$, right? Well, sort of. First, we need to find h! Let's draw a right triangle *inside* this pyramid, which has 8 as its hypotenuse and h as a side. Also notice that its bottom side is 4. (This is because regular hexagons are made up of six equilateral triangles, so *all* those little lengths down there equal 4.) Great! Pythagoras tells us we can find h like this: $4^2 + h^2 = 8^2 \rightarrow 16 + h^2 = 64 \rightarrow h^2 = 48 \rightarrow h = \sqrt{48} = $ **$4\sqrt{3}$**. And now we're ready for the formula! $V = \frac{1}{3}Bh = \frac{1}{3}(24\sqrt{3})(4\sqrt{3}) = 8 \cdot 4 \cdot \sqrt{3} \cdot \sqrt{3} = 32 \cdot 3 = $ **96.**[*]

Answer: 96 in.³

2. What is the volume of a pyramid whose Base is a square with sides = 6 cm each and whose height is $3\sqrt{3}$ cm?

3. What is the volume of a pyramid whose Base is a square with sides = 6 cm each and whose *slant height* is 6 cm?

4. The Bases of a prism are rhombuses with *diagonals* equal to 4 inches and 6 inches. If the prism is 5 inches tall, what is the volume of the prism? *(Hint: Draw a picture, and see pp. 352–3 for how to find the area of a rhombus.)*

. . . .

[*] To brush up on multiplying square roots, see pp. 285–7 in *Hot X: Algebra Exposed.*

5. A solid ball of ice cream with radius 5 cm is placed in a small can, which was already filled to the rim with root beer. Of course, root beer spills everywhere. The can is a cylinder with a base of radius 5 cm and a height of 10 cm.

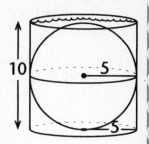

Part a. What was the original volume of root beer?

Part b. What is the volume of the ice-cream ball?

Part c. And finally, after the ice cream is put in and root beer spills everywhere, how much root beer is actually still left inside the can?

6. A chef wants to shape a small serving of risotto into a perfect, half-sphere-shaped dome.

Part a: If the volume of the risotto is $\frac{16}{3}\pi$ in.^{3}, then what is the radius of the resulting dome? *(Hint: What would be the formula for a half-sphere? Set that equal to the volume, and solve for r by isolating it.)*

Part b: If coconut is to be sprinkled over the risotto, what is the surface area that will be covered with coconut?

7. A birthday cake has a radius of 7 inches and height of 3 inches. A piece of cake is cut, creating a 60° angle. What was the volume of the whole cake before the piece was cut? What is the remaining volume of the cake *without* this piece?

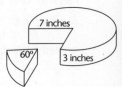

8. A cone-shaped cup is 6 inches tall with a radius of 4 inches.

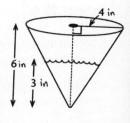

Part a: How much <u>total</u> water could the cup hold?

Part b: If water is poured out until the water is only 3 inches high, what is the new radius that the top of the water makes? What is the new volume of water in the cup? *(Hint: See drawing to the right. Use similar triangles and proportions to find the new radius, and then use that radius and the height of 3 inches to find the new volume!)*

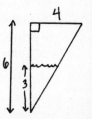

(Answers on p. 402)

Takeaway Tips

 The **surface area** of a 3-dimensional solid is the <u>sum of the areas</u> that enclose the solid—the part that would be covered by wrapping paper! It uses square units, and it's made up of base areas and lateral areas. Slant height is very helpful for finding the surface area of pyramids and cones.

 The surface area of a sphere is S.A. = $4\pi r^2$—four times the area of a circle with the same radius.

 The **volume** of a 3-dimensional solid is the space *inside* the solid and is measured in cubic units.

 The volume of flat-top solids like prisms and cylinders is just $V = Bh$, where B is the area of the Base. The volume of pointy-top solids, like pyramids and cones, is $V = \frac{1}{3}Bh$.

 For the volume of a sphere, $\frac{4}{3}\pi r^3$, remember to cut up some pie into little cubes with bad grammar: "For birds, pie are cubed!"

 When in doubt, <u>draw a picture</u> and look for what missing values might be found using right triangles or anything else we know about the existing shapes. You can do it!

A Final Word

Geometry is a whole new way of thinking, so if you're feeling confused, don't be too hard on yourself, okay? For most people, it takes tons of practice to even *start* to feel comfortable with geometry. See, unlike algebra, there aren't "set" ways to do things, and sometimes each problem can seem like a whole new topic! We often have to be creative and think outside the box.

Proofs can be a particular challenge, but if you stick with it, they will sharpen your mind in ways you can't even imagine. After reading through a challenging proof, it's totally normal to freak out a little and think, "But I don't know how I'd ever apply what I just saw to other proofs. I just barely followed this one!" The good news is, you don't *have* to know. Just keep following along, and then see if you can re-create the proof without looking. Keep practicing: Patterns will start to emerge, and soon this stuff won't seem so foreign. Before you know it, you'll find yourself sitting down to tackle a proof, saying "Bring it on!" because you've learned the skill of strategizing and completing a proof. It really is a skill unto itself, and it will serve you in *all* areas of life.

We all have the power to shape our minds and our bodies in healthy ways. Feed your body healthy food and feed your mind hearty challenges like geometry, and you'll become the young woman you've always imagined yourself to be. There will be no stopping you!

I'm <u>so</u> proud of you.

PROOFS:
The Troubleshooting Guide!

*T*ons of strategies for proofs can be found throughout this entire book, and of course the more theorems you're familiar with, the easier proofs will be, right? But sometimes we all get stuck, so if you find you're still not sure what to do, try these tricks!

Proof Strategy #1: Find the Hidden Information on the Diagram

Vertical Angles: Chopstick math! Are there vertical angles on the diagram? If so, mark 'em congruent with kitty scratches! (See p. 97 for an example.)

Supplementary Angles: Notice if there are supplementary angles. If there are two sets of supplementary angles, then there could be a set of congruent angles hiding in plain sight. (See p. 81.)

Radii of Circles: If there are circles, then mark all radii as congruent segments. Also, try *drawing in* radii from the center to key points like points of tangency, or any other "special" point, like maybe the vertex of an inscribed polygon. They will all be congruent, so mark 'em with kitty scratches! (See p. 304.)

Parallel Lines: Is there a set of parallel lines? If so, remember the mall and escalator from Chapters 13 and 14, and look for pairs of congruent angles—and vice versa. You might need to tilt your head!

Proof Strategy #2: Hidden Information in the Wording of the Problem

Look up the **definitions** of key words in the problem—words in the Givens and in the thing we're supposed to prove. Very often, the actual definitions will give us the links we need to get from Point A to Point B!

For example, *midpoint* gives us congruent segments, and *angle bisector* gives us congruent angles. See pp. 134–7 for more details on how key words in the Givens expose hidden congruencies. And then, of course, mark 'em congruent with kitty scratches!

Proof Strategy #3: Triangles Are Our Friends

Right Triangles: Even within polygons or 3-D objects, if we find a right triangle (which we might create by drawing in an altitude), we might be able to use the Pythagorean Theorem to find a missing side, and that could help us find what we're being asked for! See Chapter 11 for more.

Isosceles Triangles: Even with a bunch of overlapping triangles within a more complicated diagram, look for a pair of congruent angles or congruent sides, and see if there's an isosceles triangle involved! If so, then we automatically know that the legs are congruent and the base angles are congruent, which could lead somewhere very helpful. See the shortcuts on p. 148.

Pairs of Congruent Triangles: Last but not least, pairs of congruent triangles are enormously helpful! We can prove them congruent in many ways—SSS, SAS, ASA, SAA, HL—and once we've proven that they're congruent, CPCTC gives us tons of information about other pairs of congruent parts. See Chapter 9 for more.

Still Not Sure of the Strategy?

Draw or Redraw: Drawing our own diagrams can be a lifesaver! And if one is already provided, then consider drawing overlapping triangles separately to better "see" what's going on like we did on p. 289. Are there triangles you haven't considered yet?

Proof by Contradiction: Remember, if the thing we're supposed to prove has "not," ≠, or ∦ in it, try doing a proof by contradiction like we did in Chapter 10.

Work Backwards: Write the last line of the proof first. Then think: Gosh, if only I knew such-and-such, I could prove that. Now write the such-and-such in the line above it, using "if . . . then" logic to create an airtight argument, and see if you can keep going from the bottom up! See p. 230–2 for an example.

Prove Something Else: Still not sure of your strategy? Well, ignoring what you're *supposed* to prove, what *could* you prove with the Givens and the extra stuff you've marked on the diagram? You might be surprised to find out that what you *can* prove will often bring you closer to your goal than you realized.

And Remember: You Can Do This!

Appendix

*H*ere are some items that support what we've learned in this book, just in case you want to check 'em out!

How to Use a Protractor

Protractors are typically clear, plastic rulers in the shape of a half-circle, with degree measurements ranging from 0°–180° all along the curved part—usually in both directions. When we measure an angle, we just place the protractor so the vertex (armpit) lines up with the center circle, and then look to see where the angle crosses the curved edge. Because the numbers go in both directions, just think to yourself, "Should this angle be less than 90° or greater than 90°?" and then use that number. That's all there is to it!

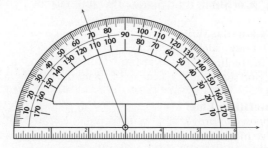

In this case, our angle crosses the curved edge of the protractor at 70° and 110°. Since we can tell that the angle is bigger than 90°, we should use the 110° Ta-da! Remember, protractors shouldn't ever be used for your proofs, because the exact size of angles in textbook diagrams can't be assumed to be accurate. (See p. 39 for a list of things we can and can't assume about a diagram.)

Angles: Consecutive vs. Adjacent

In the footnote on p. 243, we learned that consecutive segments and adjacent segments are the same thing: They're segments that are next to each other in a shape, like for example, two distinct sides of a polygon that share an endpoint.

For *angles*, however, there is a difference. In a polygon, two neighboring angles are called *consecutive angles* like ∠1 & ∠P, or ∠P and ∠PST below. On the other hand, *adjacent angles* are nonoverlapping angles that share a vertex (armpit) and a side, like ∠1 & ∠2. However, ∠1 & ∠PIT are not adjacent, because even though they share a side and the same vertex, they overlap.

Adjacent is pronounced "uh, Jason's . . .," which is a sound you might make if your little brother Jason were pretending to be a dog sniffing your armpits and you wanted to say, "Uh, Jason, don't" through clenched teeth because a really cute guy was standing right there, staring at you.

What's It Called?

Parallel planes, Perpendicular to a plane

Remember that planes extend infinitely in all 2-D directions. Planes are said to be **parallel planes** if they never intersect.

We say that a line is **perpendicular to a plane** if the line is ⊥ to *every* line in that plane that passes through the point of intersection.* For example, the line segments c and d both pass through e's point of intersection with plane P, right? (Look for point A in the diagram.) Well, if we're told that c and d lie in the plane P, then $e \perp P$ automatically means that $e \perp c$ and $e \perp d$, too.

In the diagram to the right, the line segment m at the Base of the pyramid lies in the plane Q, and we have perpendicular line segments: $l \perp m$. But m is just *one* line in the plane Q. We can't say that l is perpendicular to the entire plane Q, can we? In fact, it probably isn't!

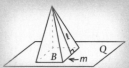

. . . .

* As it turns out, if we know that a line is ⊥ to at least two distinct lines in the plane passing through its point of intersection (like c and d), then we've proven that the line is ⊥ to the entire plane. This is like the "converse" of the definition of a line ⊥ to a plane.

Postulates About Drawing Lines

These next three postulates have a common thread: They tell us that if we draw a line according to a specific instruction, and someone else follows that same instruction, our lines will have to be exactly the same!

Two Points Determine a Line (postulate): This could also be called "Two points *define* a line." If someone tells us about two points, we can draw a line through them, and we know we drew the "correct" line because it's unique! So if somebody else drew a line through those same two points, it would have to be the exact same line as ours. Well, unless they have messy handwriting. Another way to think about it is that if we know two points of a line, that's enough to "fix" or "lock" the line into position. <u>No other line in the whole world exists that also contains these two points.</u> Wild, huh?

Similarly . . .

Perpendicular line postulate: Given any straight line and a point not on it, there exists one and only one straight line that passes through that point and is perpendicular to that line.

Parallel line postulate: Given any straight line and a point not on it, there exists one and only one straight line that passes through that point and is parallel to that line.

The reason people care about these postulates is because in proofs, sometimes it helps to draw a line, like a diagonal between two vertices (see p. 251 for an example), and the reason we're "allowed" to draw a line or segment in the middle of a proof is because it's unique. In the Statements column, we can write "Draw $\overline{AB}$," and in the Reasons column, we can write "Two points determine a line."

Definitions, Postulates, Properties, and Theorems—What Gives?

Each of the Rules we've been using in this book so far could also be called one of these: a definition, a postulate, a property, and/or a theorem. What's the difference?

Well, a **definition** is something that somebody came up with to make an item easier to talk about by giving it a single word. So instead of saying "the point that divides a segment in half," we can just say *midpoint*. Faster, right?

"Properties" is a general term that can be divided up into two categories: **postulates** and **theorems**. Generally speaking, postulates are rules that aren't proven in the book—but they often *could* be.* Theorems are usually proven in the book. But it all depends on the book! So in any given textbook, depending on whether or not the textbook writer decided to actually prove a property, it might be called a postulate and it might be called a theorem. Crazy, right? In this book, there are some "theorems" we didn't actually prove within the book, so technically they could have been called postulates. But if something is commonly known as a theorem, then that's what it's called in this book. No biggie.

Here are some of those Rules now!

. . . .

* Some postulates cannot be proven at all—these are called *axioms*.

List of Rules Appearing in This Book

*If the sum of two ∠'s is 90°, then they are comp. (p. 24)

*If the sum of two ∠'s is a right ∠, then they are comp. (p. 24)

*If the sum of two ∠'s is 180°, then they are supp. (p. 25)

*If the sum of two ∠'s is a straight ∠, then they are supp. (p. 25)

*If ∠'s are vertical, then ∠'s are ≅. (p. 29)

*If a point divides a segment into two ≅ segments, then the point is the midpoint. (p. 36 & p. 63)

*If a line (or seg) is a ⊥ bisector, then it is ⊥ to the seg and passes through the seg's midpoint. (p. 37)

*If a line (or seg) is a ⊥ bisector, then it is ⊥ to the seg and divides the seg into two ≅ seg's. (p. 37)

*If a ray is an ∠ bisector, then it divides the ∠ into two ≅ ∠'s. (p. 37)

*If two lines are ⊥, then they create right ∠'s. (p. 52)

*If two lines create a right ∠, then they are ⊥. (p. 52)

*Transitive Property for Segments: If two segments are ≅ to the same segment, then they are ≅ to each other. (p. 52)

*Transitive Property for Angles: If two ∠'s are ≅ to the same ∠, then they are ≅ to each other. (p. 52)

*Transitive Property: If $a ≅ b$ and $b ≅ c$, then $a ≅ c$. (p. 52 & p. 96)

*Addition Property #1: If a segment (or an angle) is added to two ≅ segments (or angles), then the sums are ≅. (p. 71)

*Addition Property #2: If two ≅ segments (or angles) are added to two other ≅ segments (or angles), then the sums are ≅. (p. 71)

*Subtraction Property #1: If a segment (or an angle) is subtracted from two ≅ segments (or angles), then the differences are ≅. (p. 72 & p. 73)

*Subtraction Property #2: If two ≅ segments (or angles) are subtracted from two other ≅ segments (or angles), then the differences are ≅. (p. 72 & p. 73)

*Multiplication Property: If segments (or angles) are ≅, then their like multiples are ≅. (p. 79)

*Division Property: If segments (or angles) are ≅, then their like divisions are ≅. (p. 80)

*If ∠'s are supp. to ≅ ∠'s, then they are ≅. (p. 81)

*If ∠'s are comp. to ≅ ∠'s, then they are ≅. (p. 81)

*The Substitution Property: If two angles are ≅ and we have a statement involving the measure of one ∠, we can replace it with the other ∠. (p. 81)

*A scalene triangle is a triangle that has no ≅ sides. (p. 88)

*An isosceles triangle is a triangle in which at least 2 sides are ≅. (p. 89)

*An equilateral triangle is a triangle in which all 3 sides are ≅. (p. 89)

*A scalene triangle is a triangle in which no *angles* are ≅. (p. 90)

*An isosceles triangle is a triangle in which at least two *angles* are ≅. (p. 90)

*An equilateral triangle is a triangle in which all three *angles* are ≅. (p. 90)

*An acute triangle is a triangle in which all angles are acute (less than 90°). (p. 91)

*An obtuse triangle is a triangle in which one angle is obtuse (more than 90°). (p. 91)

*A right triangle is a triangle that has one right angle (90°). (p. 91)

*Triangle Inequality: The sum of the lengths of any two sides of a triangle is always greater than the length of the third side. (p. 99)

*If the angles of a triangle are added together, the sum will equal 180°. (p. 104)

*The three angles in every equilateral triangle each measure 60°. (p. 104)

*Side-Side-Side Rule (SSS): If the three corresponding sides of two triangles are ≅, then the two (entire) triangles are ≅. (p. 121)

*Side-Angle-Side Rule (SAS): For any two triangles, if two pairs of corresponding sides and the angle in between are ≅, then the two (entire) triangles are ≅. (p. 121)

*Angle-Side-Angle Rule (ASA): For any two triangles, if two pairs of corresponding angles and the side in between are ≅, then the two (entire) triangles are ≅. (p. 121)

*Angle-Angle-Side Rule (AAS or SAA): For any two triangles, if two pairs of corresponding angles and a side *not* in between are ≅, then the two (entire) triangles are ≅. (p. 122)

*The Reflexive Property: Every segment (or angle) is ≅ to itself. (p. 124)

*The Hypotenuse-Leg Theorem (HL): If the hypotenuse and leg of one right triangle are ≅ to the hypotenuse and leg of another right triangle, then the two triangles must be ≅. (p. 143)

*CPCTC: Corresponding Parts of Congruent Triangles are Congruent. (p. 144)

*If a triangle has two ≅ sides, then the angles opposite those sides are ≅. (If sides, then angles.) (p. 148)

*If a triangle has two ≅ angles, then the sides opposite those angles are also ≅. (If angles, then sides.) (p. 148)

*If a segment connects a vertex of a triangle to the midpoint of the opposite side, then it's the median. (p. 155)

*The Pythagorean Theorem: In a right triangle, $a^2 + b^2 = c^2$, where a and b are the two legs (the shorter sides) and c is the hypotenuse. (p. 176)

*Midpoint Theorem: If a segment connects two sides of a triangle at their midpoints, then that segment is parallel to the third side and is half the length of the third side. (p. 298)

*Side-Splitter Theorem: A parallel line splits the sides of a triangle proportionally. (Footnote on p. 298)

*All radii of a circle are ≅. (p. 304)

The Equidistant Rules:

*If a line is the perpendicular bisector of a segment, then every point on the line is equidistant (the same distance) to the segment's endpoints. (p. 158)

*If two points are each equidistant to the endpoints of a segment, then those two points determine the segment's perpendicular bisector. (p. 158)

Transversal Rules (p. 216 & p. 218):

*If ∥ lines, then corr. ∠'s are ≅.

*If ∥ lines, then alt. int. ∠'s are ≅.

*If ∥ lines, then alt. ext. ∠'s are ≅.

*If ∥ lines, then same side int. ∠'s are supp.

*If ∥ lines, then same side ext. ∠'s are supp.

Converse of Transversal Rules (p. 230):

*If corr. ∠'s are ≅, then lines are ∥.

*If alt. int. ∠'s are ≅, then lines are ∥.

*If alt. ext. ∠'s are ≅, then lines are ∥.

*If same side int. ∠'s are supp., then lines are ∥.

*If same side ext. ∠'s are supp., then lines are ∥.

Quadrilateral Rules:

*If a quad is a parallelogram, then *two* pairs of opposite sides are ∥. (p. 259)

*If a quad is a parallelogram, then two pairs of opposite sides are ≅. (p. 259)

*If a quad is a parallelogram, then *two* pairs of opposite ∠'s are ≅. (p. 260)

*If a quad is a parallelogram, then the *two* diagonals bisect each other. (p. 260)

*If a quad is a parallelogram, then consecutive ∠'s are supp. (p. 260)

*If two pairs of opposite sides of a quad are ∥, then it's a parallelogram. (p. 262)

*If two pairs of opposite sides of a quad are ≅, then it's a parallelogram. (p. 262)

*If two pairs of opposite ∠'s of a quad are ≅, then it's a parallelogram. (p. 262)

*If the *two* diagonals of a quad bisect each other, then it's a parallelogram. (p. 262)

*If *one* pair of opposite sides of a quad are ∥ *and* ≅, then it's a parallelogram. (p. 262)

*If a quad is a rectangle, then it has all the properties of parallelograms. (p. 268)

*If a quad is a rectangle, then all four ∠'s are right ∠'s. (p. 268)

*If a quad is a rectangle, then its diagonals are congruent. (p. 268)

*If a quad is a kite, then two disjoint pairs of consecutive sides are ≅. (p. 269)

*If a quad is a kite, then the diagonals are ⊥. (p. 269)

*If a quad is a kite, then one pair of opposite ∠'s is ≅. (p. 269)

*If a quad is a kite, then one diagonal bisects the other. (p. 269)

*If a quad is a kite, then one diagonal bisects opposite ∠'s. (p. 269)

*If a quad is a rhombus, then it has all the properties of parallelograms and kites. (p. 270)

*If a quad is a rhombus, then all four sides are ≅. (p. 271)

*If a quad is a rhombus, then both diagonals are ⊥ bisectors of each other. (p. 271)

*If a quad is a rhombus, then both diagonals bisect their angles. (p. 271)

*If a quad is a square, then it has all the properties of rectangles and rhombuses. (p. 272)

*If a quad is a square, then the diagonals divide it into four ≅ right isosceles △'s. (p. 272)

*If a quad is a trapezoid, then it has exactly one pair of opposite, ∥ sides. (p. 273)

*If a quad is an isosceles trapezoid, then it has *exactly one* pair of opposite, ∥ sides (bases). The legs are ≅ and *not* ∥. (p. 274)

*If a quad is an isosceles trapezoid, then the lower base angles are ≅, and the upper base angles are ≅. (p. 274)

*If a quad is an isosceles trapezoid, then any lower base angle is supplementary to any upper base angle. (p. 274)

*If a quad is an isosceles trapezoid, then the diagonals are ≅. (p. 274)

Similarity Rules:

*If two triangles are similar, then their corr. ∠'s are ≅ and the ratio of their corr. sides are equal. (p. 286)

*AA Similarity: If two triangles share two sets of ≅ angles (AA), then the two triangles are similar. (p. 289)

Circle Rules:

*Symmetrical Bulldog Puppy Mouth Rule: If a radius is ⊥ to a chord, then it bisects the chord. (p. 306)

*Converse of Symmetrical Bulldog Puppy Mouth Rule: If a radius bisects a chord (that is not a diameter), then it's ⊥ to the chord. (p. 306)

*Congruent-Chord (Twin Puppy Dog) Theorem: If two chords are equidistant to the center, then the two chords are ≅. (p. 308)

*Converse of Congruent-Chord (Twin Puppy Dog) Theorem: If two chords are ≅, then the two chords are equidistant to the center. (p. 308)

*Bicycle Wheel Rule: A tangent line to a circle will always be perpendicular to the radius that touches the tangent point. (p. 313)

*Two-Tangent "Bird Beak" Theorem: If a circle has two tangent segments drawn from a single point, they will always be ≅. (p. 314)

Angle-Arc-Chord Congruent Rules: (p. 328)

*If two central ∠'s are ≅, then the intercepted arcs are ≅.

*If two central ∠'s are ≅, then the corresponding chords are ≅.

*If two intercepted arcs are ≅, then the corresponding chords are ≅.

*If two intercepted arcs are ≅, then the central ∠'s are ≅.

*If two corresponding chords are ≅, then the central ∠'s are ≅.

*If two corresponding chords are ≅, then intercepted arcs are ≅.

Answer Key

For the fully worked-out solutions—including all of the proofs—visit the "Solutions Guides" page at GirlsGetCurves.com.

Chapter 1, pp. 2–3

2. <u>If</u> my sister takes forever in the shower, <u>then</u> I'm late for school.
3. <u>If</u> I hear that song one more time, <u>then</u> I will scream.
4. <u>If</u> I eat more vegetables, <u>then</u> I feel healthy.
5. <u>If</u> a number is odd, <u>then</u> it's not divisible by 2. (Or: <u>If</u> numbers are odd, <u>then</u> they are not divisible by 2.)

Chapter 1, p. 5

2. Therefore: all Barbies are creepy.
3. No conclusion possible. Sparky could be a bird, after all!
4. Therefore: All apples grow on trees.
5. No conclusion possible. Debbie might be a human that speaks Martian!

Chapter 1, pp. 8–9

(Note: Verb tenses—past, present, future—may vary in these. That's totally fine!)
2. **Converse:** If you're bored, then you tweet. **Inverse:** If you don't tweet, then you're not bored. **Contrapositive:** If you're not bored, then you don't tweet.
3. **Converse:** If she's not in a good mood, then she misses the bus. **Inverse:** If she doesn't miss the bus, then she's in a good mood. **Contrapositive:** If she's in a good mood, then she doesn't miss the bus.
4. **Converse**: If I'm tired on Thursday, then I stayed up on Wednesday night. **Inverse:** If I didn't stay up on Wednesday night, then I'm not tired on Thursday. **Contrapositive**: If I'm not tired on Thursday, then I didn't stay up on Wednesday night.

Chapter 1, pp. 15–6

2. Given: b
   If b, then a
      If a, then c
      "c" is proven!

4. Given: not L
   If not L, then K
      If K, then M
         If M, then N
         "N" is proven!
         So Natalie wears pink!

3. Given: not R
   If not R, then not P
      If not P, then not Q
      "not Q" is proven!
      So Quinn isn't happy.

Chapter 2, p. 26

2. 144° and 36°
3. 92.5° and 87.5°

4. 66° and 24°
5. 9° and 81°

Chapter 2, pp. 31–2* (see footnote below)

2. $\overline{CB} \cong \overline{RT}$, $\overline{RA} \cong \overline{AB}$
3. ∠C, ∠ACB, ∠BCA
4. ∠CAR and ∠BAT
5. ∠CAT and ∠BAR
6. five

7. no
8. no
9. $\overline{CB} \perp \overline{CT}$
10. $\overrightarrow{QC}$

11. ∠BAQ, ∠QAB, ∠TAB
12. ∠BAC is supp. to ∠CAR;
    ∠BAT is supp. to ∠RAT;
    ∠BAC is supp. to ∠BAT;
    ∠RAT is supp. to ∠CAR.
13. ∠CAB is comp. to ∠B.

Chapter 3, p. 42

2. True
3. Not enough
   information
4. True

5. False
6. False
7. Not enough
   information

8. True
9. Not enough
   information
10. True

Chapter 4, pp. 63–5

2.

Statements	Reasons
1. $\overline{HO} \cong \overline{OP}$	1. Given
2. ∴ O is the midpoint of $\overline{HP}$.	2. If a point divides a seg into two ≅ seg's, then the point is the midpoint.

....

* Note: All segments can be written with the letters swapped; so $\overline{CB}$ is the same as $\overline{BC}$. For all angles using 3 letters, the first and third letter can be swapped, so ∠BAT is the same as ∠TAB.

3.

Statements	Reasons
1. $\overrightarrow{SI}$ is the angle bisector of $\angle KSP$.	1. Given
2. $\therefore \angle KSI \cong \angle ISP$	2. If a ray is an $\angle$ bisector, then it divides the $\angle$ into two $\cong \angle$'s.

4.

Statements	Reasons
1. $\overline{JM} \perp \overline{UP}$	1. Given
2. $\therefore \angle JMP$ is a right angle.	2. If two seg's are $\perp$, then they create right $\angle$'s.

5. Given

6. Given

7. If a point is a midpoint, _then_ it divides a segment into two $\cong$ segments.

8. Transitive Property: If $\overline{JM} \cong \overline{UM}$ and $\overline{UM} \cong \overline{MP}$, then $\overline{JM} \cong \overline{MP}$.

9. Given

10. $\angle 1 \cong \angle 2$ (or $\angle WTI \cong \angle ITR$)

11. $\angle 1 \cong \angle 3$

12. Transitive Property: If $\angle 1 \cong \angle 2$ and $\angle 1 \cong \angle 3$, then $\angle 2 \cong \angle 3$.

13. If a ray ($\overrightarrow{TR}$) divides an $\angle$ into two $\cong \angle$'s, then it's the $\angle$ bisector.

Chapter 5 p. 74–5

2. $127°$
3. 6 cm
4. $\overline{WA} \cong \overline{VE}$
5a. $\angle IRC \cong \angle MRA$, $\angle IRA \cong \angle MRC$
5b. $65°$

Chapter 5, pp. 75–7 (proof solutions at GirlsGetCurves.com)

4. If a point is a midpoint, _then_ it divides the segment into two $\cong$ segments.

5. $\overline{LU} \cong \overline{UE}$

6. Transitive Property: If $\overline{BL} \cong \overline{LU}$ and $\overline{LU} \cong \overline{UE}$, then $\overline{BL} \cong \overline{UE}$.

7. Addition Property: _If_ a segment ($\overline{LU}$) is added to two $\cong$ segments ($\overline{BL}$ and $\overline{UE}$), _then the sums are_ $\cong$.

Chapter 5, pp. 83–4 (proof solutions at GirlsGetCurves.com)

2. $\angle FSU = 120°$
3. $HY = 12$

Chapter 6, p. 94

2. sometimes
3. sometimes
4. never
5. never

6. sometimes
7. never
8. sometimes
9. always

Chapter 6, p. 97–8

2. Given
3. Given
4. $\overline{AT} \cong \overline{TS}$
5. Transitive Property (If $\overline{LT} \cong \overline{TS}$ and $\overline{AT} \cong \overline{TS}$, then $\overline{AT} \cong \overline{LT}$.)
6. If a $\triangle$ has two $\cong$ sides, then it's isosceles.
7. $\angle LTA$ is a right angle.
8. $\triangle ATL$ is a right isosceles triangle.

Chapter 6, pp. 102–3

2. $3 < x < 17$
3. $0 < b < 8$
4. $UE = 20$ (Those $\cong \angle$'s tell us that $\triangle CUE$ is isoceles!); $11 < y < 29$
5. $3a < x < 17a$

Chapter 6, pp. 105–6

2. right
3. 21°, 21°, 138°
4. 95°
5. 30°

Chapter 7, pp. 114–5

2. G ᗡ

3. ⅃

4. Any 5 of these: A, H, I, M, O, T, U, V, W, X, Y
5. Any 5 of these: B, C, D, E, H, I, K, O, X
6. Any 5 of these: H, I, N, O, S, X, Z
7. Any 2 of these: N, S, Z
8. Reflectional symmetry across a vertical line
9. 90°, 180°, 270°, 360°
10. 60°, 120°, 180°, 240°, 300°, 360°

Chapter 8, pp. 124–5

2. $\triangle TAR$
3. ASA or AAS (SAA)
4. B
5. G
6. $\overline{BG}$
7. $\angle BGU$
8. C
9. B and C
10. B and D (Notice B has vertical $\angle$'s)
11. B (and technically, D. Since $\triangle$'s $\angle$'s sums always = 180°, then we know the third $\angle$'s in D must also be $\cong$.)
12. A

Chapter 8, pp. 138–40 (all proofs—see GirlsGetCurves.com)

Chapter 9, pp. 151–3 (all proofs—see GirlsGetCurves.com)

Chapter 9, pp. 160–1 (proof solutions at GirlsGetCurves.com)

2. $\overline{IY}$, $\overline{RG}$, $\overline{BV}$
3. $\overline{RS}$, $\overline{IR}$, $\overline{KR}$, $\overline{NA}$, $\overline{BR}$ (not $\overline{IN}$—the dotted line doesn't actually extend a side, does it? So $\overline{IN}$ isn't an altitude!)
4. Yes: $\angle H = 90°$, so $\overline{HY}$ is $\perp$ to $\overline{HW}$.

Chapter 10 (all proofs—see GirlsGetCurves.com)

Chapter 11, pp. 184–5

2. 4
3. $\sqrt{34}$
4. 24
5. $\sqrt{5}$
6. $12\sqrt{13}$
7. $\frac{17}{71}$
8. 20
9. 0.013
10. 24
11. $15x$
12. $4\sqrt{5}$
13. 24π
14. 12 ft; perimeter = 90 ft
15. 6
16. $4\sqrt{2}$; perimeter = $32 + 4\sqrt{2}$
17. Yes, it's a member of the 7-24-25 family.

Chapter 11, pp. 192–3

2. l.leg = $7\sqrt{3}$; hyp. = 14
3. s.leg = 10; l.leg = $10\sqrt{3}$
4. s.leg = $\frac{5}{2}$; l.leg = $\frac{5\sqrt{3}}{2}$
5. s.leg = 3; hyp. = 6
6. s.leg = $6\sqrt{3}$; hyp. = $12\sqrt{3}$
   ($\frac{18}{\sqrt{3}}$ & $\frac{36}{\sqrt{3}}$ before simplifying)
7. perimeter = $10\sqrt{2}$ inches (before simplifying $\frac{20}{\sqrt{2}}$)
8. $\sqrt{3}$ feet
9. l.leg = $x\sqrt{3}$; hyp. = $2x$
10. $BO = \sqrt{3}$; not a square
11. $HK = \sqrt{5}$; not equilateral
12. $s = 4$ in.; perimeter = 24 in.

Chapter 12, pp. 210–2

2. Here's an example; answers will vary.

8. 50

9. 360°

3. 7; 8; 35

4. 498,500

5. 8; 1,080°

6. 30; 5,040°

7. $1,620°; 147\frac{3}{11}°$

10. heptagon (7-gon), octagon (8-gon), decagon (10-gon)

11a. $180° - \dfrac{360°}{n}$

11b. $\dfrac{180(n-2)°}{n}$

12. 190 hugs

Chapter 13, pp. 219–20

2. $\angle ADB$ & $\angle 4$ and $\angle BDC$ & $\angle 3$

3. $\angle ADE$ & $\angle 1$ and $\angle CDE$ & $\angle 2$

4. $\angle BDA$ & $\angle 3$ and $\angle BDC$ & $\angle 4$

5. $\angle 1$ & $\angle 4$ and $\angle 2$ & $\angle 3$

6. $\angle 4$

7. $\angle 1$ & $\angle 4$

8. 98°

9. $x° = 75°$

Chapter 13, pp. 224–5

2. 75°

3. 50°

4. 23°; if ‖, then alt. int. ∠'s are ≅.

5. 50°

6. 40°

7. 68°

8a. $\angle SOP = c°$

8b. 180°

8c. 180°

8d. $c° = 60°$; yes

Chapter 14, pp. 234–5 (proof solutions at GirlsGetCurves.com)

2. Yes, $\overleftrightarrow{AH} \parallel \overleftrightarrow{SN}$

3. Cannot prove any lines are ‖; Yes, $\triangle SFN$ is isosceles and $\angle F = 100°$

4. $\overline{LO} \parallel \overline{VE}$; If corr. ∠'s are ≅, then lines are ‖.

5. Yes, $\angle OVE = 90°$

Chapter 15, p. 248

2. never

3. sometimes

4. always

5. sometimes

6. sometimes

7. always

8. always

9. never

10. sometimes

11. rectangle (or parallelogram!)

12. square

13. a rhombus (that isn't a square)

Chapter 15, p. 255 (all proofs—see GirlsGetCurves.com)

Chapter 16, p. 261

2. yes

3. no

4. yes

5. yes

6. no

7. yes

8. no

9. no

10. yes

11. yes

12. yes

13. no

Chapter 16, p. 266 (proof solutions at GirlsGetCurves.com)

2. $a = 80°$, $b = 70°$

Chapter 16, pp. 275–7

2. kite, $\angle I$ and $\angle E$
3. rhombus, square
4. kite, rhombus, square
5. yes
6. no
7. $\overline{RE} \cong \overline{NG}$, $\overline{SU} \cong \overline{QA}$,
   $\overline{IO} \cong \overline{SZ}$

8. parallelogram, rectangle, rhombus, square
9. trapezoid
10. congruent diagonals
11. rhombus, square
12. rectangle

13. square
14. kite; no ∥ lines
15. congruent
16. angles
17. congruent and ∥
18. are congruent to
19. their angles
20. ⊥ to

Chapter 16, pp. 277–8 (proof solutions at GirlsGetCurves.com)

2. 120°, 60°, 60°; two equilateral △'s
3. $x° = 50°$; $y° = 15°$

Chapter 17, p. 291

2. $BG = 12$
3. 9 inches

4. $LA = 3\frac{1}{5}$
5. $LS = 9$

Chapter 17, pp. 297–8 (all proofs—see GirlsGetCurves.com)

Chapter 18, pp. 310–1 (proof solutions at GirlsGetCurves.com)

2. $\sqrt{5}$ ft.
3. 16 in.
4. 6.5

Chapter 18, pp. 315–6 (proof solutions at GirlsGetCurves.com)

2. $TN = 8$ cm, $TA = 10$ cm, $TC = 4$ cm

Chapter 18, pp. 320–1 (proof solutions at GirlsGetCurves.com)

2. $2\sqrt{7}$
3. $4\sqrt{3}$

Chapter 19, pp. 329–30 (proof solutions at GirlsGetCurves.com)

2. Both are 100°.
3. Both are 80°.
4. $\widehat{AUI} = 260°$ or $\widehat{AIU} = 280°$
5. True
6. False. Their measures are equal, but their lengths are not!

7. $\overline{SM}, \overline{SC}$, and $\overline{SL}$
8. 50°
9. $\frac{5}{18}$

10. $6\frac{2}{3}$ feet or 6 feet, 8 inches

11. 9 feet

12. $\frac{1}{5}$; 72°

13. 60°

14. $\frac{360°}{n}$

Chapter 19, pp. 337–8 (proof solutions at GirlsGetCurves.com)

2. $\overset{\frown}{CE} = 120°$

3. 66°

4. $\angle 1 = 26°$ and $\overset{\frown}{ZY} = 26\frac{1}{2}°$

5. 90°

Chapter 20, pp. 351–2

2. 18 mm²; 4 mm

3. $\frac{8\sqrt{\pi}}{\pi}$ in.² (before simplifying: $\frac{8}{\sqrt{\pi}}$ in.²)

4. 9 : 1

5. 4.9π in.²

6. 150$\sqrt{3}$ ft²

7. 25 cm²

8. perimeter = 84 miles; area = 294$\sqrt{3}$ mi²

9. 6 inches

10. total = 9π in.²; slice = $\frac{3\pi}{2}$ in.²

Chapter 20, pp. 358–9

2. $217\frac{1}{2}$ cm²

3. $(4 - \pi)$ in.²

4. 5 cm and 10 cm

5. $(25\pi - 15)$ in.²

6. 108$\sqrt{3}$ ft²

7. 390 ft²

8. dist = 2$\sqrt{3}$ cm (before simplifying: $\frac{6}{\sqrt{3}}$ cm); Area = 18$\sqrt{3}$ − 9π cm²

Chapter 21, pp. 374–5

2. 75 in.²

3. 96 m²

4. 100π mm²

5. 64 in.²

6. 72.5π m²

7. 66π ft²

Chapter 21, pp. 380–1

2. 36$\sqrt{3}$ cm³

3. 36$\sqrt{3}$ cm³ (yep, same answer!)

4. 60 in.³

5a. 250π cm³

5b. $\frac{500\pi}{3}$ cm³

5c. $\frac{250\pi}{3}$ cm³ (before simplifying: $250\pi - \frac{500\pi}{3}$ cm³)

6a. 2 in.

6b. 8π in.²

7a. 147π in.³

7b. $\frac{245\pi}{2}$ in.³

8a. 32π in.³

8b. 2 in.

8c. 4π in.³

Symbol Glossary

$\overline{AB}$ segment AB	= is equal to	∥ parallel
$\overleftrightarrow{AB}$ line AB	≠ is not equal to	∦ not parallel
$\overrightarrow{AB}$ ray AB	< is less than	⊥ perpendicular
$\overgroup{AB}$ arc AB	> is greater than	⊥̸ not perpendicular
AB the length of $\overline{AB}$	≤ is less than or equal to	△ triangle
∠A angle A	≥ is greater than or equal to	▱ parallelogram
∠, ∠'s angle, angles	≈ is approx. equal to	○ circle
≅ congruent	⇔ is equivalent to	° degrees
≇ not congruent	⇒ implies	∴ therefore
∼ similar	π pi	

Index

Angles (*cont.*)

alternate interior, 215, 218, 220, 230–233, 251–254, 295–296, 356n

base, 88–89, 273–274, 386

central, 324–328, 339

complementary, 24–26, 32, 60, 81, 84, 137, 224, 297

congruent, 27–32, 37–42, 51–52, 58–59, 71–81, 84, 90, 136–137, 145, 216–220, 225, 227–235, 260, 273, 293

consecutive, 216n, 260, 260n, 263, 268n, 388

corresponding, 121–122, 145–146, 152, 215–219, 222, 227–232, 265, 286, 289–300, 336

defined, 22

in a polygon (exterior), 207–208, 211, 213

in a polygon (interior), 204–206, 208, 210–213

in a triangle, 103–106

obtuse, 23, 31, 91, 94, 170

right, 23–24, 32, 39–40, 52, 63–64, 67, 81, 91, 131–132, 132n, 133n, 137, 143–144, 146, 151–153, 156–159, 183, 184n, 194, 241–242, 258, 268, 268n, 271, 305–313, 319–320, 347–353, 366–369, 372

same side exterior, 216, 218–220, 222, 230

same side interior, 216, 218–220, 230, 255, 260n, 273, 356n

sides of, 22

straight (*see also* 180°), 23, 31, 103, 139, 139n, 217

supplementary, 24–30, 32, 60, 81, 83, 96, 137–139, 149, 207, 213, 217–225, 228, 230, 235, 255, 260, 263, 268n, 273–274, 279, 334, 356n, 385

transitive property for (*see also* Transitive Property), 52–53, 58, 63

vertex/vertices of, 22, 40, 90, 120, 155–159, 201–213, 250, 294n, 324, 331–339, 348, 367

vertical, 28–29, 31, 32, 39, 51–52, 58–59, 63, 76, 97, 126, 129, 129n, 136, 217, 218, 221–225, 231–232, 254, 304, 334, 336, 385

Apothem, 348, 350, 360

Arcs, 324–339, 344, 352

Area (*see also* Surface Area), 342–360, 379

of circles, 342

of circle sector (slice), 344

of compound shapes, 354–359

of quadrilaterals, 346, 352–354

of regular hexagons, 350

of regular polygons, 348

of triangles, 347, 360

ASA (*see* Angle-Side-Angle Rule)

ASS, 133, 140, 143

"Assumed from diagram," 59, 62, 76, 76n, 97, 129, 129n, 139n

Assumptions in life, 38, 40–45, 165

Assumptions in math, 35, 39–45, 165

Augmentation, 284

Axioms, 18n, 389n

Base angles, 88–89, 273–274, 386

Bases of isosceles triangles, 89, 102, 105, 135, 136, 157, 188n

Bases of quadrilaterals, 346, 352–354, 360

Bases of solids, 365–372, 374–382

Bases of trapezoids, 242, 273–274, 353, 354, 360

Bases of triangles, 347, 350

Beauty tips, 14, 147, 154, 364n

Betweenness, 21, 31, 39, 41

Biconditional rules, 10–15

Bicycle Wheel (for rotations), 111–114, 118, 128, 146

Bicycle Wheel Rule (*see* Tangent "Bicycle Wheel" Theorem)

Bisect, 36–38, 43, 51, 75, 79, 126, 136, 155, 159n, 170, 187, 188, 254, 260, 264, 266, 269–270

Bisected Radii "Symmetrical Bulldog Puppy Face" Theorem, 305–306, 321

Blinking, 14

Body image, 17, 46–49, 174, 236–239, 292, 300–302, 339–341

Brady, Fernmarie, 361–362

Breathing, 14

204*n*, 211, 211*n*, 212*n*, 249*n*, 267,
283, 291*n*, 320*n*, 344*n*, 377*n*, 380*n*
Hypotenuse (*see also* HL), 91, 173,
176–185, 188–195, 311
Hypotenuse-Leg Theorem (HL), 126,
133*n*, 143–146, 152, 162, 317, 350,
356, 380

"If angles, then sides," 139*n*, 148–152,
159, 162, 167, 251
"If sides, then angles," 148, 162, 251,
326
"If . . . then" statements, 1–16, 51,
54–57, 60, 62, 66–67, 82, 92–93,
96, 122–123, 132
Image (*see also* Body image), 109–112,
117–118
Included angles, 121*n*
Indirect proofs, 165–171, 227–229,
235, 278, 278*n*, 338, 386
Induction, 3, 4, 16
Inductive reasoning (*see* Induction)
Inequalities (*see also* Triangle
Inequality), 99–102
Insecurities, 33, 40, 48, 78, 85,
141–142, 174–175, 282
Intercepted arcs, 325–328, 331–339,
334
Interior angles, 103–106, 204–213
Inverse, 6–9, 13, 16
Irrational emotions, xii
Irrational numbers, 343–344, 344*n*
Isometry (*see* Congruence
transformations)
Isosceles trapezoids, 243–247, 256,
268*n*, 273–278
area of, 353
properties of, 274
Isosceles triangles, 89, 90, 92–98, 101,
105–106, 126, 135, 136, 147–150,
157, 162, 167, 182, 188*n*, 189,
250–251, 256, 263, 272, 273,
308–309, 326, 386

Keramas, Jessica, 300–302
Kiss My Math (McKellar), xiii, 8*n*, 25*n*,
99*n*, 177*n*, 194*n*, 199*n*, 226, 323*n*,
329*n*

Kites, 243–248, 269–270, 275–277,
279, 354*n*
area of, 353
properties of, 269

Landow, Robyn, 46
Lateral faces, 365–369
Lateral surface area, 368–374, 382
Legs of triangle, defined, 89
Line of reflection, 110, 117, 118
Line of symmetry, 112
Lines, 18, 19
coplanar, 35–36, 41, 43, 214
parallel, 36, 39–43, 60, 214,
216–222, 225–235, 241–243, 247,
250–256, 262, 267, 272, 273,
290–295, 298*n*, 299, 300, 385
perpendicular (*see also* 90°,
Altitudes, Perpendicular
bisectors, Perpendicular
diagonals, Right angles), 23, 32,
63, 156, 305–310, 317
transversal, 214–222, 225–235,
250–256, 263, 293, 300, 311,
392
Lipstick, 143, 147
Little *m*'s (*see* Measure notation)
Logic in life, xi–xii, 1, 2, 58, 68–69, 72,
95, 102, 165, 170
Logic in math, 1–17, 57–58, 62, 67,
82, 92, 96, 122–123, 132, 165,
232

m (*see* Measure notation)
Major arcs, 324–326, 329, 333–334
Math Doesn't Suck (McKellar), xiii, 95*n*,
179*n*, 186*n*, 249*n*, 285*n*, 346*n*
Measure notation (*m*), 21, 23, 23*n*,
28*n*, 325, 332*n*, 335–336, 336*n*
Median
of trapezoids, 353, 353*n*
of triangles, 134, 155–157, 160–162,
182, 188*n*, 353*n*
Midpoint, 36–38, 41, 61, 63–65, 77,
79–80, 84, 126, 134, 148–150,
155, 298, 385
Midpoint Theorem, 298
Mini-proofs, 60–64, 67, 77

About the Author

Well known for her role as Winnie Cooper on *The Wonder Years* and her recurring roles on *The West Wing* and *How I Met Your Mother*, Danica McKellar is an internationally recognized mathematician and advocate for math education, and is now a three-time *New York Times* bestselling author.

Upon the release of her first groundbreaking bestseller, *Math Doesn't Suck*, Danica made headlines and was named "Person of the Week" by *ABC World News with Charles Gibson*. Her next books, *Kiss My Math: Showing Pre-Algebra Who's Boss* and *Hot X: Algebra Exposed,* followed suit on the *New York Times* bestseller list, and have inspired countless emails from happy young students. McKellar has also been honored in Britain's esteemed *Journal of Physics* and the *New York Times* for her prior work in mathematics, most notably for her role as coauthor of a mathematical physics theorem that bears her name (The Chayes-McKellar-Winn Theorem).

A summa cum laude graduate of UCLA with a degree in mathematics, McKellar's passion for promoting girls' math education earned her an invitation to speak before Congress on the importance of women in math and science. Amidst her busy acting schedule, Danica continues to make math education a priority as a featured guest and speaker at mathematics conferences nationwide.

McKellar is also spokesperson for the Math-A-Thon program at the St. Jude Children's Research Hospital, which raises millions of dollars every month for cancer research and to provide free care for young cancer patients.

McKellar lives in Los Angeles, California, and this is her fourth book.